ANIMAL SCIENCE, ISSUES AND PROFESSIONS

VERTEBRATE AND INVERTEBRATE ANIMAL APPROACHES

ANIMAL SCIENCE, ISSUES AND PROFESSIONS

Additional books in this series can be found on Nova's website
under the Series tab.

Additional e-books in this series can be found on Nova's website
under the e-book tab.

ANIMAL SCIENCE, ISSUES AND PROFESSIONS

VERTEBRATE AND INVERTEBRATE ANIMAL APPROACHES

JOSÉ ROSA GOMES
EDITOR

Nova Publishers
New York

Library of Congress Cataloging-in-Publication Data

Vertebrate and invertebrate animal approaches / editor, Josi Rosa Gomes.
pages cm
Includes index.
ISBN 978-1-63117-102-4 (hardcover)
1. Physiology. 2. Cytology. 3. Animal genetics.. I. Gomes, Josi Rosa.
QP81.V47 2014
570--dc23

2014001922

Published by Nova Science Publishers, Inc. † New York

CONTENTS

Preface **ix**

Section I: Studies on Cells, Tissues, and Biochemistry in Vertebrate and Invertebrate Animals **1**

Chapter 1 Molecular Conservation of the Telomerase Protein in Vertebrate and Invertebrate Species Using Computational Biology Analysis **3**
Jane Eyre Gabriel

Chapter 2 The Somatotropin Protein and Its Conservation in Primates by Using Comparative Analyses of Multiple Sequence Alignments **9**
Jane Eyre Gabriel

Chapter 3 Colchicine-Induced Apoptosis Daily Rhythms in Male and Female Suckling Mice **15**
Norma. V. González and Claudio G. Barbeito

Chapter 4 Thyroid Deregulation Causes Cellular Damage in Pancreatic Islets and Spleens in Rats **25**
Srikanta Guria, Nisha Balmiki, Sudipta Chakrabarti and Madhusudan Das

Chapter 5 Hypothyroidism Causes Cellular Damage in Retinal Tissue and Biceps in Rats **33**
Manalee Guha, Preeta Sen, Sudakshina Ghosh, Soumee Ghosh, Anisha Chaudhuri, Pritha Gupta, Somtirtha Das, Sweta Mukherjee, Reni Ghosh, Rini Roychowdhury, Srikanta Guria and Madhusudan Das

Chapter 6 Methimazole-induced Hypothyroidism Modulates Cytomorphology of Skin Epidermis and Collagen Synthesis in Rats **39**
Srikanta Guria, Nisha Balmiki, Sudipta Chakrabarti and Madhusudan Das

Chapter 7 Hypothyroidism Reduces Phagocytic Activity of Rat Macrophages **45**
Srikanta Guria and Madhusudan Das

Chapter 8 Phagocytosis of Yeast Particles by Peritoneal Macrophage in Normal and Hyperthyroid Rats **53**
Srikanta Guria and Madhusudan Das

Chapter 9 Pulmonary Involvement of Arsenic Poisoning and Effect of Arsenic Toxicity on Peritoneal Macrophages and Spleen Cells in Rats **61**
Srikanta Guria, Swapna Chhetri, Sayantan Saha, Nikita Chetri, Gitanjali Singh, Piali Bala Saha, Bilkis Sultana Sarkar and Madhusudan Das

Chapter 10 Study of Cytomorphology of Pancreatic Islets and Peritoneal Macrophage in Alloxan-Induced Diabetic Rats: A Mechanistic Insight **71**
Srikanta Guria, Swapna Chhetri, Sayantan Saha, Gitanjali Singh, Piali Bala Saha, Nikita Chetri, Bilkis Sultana Sarkar and Madhusudan Das

Chapter 11 Antioxidant Enzyme Activity and Lipid Peroxidation in Liver of Wistar Rats Exposed to Arabian Incense **81**
Majed S. Alokail, Mohammad A. Ismael and Saud A. Alarifi

Chapter 12 Betaine Reduces Hyperhomocysteinemia and Enhances 5-Hydroxyindoleacetic Acid in Ethanol-Induced Hyperhomocysteinemia in Rabbits **91**
Masoud Alirezaei, Mehdi Saeb, Katayoun Javidnia, Saeed Nazifi and Najmeh Khalighyan

Chapter 13 Volatile Substances of Mandibular Gland Secretion of a Stingless Bee: Scaptotrigona postica Latreille **105**
Luciana Fioretti Gracioli -Vitti, Carminda da Cruz-Landim and Fábio Camargo Abdalla

Chapter 14 Effects of Dimethonate on the Survival and Growth of (Bufo himalayanas) Tadpoles in Laboratory Conditions **117**
Rujas Yonle, Arpita Chaudhury, Deependra Sharma, Santosh Basnet, Somyadip Ghosh and Tashi T. Bhutia

Chapter 15 Developmental Stages of Tetranychus ludeni Zacher (Acari: Tetranychidae) and Brevipalpus deleoni Pritchard and Baker (Acari: Tenuipalpidae) on New Host Plant **123**
Indranil Roy and Goutam Kumar Saha

Chapter 16 Antioxidant and Detoxification Activities of Gill and Digestive Gland of Lamellidens marginalis (Mollusca: Bivalvia) Exposed to Azadirachtin **129**
Suman Mukherjee, Mitali Ray and Sajal Ray

Chapter 17 Cypermethrin Induced Dynamics of Hemocyte Density of Indian Mollusc Lamellidens marginalis **139**
Krishnendu Das, Mitali Ray and Sajal Ray

Chapter 18 Histopathology of Gill, Digestive Gland and Antenna of Bellamya Bengalensis Exposed to Fenvalerate **151**
Chiranjib Mandal, Mitali Ray and Sajal Ray

Chapter 19 Structural Impairment of Gill and Digestive Gland of Lamellidens marginalis Exposed to Cypermethrin **159**
Krishnendu Das, Mitali Ray and Sajal Ray

Chapter 20 Morphological Features of Three Endemic Pill-Millipedes of the Genus Arthrosphaera in the Western Ghats of India **171**
B. S. Kadamannaya, C. N. Ambarish and K. R. Sridhar

Chapter 21 Impacts of Ammonia Toxicity on Some Enzymes of a Freshwater Fish at Different pH and Temperatures **185**
A. Shiwanand and G. Tripathi

Section II: Studies on Morphology and Reproduction in Vertebrate and Invertebrate Animals **201**

Chapter 22 Ultrastructure and Distribution of Immunopotent Leucocytes in an Air-Breathing Fish, Heteropneustes fossilis (Bloch) **203**
Rajarshi Ghosh and Sumit Homechaudhuri

Chapter 23 Are Sex and Reproductive Phase-Specific Changes in Catalase and Protein Profile of a Freshwater Catfish Season-Dependent? **213**
G. Tripathi and N. Bandooni

Chapter 24 Fecundity, Sex Ratio and Gonado-Somatic Index of Epinephelus chlorostigma and Epinephelus areolatus sampled from Arabian Gulf **225**
Zubair Ahmad, Hmoud Fares Alkahem and Amir A. A. Aldhahi

Chapter 25 Laboratory Studies on the Terrestrial Marsh Slug Deroceras laeve (Müller) (Agriolimacidae: Mollusca) **235**
M. I. Mohamed and R. F. Ali

Chapter 26 Life Cycle and Growth Rates of the Conical Snail Cochlicella acuta (Müller, 1774) (Gastropoda: Cochlicellidae) **245**
Mahmoud I. Mohamed and Reham F. Ali

Section III: Studies on Genetic Aspects in Vertebrate and Invertebrate Animals **255**

Chapter 27 Karyotypic Data Detect Interpopulational Variation in Physalaemus olfersii and the First Case of Supernumerary Chromosome in the Genus **257**
Mateus Milani, Carla Santana Cassini, Shirlei M. Recco-Pimentel and Luciana Bolsoni Lourenço

Chapter 28 Genetic Diversity and Different Ecological Conditions in Salamandra infraimmaculata Larvae from Various Breeding Sites **265**
Tali Goldberg, Eviatar Nevo and Gad Degani

Section IV: Reviews Related to Vertebrate and Invertebrate Animals and Humans **279**

Chapter 29 Historical Epidemiological Aspects of American Visceral Leishmaniasis in Brazil **281**
Kárita Cláudia Freitas-Lidani, Jhulie Mirandola, Jimena Ferreira, Raphaela Mordaski and Edna Aoba Yassui Ishikawa

Chapter 30 Black-tailed Prairie Dog Affinity for Sunlight and its Potential Animal Welfare Management Implications **291**
Adam E. M. Eltorai and Robert W. Sussman

Chapter 31 Clinical Spectrum of Hypothyroidism in West Bengal, India **295**
Nisha Balmiki, Srikanta Guria, Biswabandhu Bankura, Arup Kumar Pattanayak, Tapas Kumar Das, Anirban Sinha, Sudipta Chakrabarti, Subhankar Chowdhury and Madhusudan Das

Index **303**

PREFACE

This book is comprised of important contributions from expert researchers around the world concerning the biology of animals from a variety of approaches. In particular are manuscripts that deal with cellular, biochemical, genetic, reproductive, and ecological themes including various manuscripts regarding invertebrate animals.

Regarding this last animal class, the science provides excellent models to study a lot of biological processes that can explain the evolution and diversity of life on the earth's surface. They are simple organisms, the studies of which can contribute to explanations of how the metabolic processes found in vertebrates and humans started and have been maintained.

This book also provides results that demonstrate some effects and interactions among environmental conditions and drugs on the morphology and biochemical processes in cells, and it contains reviews concerning the interaction between pathogenic invertebrates and human diseases. The aim of these studies was to provide important results that are not commonly treated in traditional and experimental models within the current scientific scene.

SECTION I: STUDIES ON CELLS, TISSUES, AND BIOCHEMISTRY IN VERTEBRATE AND INVERTEBRATE ANIMALS

In: Vertebrate and Invertebrate Animal Approaches
Editor: José Rosa Gomes
ISBN: 978-1-63117-102-4

Chapter 1

MOLECULAR CONSERVATION OF THE TELOMERASE PROTEIN IN VERTEBRATE AND INVERTEBRATE SPECIES USING COMPUTATIONAL BIOLOGY ANALYSIS

Jane Eyre Gabriel*

Colegiado de Ciências Biológicas, Centro de Ciências Agrárias,
Universidade Federal do Vale do São Francisco UNIVASF, Petrolina, PE, Brazil

ABSTRACT

In order to assess the conservation of the telomerase protein among distinct animal species were employed comparative analysis of multiple sequence alignments using computational biology tools. The search of amino acid sequences of the telomerase enzyme and their alignments were performed using the non-redundant and highly annotated protein sequence database UniProtKB/Swiss-Prot that provides a wealth of information regarding protein variety, function and associated disorders. An identity value corresponding to 3.249% was observed in the alignments of the telomerase protein between vertebrates and invertebrates, presenting 49 and 184 identical and similar amino acids, respectively. Nevertheless, alignments of this protein exclusively generated from vertebrate or invertebrate species revealed identity values of 16.435% and 22.489%, respectively. The findings presented in this study describing the identity degree of the telomerase protein among distinct animal species using computational biology tools reinforce relevant results about evolutionary biology and biological molecule conservation in animal species.

Keywords: Protein conservation, bioinformatics, telomerase

* Corresponding author: Jane Eyre Gabriel, Caixa postal 91, CEP: 56.302-970, Petrolina/PE, e-mail: jane.gabriel@univasf.edu.br, telephone: 87 3861 8289.

Introduction

With the advent of computational biology a large number of important discoveries have reported crescent approaches regarding conservation and molecular characterization of biological sequences in different species along the evolutionary process (Wasserman and Sandelin, 2004, Li et al., 2006). In particular, several studies have been conducted in order to exploit collections of complete genome sequences to detect biological patterns that explain how genotype leads to phenotype using molecular evolutionary theory and computational biology, as well as to ask questions about how sequences evolve at the molecular level.

Bioinformatics has assisted the evolutionary biology in several key ways; it has enabled researchers to trace the evolution of numerous organisms by measuring changes in their DNA or proteins, rather than through physical taxonomy or physiological observations alone (Lesk, 2002). More recently, compare entire genomes which permits the study of more complex evolutionary events, such as gene duplication, horizontal gene transfer, and the prediction of important factors in bacterial speciation, and it has become possible by building complex computational models of populations to predict the outcome of the system over time, track and share information on an increasingly large number of species and organisms (Li et al., 2006).

The telomerase, protein of interest in this study, is a ribonucleoprotein enzyme essential for the replication of chromosome termini in most eukaryotes and catalytic component of the teleromerase holoenzyme complex whose main activity is the elongation of telomeres by acting as a reverse transcriptase that adds simple sequence repeats to chromosome ends by copying a template sequence within the RNA component of the enzyme (Shay and Wright, 2001, Cifuentes-Rojas and Shippen, 2012). This enzyme catalyzes the RNA-dependent extension of 3'-chromosomal termini with the 6-nucleotide telomeric repeat unit, 5'-TTAGGG-3' (Podlevsky and Chen, 2012). Although telomere length reduction and following initiation of senescence is the usual fate of somatic cells, it is possible that the lack of certain telomerase components results in early phenotypic senescence with loss of function at the cellular and system levels (Shay and Wright, 2001). Based on the unlimited potential of biology computational tools, the purpose of the present study was to assess the conservation degree of the amino acid sequences in the telomerase protein between vertebrates and invertebrates using comparative analysis of multiple sequence alignments.

Material and Methods

In the present study the amino acid sequences corresponding to telomerase protein in distinct animals were searched and aligned using the non-redundant and highly annotated protein sequence database, named UniProtKB/Swiss-Prot protein knowledgebase (Boeckmann et al., 2003). Such database provides a wealth of information regarding protein variety, function and associated disorders. The Swiss-prot protein knowledgebase (http://www.expasy.org/sprot/ and http://www.ebi.ac.uk/swissprot/) connects amino acid sequences and provides an interdisciplinary overview of relevant information by bringing together experimental results and computed features, ensuring the presence of high quality annotation for representative members of all protein families (Jungo et al., 2012).

The telomerase protein sequences with their respective access numbers selected during computational search in the Swiss-prot database were five vertebrate animals [human *Homo sapiens* (O14746), mouse *Mus musculus* (O70372), chicken *Gallus gallus* (Q537V4), Medaka fish *Oryzias latipes* (Q1PS67) and Western clawed frog *Xenopus tropicalis* (F6TTE7)], besides two invertebrate species represented by two ciliated protozoa *Tetrahymena thermophila* (O77448) and *Euplotes aediculatus* (O00939). The vertebrate species selected for these alignments have been grouped in distinct taxonomic classes through traditional procedure, demonstrating great divergence morphology and physiological characteristics between them.

RESULTS

The multiple alignment analyses proportionate relevant statistical parameters concerning conservation of the telomerase protein among distinct animal species using Swiss-prot tools (Figure 1). The identity value observed in the alignments of the telomerase protein in all species selected was corresponding to 3.249%, presenting 49 and 184 identical and similar amino acids, respectively. Nevertheless, multiple alignments exclusively among vertebrates demonstrated 16.435% identity value with identical and similar amino acid residues corresponding to 225 and 259 positions, respectively. Additionally, alignments of the telomerase protein between two invertebrate species (the ciliated protozoa *Tetrahymena thermophila* and *Euplotes aediculatus*) revealed identity of 22.489% with identical and similar amino acid residues in 262 and 424 positions, respectively.

```
774  GLNISGVLCSFYFGKLEEEYTQFLKNAE--QVNGSINLLMRLTDDYLFISDSQQNALNLI  831  077448   TERT_TETTH
827  GSSLSTLLCSLCFGDMENKLFAEVQR---------DGLLLRFVDDFLLVTPHLDQAKTFL  877  070372   TERT_MOUSE
834  GSILSTLLCSLCYGDMENKLFAGIRR---------DGLLLRLVDDFLLVTPHLTHAKTFL  884  014746   TERT_HUMAN
738  GLCVSSILSSFYYATLEESSLGFLRDESMNPENPNVNLLMRLTDDYLLITTQENNAVLFI  797  000939   TERT_EUPAE
1049 GSILSTLLCSLCYGDMENKLLCGIQK---------DGVLIRLIDDFLLVTPHLMQARTFL 1099  Q537V4   Q537V4_CHICK
792  GSVVSSLLCCLCYGHMENVLFRDIKN---------KGCLMRLVDDFLLITPDRNQAQSFL  842  Q1PS67   Q1PS67_ORYLA
894  -STGNTVKMLLKWWVTTNILSVVLNY---------NLVLMRLIDDFLFVTPHLNQAKTFL  943  F6TTE7   F6TTE7_XENTR
         . :   : :    :    :.             *:*: **:*:::   .*  ::

832  VQLQNCANNNGFMFNDQKITTNFQFPQEDYNLE-----HFKISVQNECQWIGKSIDMNTL  886  077448   TERT_TETTH
878  STLVHGVPEYGCMINLQKTVVNFPVEPGTLGGAA----PYQLPAHCLFPWCGLLLDTQTL  933  070372   TERT_MOUSE
885  RTLVRGVPEYGCVVNLRKTVVNFPVEDEALGGTA----FVQMPAHGLFPWCGLLLDTRTL  940  014746   TERT_HUMAN
798  EKLINVSRENGFKFNMKKLQTSFPLSPSKFAKYGMDSVEEQNIVQDYCDWIGISIDMKTL  857  000939   TERT_EUPAE
1100 RTIAAGIPEYGFLINAKKTVVNFPVDDI-PGCSK----FKHLPDCRLISWCGLLLDVQTL 1154  Q537V4   Q537V4_CHICK
843  SILLAGVPQYGVVANPQKVVVNFQGSEGGGAFPD----IRVLPPHCLFPWCGLLLDTRSL  898  Q1PS67   Q1PS67_ORYLA
944  RTLAEGIPQYGCSISPQKTVVNFPVDDI-PACSE----VEQLPVHCLFRWCGLLLDTQTL  998  F6TTE7   F6TTE7_XENTR
       :     : *   . :*  ..*                         * *  :* .:*

887  EIKSIQKQTQ-QEINQTINVAISI-------KNLKSQLKNKLRSLFLNQLIDYFNPNINS  938  077448   TERT_TETTH
934  EVFCDYSGYAQTSIKTSLTFQSVFKAGKTMRNKLLSVLRLKCHGLFL-------DLQVNS  986  070372   TERT_MOUSE
941  EVQSDYSSYARTSIRASLTFNRGFKAGRNMRRKLFGVLRLKCHSLFL-------DLQVNS  993  014746   TERT_HUMAN
858  ALMPNINLRI-EGILCTLNLNMQT-------KKASMWLKKKLKSFLMNNITHYFRKTITT  909  000939   TERT_EUPAE
1155 EVYCDYSSYAFTSIRSSLSFNSSRIAGKNMKCKLTAVLKLKCHPLLL-------DLKINS 1207  Q537V4   Q537V4_CHICK
899  DVCKDYSSYAGLSLRYSLTLGSAHSAGQQMRRKLMSILRIKCHPLFL-------DLKTNS  951  Q1PS67   Q1PS67_ORYLA
999  DVYYDYSSYACTSIRSSMTFCHSSAAGKNMKQKLLRVLKLKCHSLFL-------DLQVNS 1051  F6TTE7   F6TTE7_XENTR
      :    .      :  ::..           :    *: * : :::          .:
```

Figure 1. Multiple alignments of amino acid residues observed in a partial segment of the telomerase protein between vertebrate and invertebrate species. Columns indicating identical amino acids in the same position in all types analyzed are exclusively signalized by asterisk (*), whereas alignments showing the presence of similar amino acids are represented by other symbols (: or .).

The presence of similar amino acids indicates conservative changes throughout telomerase sequence and such changes could represent substitutions of both amino acids with identical chemical proprieties (for example, the substitutions could occur exclusively among hydrophobic, nonpolar amino acids). In the biology computational analysis similar positions indicate that an amino acid might be substituted for another with similar chemical properties, which is more likely to observe, or for one with different chemical characteristics (Jungo et al., 2012). In the regions where an amino acid has no correspondence in the other in the same position, it would be interpreted as the occurrence of an insertion or deletion of one residue in the peptide sequence. Thus, in the present study the identification of an expressive number of identical amino acids could demonstrate the higher conservation of the telomerase protein throughout evolutionary process in animals belonging to same taxonomic category as observed in the multiple alignments exclusively obtained from ciliate protozoa.

Although certain number of identical amino acids is observed in the sequences analyzed, the great diversity of morphology and physical characteristics in vertebrates belonging to distinct taxonomic categories is accompanied for the occurrence of low identity in the amino acid sequences of the telomerase protein. On the other hand, as observed in this study, comparative analysis of the telomerase enzyme in the ciliated protozoa grouped in the same taxonomic classes generated higher values of identity in their amino acid residues. The structural and molecular characterization of biological molecules has proportioned important contributions for the taxonomy, identifying traces of ancestry among species during evolutionary process. Within this context, the computational biology has enriched new insights to assess the phylogenetic relationships of vertebrates and invertebrates using comparative analysis of complete genomes. Since the unlimited potential of computational biology tools in order to generate high quality and non redundant annotation from biological molecule sequences, the findings presented in this study describing the identity degree among distinct animal species using computational biology tools reinforce relevant results about evolutionary biology and biological molecule conservation between vertebrates and invertebrates.

REFERENCES

Boeckmann, B; Bairoch, A; Apweiler, R; Blatter, MC; Estreicher, A; Elisabeth Gasteiger, E; Martin, MJ; Michoud, K; O'Donovan, C; Phan, I; Pilbout; S; Schneider, M. The SWISS-PROT protein knowledgebase and its supplement TrEMBL in 2003. *Nucleic Acids Research*, 2003, 31, 365–370.

Cifuentes-Rojas, C; Shippen, DE. Telomerase regulation. *Mutation Research,* 2012, 730, 20-7.

Jungo, F; Bougueleret, L; Xenarios I; Poux, S. The UniProtKB/Swiss-Prot Tox-Prot program: a central hub of integrated venom protein data. *Toxicon,* 2012, 60, 230-243.

Lesk, AM. *Introduction to Bioinformatics.* 1st ed. London: Oxford University Press. 2002, 283 p.

Li, YY; Yu, H; Guo, ZM; Guo, TQ; Tu, K; Li, YX. Systematic analysis of head-to-head Gene organization: evolutionary conservation and potential biological relevance. *PLoS Computational Biology*, 2006, 2, 687-697.

Podlevsky, JD; Chen, JJ. It all comes together at the ends: telomerase structure, function, and biogenesis. *Mutation Research*, 2012, 730, 3-11.

Shay, JW; Wright, WE. Telomeres and telomerase: implications for cancer and aging. *Radiation Research,* 2001, 155, 188-193.

Wasserman, WW; Sandelin, A. Applied bioinformatics for the identification of regulatory elements. *Nature Reviews Genetics*, 2004, 5, 276-287.

In: Vertebrate and Invertebrate Animal Approaches
Editor: José Rosa Gomes
ISBN: 978-1-63117-102-4

Chapter 2

THE SOMATOTROPIN PROTEIN AND ITS CONSERVATION IN PRIMATES BY USING COMPARATIVE ANALYSES OF MULTIPLE SEQUENCE ALIGNMENTS

Jane Eyre Gabriel*

Colegiado de Ciências Biológicas, Centro de Ciências Agrárias, Universidade Federal do Vale do São Francisco UNIVASF Petrolina, PE, Brasil

ABSTRACT

The present study aimed to assess comparatively sequences of the somatotropin protein in distinct primates by using computational biology tools. The sequences of interest selected in Bolivian squirrel monkey (*Saimiri boliviensis*, P58343), Senegal bushbaby (*Galago senegalensis*, Q9GKA1), human (*Homo sapiens*, P01241), Rhesus macaque (*Macaca mulatta*, P33093) and chimpanzee (*Pan troglodytes*, P58756) were randomly searched and aligned from Uni-prot/Swiss-Prot database and COBALT program.

Comparative alignments of the multiple sequences of the somatotropin protein in all selected primates revealed an identity value corresponding to 60.731% with 55 amino acid residues characterized in similar positions. Nevertheless, matches of these sequences carried out exclusively among the three sequences belonging to Catarhini taxonomic category (human, Rhesus macaque and chimpanzee) yielded higher identity value corresponding to 96.313%, containing 5 and 209 amino acids in similar and identical positions, respectively.

The findings noticed in this study demonstrate an accentuated identity and remarkable presence of conservative changes in the sequences of the somatotropin peptide in evolutionarily closest species of primates, reinforcing the powerful application

* Corresponding author: Jane Eyre Gabriel, Caixa postal 91, CEP: 56.302-970, Petrolina/PE, e-mail: jane.gabriel@univasf.edu.br, telephone: 87 21014836.

of the computational biology tools in the studies of protein conservation and evolutionary biology in vertebrates.

Keywords: Computational biology, protein identity, somatotropin

INTRODUCTION

The somatotropin, or growth hormone, is a peptide hormone which stimulates the growth, the cell reproduction and the tissue regeneration in humans and other animals. Characterized as a single-chain polypeptide containing approximately 210-amino acids, it is already well known that its synthesis and secretion is modulated in a pulsatile manner from somatotropic cells within the lateral wings of the anterior pituitary gland by activating neurosecretory nuclei from hypothalamus. Biological roles of the somatotropin generally promote anabolic effects on the tissues of the body and its functions have been involved in the calcium retention, the mineralization of bone, the muscle mass hyperplasia, the stimulus of protein synthesis and the immune system (Varela-Nieto and Chowen, 2005). Surprisingly, few studies have been conducted in order to investigate in detail how the sequences of the somatotropin hormone have remained conserved over evolutionary process in mammals. Considering the powerful potential of the computational biology tools in the structural and functional characterization of biological molecules and in the establishment of kinship relationships in distinct species, the purpose of the present study was to investigate the identity degree of multiple sequences of the somatotropin protein in primates in order to clarify new evidences into molecular conservation of biological molecules and evolutionary closeness among vertebrates.

MATERIALS AND METHODS

Under these experimental conditions, the advanced search and selection of the sequences corresponding to somatotropin protein were performed by using Uni-prot/Swiss-Prot database which contains a high quality annotated and non-redundant protein sequence database, as well as provides relevant evaluative data concerning significance and identity of the matches (Boeckmann et al., 2003). Five sequences of the somatotropin protein were randomly selected in following primates: Bolivian squirrel monkey (*Saimiri boliviensis*, P58343), Senegal bushbaby (*Galago senegalensis*, Q9GKA1), human (*Homo sapiens*, P01241), Rhesus macaque (*Macaca mulatta*, P33093) and chimpanzee (*Pan troglodytes*, P58756). Human, Rhesus macaque and chimpanzee species are vertebrates ordered taxonomically in Haplorhini suborder and Catarrhini parvorder which comprises the higher primates (Old World monkeys), whereas Bolivian squirrel monkey and Senegal bushbaby are grouped in Haplorhini suborder and Platyrrhini parvorder (tarsiers, monkeys and apes) and in Strepsirrhini suborder (lemurs, lorises and galagos), respectively (Poux and Douzery, 2004).

Firstly, the sequences of the peptides of interest were comparatively aligned by using Uni-prot/Swiss-Prot program, generating automatically relevant evaluation parameters, such as identity and similarity. Secondly, additional comparative alignments were conducted through COBALT program, a constraint based alignment tool which implements a general framework for multiple alignments of protein sequences concerning conserved domain and local sequence similarity information (Papadopoulos and Agarwala, 2007). The COBALT alignments of the mentioned-above sequences generated complementary evaluative parameters concerning significance of these matches, such as expectation-values (E-values), able to estimate the number of possible matches that can be expected by chance during a database scan in the computational biology analyses (Lambert et al., 2005).

RESULTS

Comparative alignments of multiple sequences of the somatotropin protein in five-selected primates revealed an identity value corresponding to 60.731% with 55 and 133 amino acids characterized as similar and identical positions in a total content of 217 residues. Higher identity value corresponding to 96.313% was observed in the Swiss-Prot results obtained exclusively in three protein sequences of primates belonging to Catarhini, containing 5 and 209 amino acid residues in similar and identical positions, respectively. In these matches E-values of 0.003 were observed in all comparative alignments by using COBALT tools, demonstrating the good significance and accuracy of these matches.

Based on the substitutions that really do or do not happen in the real protein sequences, the ability of a protein to tolerate those replacements is related to the chemical properties of the amino acids in the pair. Some replacements of amino acid residues are frequently detected specially among those that maintain the same biological proprieties and such variations were noticed in the sequences of the somatotropin hormone, as observed in Figure 1 (for example, of isoleucine (I) for methionine (M), as indicated in the arrow A; of phenylalaline (F) for leucine (L), as indicated in the arrow B; of methionine (M) for leucine (L), as indicated in the arrow C).

Within this context, such replacements may indicate conservative changes throughout mammalian somatotropin protein, which could represent substitutions of both amino acids with identical chemical proprieties (for example, the replacements pointed in the Figure 1 occurred exclusively among neutral, nonpolar and very hydrophobic amino acids). In the biology computational analyses similar positions indicate that an amino acid may be substituted for another with similar chemical properties, which is more likely to observe, or for one with different chemical characteristics. In the regions where an amino acid has no correspondence in the other in the same position, it would be interpreted as the occurrence of an insertion or deletion of one residue in the peptide sequence.

Recent studies conducted in order to investigate the conservation degree of the somatotropin peptide and of its receptor have created excellent opportunities to better elucidate the coevolution of these proteins in distinct animals (Li, 1997; Liu et al., 2001, Manjou et al., 2004). It is already well established that the mammalian somatotropin peptide has rapidly evolved in primates, having undergone approximately 30% amino acid replacements from the inferred ancestral eutherian sequence (Li, 1997; Liu et al., 2001, Manjou et al., 2004). Wallis (1994), compiling the evolutionary rates of the somatotropins from various mammalian taxa, showed that, in general, this peptide hormone evolves at a slow basal rate, but significant acceleration in evolutionary rate occurred in two lineages, one leading to the primates and the other to the artiodactyls.

```
                                  A     B
                                  ↓     ↓
1    MAAGSR-TSLLLAFALLCLPWLQEGSAFPTIPLSRLFDNAMLRAHRLHQLAFDTYQEFEE  59  P33093  SOMA_MACMU
1    MAPGSR-TSLLLAFGLLCLPWLQEGSAFPTIPLSRLFDNAMLRAHRLHQLAFDTYQEFEE  59  P58756  SOMA_PANTR
1    MATGSR-TSLLLAFGLLCLPWLQEGSAFPTIPLSRLFDNAMLRAHRLHQLAFDTYQEFEE  59  P01241  SOMA_HUMAN
1    MATGSR-TSLLLAFTLLCLPQLKEAGAFPTIPLSRLLDNAMLRAHRLHQLAFDTYQEFEE  59  P58343  SOMA_SAIBB
1    MATGSHTTTLLLAVALLGLPWPQEAGAFPAMPLSSLFANAVLRAQHLHQLAADTYKEFER  60  Q9GKA1  SOMA_GALSE
     ** **: *:****. ** **  :*..***::*** *: **:***::***** ***:***.

60   AYIPKEQKYSFLQNPQTSLCFSESIPTPSNREETQQKSNLELLRISLLLIQSWLEPVQFL  119  P33093  SOMA_MACMU
60   AYIPKEQKYSFLQNPQTSLCFSESIPTPSNREETQQKSNLELLRISLLLIQSWLEPVQFL  119  P58756  SOMA_PANTR
60   AYIPKEQKYSFLQNPQTSLCFSESIPTPSNREETQQKSNLELLRISLLLIQSWLEPVQFL  119  P01241  SOMA_HUMAN
60   AYIPKEQKYSFLQNPQTSLCFSESIPTPASKKETQQKSNLELLRISLILIQSWFEPVQLL  119  P58343  SOMA_SAIBB
61   AYIPEGQRYSI-QNTQAAFCFSETIPAPTGKDEAQQRSDMELLRFSLLLIQSWLGPVQLL  119  Q9GKA1  SOMA_GALSE
     ****: *:**: ** *:::****:**:*: :.*:**:*::****:**:*****: ***:*

120  RSVFANSLVYGTSYSDVYDLLKDLEEGIQTLMGRLEDGSSRTGQIFKQTYSKFDTNSHNN  179  P33093  SOMA_MACMU
120  RSVFANSLVYGASDSNVYDLLKDLEEGIQTLMGRLEDGSPRTGQIFKQTYSKFDTNSHND  179  P58756  SOMA_PANTR
120  RSVFANSLVYGASDSNVYDLLKDLEEGIQTLMGRLEDGSPRTGQIFKQTYSKFDTNSHND  179  P01241  SOMA_HUMAN
120  RSVFANSLLYGVSDSDVYEYLKDLEEGIQTLMERLEDGSPRTGAIFRQTYSKFDINSQND  179  P58343  SOMA_SAIBB
120  SRVFTNSLVLGTLD-RVYEKLKDLEEGIQALMRELEDGSPRVGQILKQTYDKFDTNLRSD  178  Q9GKA1  SOMA_GALSE
       **:***: *.     **: *********:** .***** *.* *::***.*** * :.:

180  DALLKNYGLLYCFRKDMDKIETFLRIVQCRS-VEGSCGF  217  P33093  SOMA_MACMU
180  DALLKNYGLLYCFRKDMDKVETFLRIVQCRS-VEGSCGF  217  P58756  SOMA_PANTR
180  DALLKNYGLLYCFRKDMDKVETFLRIVQCRS-VEGSCGF  217  P01241  SOMA_HUMAN
180  DALLKNYGLLYCFRKDMDKVETFLRIVQCRS-VEGSCGF  217  P58343  SOMA_SAIBB
179  DALLKNYGLLSCFKKDLHKAETYLRVMKCRRFVESSCAF  217  Q9GKA1  SOMA_GALSE
     ********** **:**:.* **:**:::**  **.**.*
                   ↑
                   C
```

Figure 1. Comparative alignments of multiple sequences of the somatotropin protein in Bolivian squirrel monkey (*Saimiri boliviensis*, P58343), Senegal bushbaby (*Galago senegalensis*, Q9GKA1), human (*Homo sapiens*, P01241), Rhesus macaque (*Macaca mulatta*, P33093) and chimpanzee (*Pan troglodytes*, P58756). Different arrows (A, B and C) indicate the presence of conservative changes along sequences of the somatotropin hormone.

The findings reported in the present study describe an accentuated identity in the sequences of the somatotropin hormone especially in the primates restricted to Catarrhini subdivision (Figure 1). Such evidence suggests certain evolutionary closeness among these species, reinforcing some discoveries described in the literature which argue the accentuated conservation of a large broad of proteins in phylogenetically closely related animals (Gabriel, 2012, Gabriel et al., 2013, Poux and Douzery, 2004). Although the sequence of the somatotropin seems to be strongly conserved in mammals (Yi et al., 2002), the apparent burst of rapid evolution of the somatotropin hormone occurred largely after the separation of the line leading to ruminants from other cetartiodactyls (Manjou et al, 2004). According to Manjou et al. (2004), the sequences of the higher primate and ruminant somatotropins differ markedly in comparison to other mammals (for example, the sequence of the giraffe somatotropin is very similar to that of other ruminants but differs from that of nonruminant cetartiodactyls at about 18 residues).

The complexity of the genome evolution has been extensively unveiled by using the computational biology tools in several key ways; enabling to trace the evolution of a large number of organisms by measuring changes in their biological molecules, rather than through physical taxonomy or physiological observations alone. Moreover, one important application of this area has assured to track and share information on an increasingly large number of species and organisms by using comparative analyses among entire genomes, which permits the study of more complex evolutionary events, such as gene duplication and transfer, and the prediction of important factors in the mammalian speciation (Hogeweg, 2011). The core of the comparative genome analyses is the establishment of the correspondence between genes or other genomic features in different organisms that make possible to trace the evolutionary processes responsible for the divergence of multiple genomes.

In conclusion, the findings reported in the present study provide new insights into increasing application of the computational biology tools in the molecular characterization of biological sequences among animal species, revealing an accentuated degree of identity and remarkable presence of conservative changes among sequences of the somatotropin hormone in distinct primates.

Acknowledgments

We are grateful to the members of the Bioinformatics and Computational Biology Group, designated “BIO in BYTES”, for cooperation dedicated by them during elaboration of this manuscript.

References

Gabriel, JE. Molecular conservation of the telomerase protein in vertebrate and invertebrate species using computational biology analysis. *Animal Biology Journal,* 2012, 3, 1-4.

Gabriel, JE; de Figueiredo, DDLG; de Farias, RP. Revealing highly conserved regions in the E6 protein among distinct human papillomavirus types using comparative analysis of multiple sequence alignments. *Brazilian Journal of Biology*, 2013, 73, in press.

Healy, MD. Using BLAST for performing sequence alignment. *Current Protocols in Human Genetics,* 2007, 52, 6.8.1–6.8.23.

Hogeweg, P. The roots of bioinformatics in theoretical biology. PLoS Computational Biology, 2011, 7 e1002021. doi:10.1371/journal.pcbi.1002021.

Jungo, F; Bougueleret, L; Xenarios, I; Poux, S. The UniProtKB/Swiss-Prot Tox-Prot program: a central hub of integrated venom protein data. *Toxicon*. 2012, 60, 230-243.

Lambert, A; Legendre, M; Fontaine, JF; Gautheret, D. Computing expectation values for RNA motifs using discrete convolutions. *BMC Bioinformatics*, 2005, 6, 118-135.

Li, WH. 1997. Molecular evolution. Sinauer Associates, Sunderland, Mass.

Liu, JC; Makova, KD; Adkins, RM; Gibson, S; Li, WH. Episodic evolution of growth hormone in primates and emergence of the species specificity of human growth hormone receptor. *Molecular Biology and Evolution*, 2001, 18, 945–953.

Papadopoulos, JS; Agarwala R. COBALT: constraint-based alignment tool for multiple protein sequences. *Bioinformatics,* 2007, 23, 1073-1079.

Wallis, M. Variable evolutionary rates in the molecular evolution of mammalian growth hormones. *Journal of Molecular Evolution*, 1994, 38, 619–627.

Yi, S; Bernat, B; Pál, G; Kossiakoff, A; Li, WH. Functional promiscuity of squirrel monkey growth hormone receptor toward both primate and nonprimate growth hormones. *Molecular Biology and Evolution,* 2002, 19, 1083–1092.

Varela-Nieto, I; Chowen, JA. 2005. The growth hormone/insulin-like growth factor axis during development: advances in experimental medicine and biology. Springer Science: Business Media Inc, vol 567.

Poux, C; Douzery, E JP. Primate phylogeny, evolutionary rate variations, and divergence times: a contribution from the nuclear gene *irbp*. *American Journal of Physical Anthropology*, 2004, 124, 1–16.

In: Vertebrate and Invertebrate Animal Approaches
Editor: José Rosa Gomes
ISBN: 978-1-63117-102-4

Chapter 3

COLCHICINE-INDUCED APOPTOSIS DAILY RHYTHMS IN MALE AND FEMALE SUCKLING MICE

***Norma. V. González* [1,2] *and Claudio G. Barbeito* [2,3*]**
[1]Cátedra de Histología y Embriología Animal, Facultad de Ciencias Naturales, Universidad Nacional de La Plata, Argentina
[2]Cátedra de Histología y Embriología, Facultad de Ciencias Veterinarias, Universidad Nacional de La Plata, Argentina
[3]Instituto de Patología. Facultad Ciencias Veterinarias, Universidad Nacional de La Plata, Argentina

ABSTRACT

Cell death in the intestinal crypts is increased in many conditions including exposure to several cytotoxic compounds, among them colchicine. In despite of *in vitro* studies´ abundance, *in vivo* investigations assessing the effect of this microtubule inhibitor on apoptosis are scarce and no data dealing specifically with colchicine-induced apoptosis in suckling mice are available. The aims were to determine the colchicine inductive effect on apoptosis in 14 d old mice intestinal crypts; to establish the existence of daily variations in the apoptotic indices and to examine whether there is sexual dimorphism in such variations. Experimental work was performed on C3H/S male and female 14 d old mice exposed to 12:12 light-dark conditions. Commencing at midnight and every 4 h thereafter animals were sacrificed having received an i.p. 1 μg colchicine dose per g body weight 4 h before. Samples of duodenum and colon were processed for hematoxylin-eosin staining and TUNEL technique. Twenty longitudinally sectioned crypts per organ and individual were monitored for apoptotic cells. The apoptotic index was expressed as the number of apoptotic cells per 1, 000 nuclei for each animal. The arithmetic mean ± SE was calculated for each study time and gender. Differences between genders were determined by the Student's t test. Within group differences were tested by one-way analysis of variance (ANOVA). Results revealed that. 1) the apoptotic daily changes phasing in the small bowel between young males and females showed sexual dimorphism; 2) the absence of daily variations in the young female large bowel cell death

* Corresponding author: Cátedra de Histología y Embriología. Facultad de Ciencias Veterinarias. 60 y 118. La Plata (1900). Buenos Aires. Argentina. barbeito@fcv.unlp.edu.ar.

indices; 3) a highly prominent induced apoptosis intensity in the small bowel; 4) the sensitivity and number of colchicine induced-apoptosis changes throughout the day are age-dependant. Colchicine is to be taken regularly as a treatment for various human diseases and poisoning results in a multi-organ dysfunction syndrome, particularly affecting tissues with rapid cell turnover among them the intestinal epithelium. The complex pattern of variables implied in the colchicine-induced intestinal apoptosis rhythms (gender, age, time of drug delivery) highlights the need of treatments design with colchicine dosing according with the patient condition.

Keywords: Apoptosis-colchicine-circadian rhythms-intestinal crypts-age-gender

INTRODUCTION

In multicellular organisms homeostasis is maintained through a balance between cell death and cell proliferation (Danial and Korsmeyer, 2004; Fan and Bergamann, 2008). Cell death spontaneously occurs in the intestinal crypts; this process is increased in many conditions including exposure to several citotoxic compounds (Kasibhatla et al., 2004), radiation (Inagaki-Ohara et al., 2002), venous congestion (Wu et al., 2004), feeding on a combination of fish oil and pectin (Sanders et al., 2004) or by hypoxia/reoxygenation (Ozen et al., 2005).

Duncan and Heddle (1984) demonstrated that colchicine stimulates the apoptosis in large intestine crypts and more recently we reported the same effect on the small intestine crypts in adult mice (González et al., 2005). Colchicine is an agent capable of altering the mitotic spindle formation by sequestering single tubulin molecules (Inoué, 1981; Salmon et al., 1984).

Several investigations have established the occurrence of daily rhythmicity in many aspects of the intestinal physiology including enzyme expression (Abolmaali et al., 2009), transcription of drug transporters (Stearns et al., 2008), DNA synthesis (Bjarnason and Jordan, 2002), mitosis (Barbeito et al., 2002) and enterocytes apoptosis at the cessation of feeding and commencement of fasting (Iwakiri et al., 2001). In a previous work we described the existence of temporal variations in the proliferation of C3H/S-mouse intestinal enterocytes at 7 and 14 days old (Barbeito et al., 2003) a much earlier developmental stage than those previously reported (Reyna et al., 1997; Barbeito et al., 2002). We also showed the existence of sex-differences in the temporal structure of mitotic index curves in these animals.

Spontaneous and radiation-induced cell death by apoptosis also displays clear circadian rhythms (Ijiri and Potten, 1986; Ijiri and Potten, 1988). We reported that colchicine-induced apoptosis in the adult male mice duodenal crypts showed daily variations (González et al., 2005). No data concerning possible differences for the circadian rhythmicity of apoptosis indices related to sex and gender in mice are available.

Due to the lack of works dealing specifically with colchicine-induced apoptosis in suckling mice, the following study was designed bearing in mind the above investigations. The aims were to determine the colchicine inductive effect on apoptosis in 14 day old mice intestinal crypts; to establish the existence of daily variations in the apoptotic indices and to examine whether there is sexual dimorphism in such variations.

MATERIALS AND METHODS

Animals and Housing

C3H/S 14 days old male and female mice were used. These suckling animals were the progeny of breeder females that had been housed under standard conditions for circadian-periodicity analysis in individual cages after conception. These conditions included: food and water available *ad libitum*, room temperature controlled at 22 ± 2 °C and light alternating in a 12 h light-dark cycle beginning at 06:00 h. Even illumination was provided by a 40W fluorescent light placed over the cages.

Experimental Design and Procedures

At 14 days of age, the mice were divided into control and experimental groups of males and females (n=4-8) for killing by decapitation and exsanguination at the following times of day: 00:00, 04:00, 08:00, 12:00, 16:00, and 20:00 h. Four h before sacrifice, each animal of the experimental lots was injected intraperitoneally with 1 µg of colchicine (Sygma Chemical Co., St. Louis, MO, USA) per g body weight. Samples of duodenum and colon were fixed in 10% buffered formalin and embedded in paraffin for subsequent sectioning at a thickness of 5 µm and staining with hematoxylin & eosin (Sygma Chemical Co., St. Louis, MO, USA).

A 10 % of the sections of the sample were nick end-labeled for DNA fragmentation using the TUNEL technique (Roche Diagnostics, Indianapolis, IN) according to manufacturer specifications.

Microscopic Observations

The preparations were keyed and examined by a single observer, under oil immersion objective (1, 500 fold magnification) in order to score the number of apoptosis in 20 crypts sections. The definition of an apoptotic cell in the H-E stained samples attended to morphological criteria: cell shrinkage, intense eosinophilic cytoplasm, and nucleus showing typically condensed chromatin (pyknosis), as described by Kerr et al. (1972), Colombo et al. (2000), Coopersmith et al. (2002) and Vyas et al. (2007). Several apoptotic bodies clustered were given a single count. Any doubtful cells were disregarded. A parallel TUNEL assay was employed to corroborate the reliability of the apoptosis morphological assessment.

Apoptotic Quantification

The scanty depth and other morphological characteristics of the immature crypts in suckling mice did not allow to divide them into three regions (tiers 1-4; 5-12 and 13-20, proceeding from bottom to top) (Barbeito et al., 2003) as in adult mice (Potten, 1991). Instead, the individual count was performed in longitudinally sectioned crypts showing a lumen that had no less than 14 cell tiers. The apoptotic index was expressed as the number of apoptotic cells per 1, 000 nuclei for each animal.

Statistical Analysis

The arithmetic mean ± standard error (SE) was calculated for each study time and gender. Differences between sexes were determined by the Student's t test. Within group differences were tested by one-way analysis of variance (ANOVA).

RESULTS

All results using H&E were in concert and confirmed as apoptotic cells by internucleosomal DNA fragmentation by TUNEL technique.

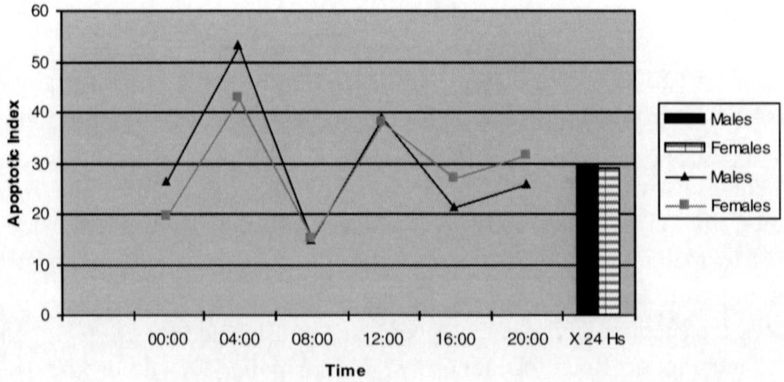

Figure 1. Apoptotic cells in duodenal crypts.

Duodenum

Males. The apoptotic indices varied significantly throughout the day (Table 1), exhibiting a peak at 04:00 h (53.06±6.45) and a trough at 08:00 h (14.5±4.75). The difference between these values was statistically significant ($p<0.001$).

Table 1. Apoptotic index every 1000 cells in duodenal crypts

	Males		Females	
Time	X±SE	n	X±SE	n
00.00	26.34±6.33	4	19.29±4.50	5
00.04	53.06±6.45	7	42.56±5.73	6
00.08	14.73±4.75	4	14.96±3.25	8
12.00	38.84±8.42	4	38.01±4.54	7
16.00	21.21±3.90	8	26.79±2.19	5
20.00	25.71±2.91	5	31.25±5.49	6
		32		37

X 24 hs 29.28 ± 5.63 X 24 hs 28.81 ± 4.34
p 04.00-08.00 < 0.001 p 04.00-08.00 < 0.001
p 12.00-00.00 < 0.001
p: statistical signification.

Females. In these animals a bimodal apoptosis daily curve was registered with peaks at 04:00 and 12:00 h (42.56±5.73; 38.01±4.54 respectively, $p < 0.001$) and minimal indices at 00:00 and 08:00 (19.29±4.50; 19.29±4.50 respectively; $p < 0.001$).

The average apoptotic indices were similar in both sexes (males: 29.28±5.63, females: 28.81±4.34) (Table 1).

Colon

Males. In this group the peak was detected at 04:00 h (18.30±2.23). This value was higher than all the other values registered for male colonic apoptosis (Table 2).

Females. In contrast to the findings in the other sector and organs examined, female mice showed no circadian variation in the apoptotic intensity in the large intestine (Table 2).

The comparison of the results obtained from males and females showed similar average apoptotic indices in this organ (males: 7.96±2.35; females: 8.71±3.38) (Table 2).

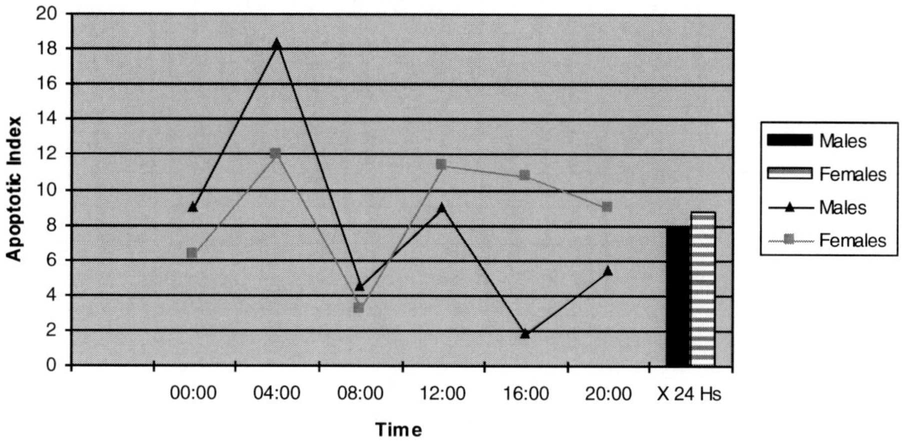

Figure 2. Apoptotic cells in colonic crypts.

Table 2. Apoptotic indices every 1000 cells in colonic crypts

	Males		Females	
Time	X±SE	n	X±SE	n
00.00	8.93±1.78	5	6.25±2.68	4
00.04	18.30±2.23	4	11.91±2.34	6
00.08	4.46±1.71	4	3.22±1.18	5
12.00	8.93±2.39	5	11.31±2.90	6
16.00	1.79±0.98	5	10.71±3.17	4
20.00	5.35±1.00	4	8.93±5.49	5
		27		30

X 24 hs 7.96 ± 2.35 X 24 hs 8.71 ± 3.38

p 04.00-16.00 < 0.001 p=NS

p: statistical signification; NS: not significant.

Discussion

It has been demonstrated that diverse cell types undergo apoptosis when they are treated with colchicine (Suzuki, 1998; Suzuki et al., 1998; Takeda et al., 2000; Cervinka et al., 2004; Kalvelyte et al., 2003; Jorda et al., 2005). In despite of *in vitro* studies´ abundance, *in vivo* investigations assessing the effect of this microtubule inhibitor on apoptosis are scarse. Our work demonstrated that colchicine induced apoptosis in the suckling mouse duodenal and colonic crypts enterocytes as it had been shown by Duncan and Heddle for adult mouse colonic crypts (1984) and in our laboratory for adult male mouse duodenal crypts (González et al., 2005).

The comparison of the apoptotic daily changes phasing in the small bowel between young males and females showed sexual dimorphism, a condition that has been reported in relation to mitotic activity in the same population (Reyna et al., 1997; Barbeito et al., 2003) or in others such as endocrine cells of the pars intermedia (Barbeito et al., 2000) in this mouse strain. The absence of daily variations in the young female large bowel cell death indices does not permit to consider a sexual dimorphism though it might be reflecting an uneven sensitivity to colchicine between genders.

In regard to the induced apoptosis intensity found for each intestinal sector, the levels achieved resulted highly prominent in the small bowel. This is in coincidence with the observations in adult mice of the same strain (Barbeito and González, unpublished data) in which the apoptotic indices values for the duodenal crypts had a many-times fold difference to rectal crypts. The dead cell index resulted higher in the small intestine; this agrees with the findings of Yamada et al. (2000) in guinea pig and 6 weeks old mouse after autumn crocus poisoning. Potten et al. (1992) reported an elevated apoptosis index for the small bowel of gamma-exposed mice and Gibson et al. (2005) found a 10-fold higher apoptosis index in large intestine than in jejunum. Differences between the induced apoptosis yield in the two intestinal sectors may be due to a differential gene expression related to cell death control like p53 and Bcl2 (Roberts et al., 1997; Gauthier et al., 2001).

It is known that apoptotic indices changes in accord to the time of day the cell insult is delivered, i.e., Ijiri and Potten (1990) found a circadian rhythm in the apoptotic cell yield indicating that the sensitivity and number of radiation-induced apoptosis varied along the day. Similar findings were reported by Duncan et al. (1983). Results from a previous study in adult male mice showed that the peak hour of day for inducing duodenal apoptosis was achieved when colchicine was injected at 04:00 (González et al., 2005). We also detected daily variations in the apoptotic indices for the duodenum from both sexes. However, the highest apoptotic index value was observed when animals received colchicine at midnight. The large bowel crypts only showed daily changes in the male individuals but in this case the maximal apoptotic index value was observed at 04:00, when the treatment was performed at 00:00 h. These results also differ from findings in adults which exhibited variations in both genders [these data have not been presented here]. The adult's male apoptotic highest values were obtained when the cells were challenged at 04:00 h, for the adult females the same result was registered when the colchicine administration was performed at 08:00 h. Thus, the sensitivity and number of colchicine-induced apoptosis changes throughout the day are age-dependant. A similar age-dependant condition has been observed in assessments concerning the proliferation activity in diverse organs including the bowel (Barbeito et al., 2000, Barbeito et al., 2003).

Advances in colchicotheraphy have shown fascinating new fields for research (Fabre, 2005). Colchicine is used in the treatment of gout, inflammatory dermatoses such as scleroderma, Sweet´s syndrome and autoinmune bullous disease (Sullivan et al., 1998). It is also the most effective treatment for familial Mediterranean fever to be taken regularly on a life-long basis (Medlej-Hashim et al., 2004) and recently its potential therapeutic value in the prevention of p25 formation for the treatment of neurodegenerative disorders has been assessed (Jorda et al., 2005). The colchicine doses employed in these diseases' treatments are lower than those administered in antineoplastic chemotherapy. In spite of this, patients with hepatic alterations or ingesting barbiturates tend to exhibit a predisposition to colchicine intoxication (Arroyo et al., 2004). The drug´s effect on the intestinal epithelium is the cause of the initial signs in colchicine intoxication. As colchicine poisoning results in a multi-organ dysfunction syndrome, particularly affecting tissues with rapid cell turnover (Brvar et al., 2004), among them the intestinal epithelium, the complex pattern of variables implied in the colchicine-induced intestinal apoptosis rhythms showed here (gender, age, time of drug delivery, enterotoxicity species-differences) highlights the need of treatment design with colchicine dosing according to the patient condition.

Acknowledgments

The authors wish to thank Mr. Ruben Mario for technical assistance. CGB is research career member of the National Council of Scientific and Technological Research (CONICET), Argentina. Conditions concerning animal management fully respected the policy and mandates of the guide for the care and use of laboratory research of the National University of La Plata and standards for quality biological rhythm research as recommended by Touitou et al. (2004).

References

[1] Danial, NN; Korsmeyer SJ. Cell Death: Critical Control Points. *Cell,* 2004 116, 205–219.

[2] Fan, Y; Bergmann, A. Apoptosis-induced compensatory proliferation. The Cell is dead. Long live the Cell! *Trends in Cell Biology*, 2008 18, 467-473.

[3] Kasibhatla, S; Gourdeau, H; Meerovitch, K; Drewe, J; Reddy, S; Qiu, L; Zhang, H; Bergeron, F; Bouffard, D; Yang, Q; Herich, J; Lamothe, S; Cai, SX; Tseng, B. Discovery and mechanism of action of a novel series of apoptosis inducers with potential vascular targeting activity. *Mol. Cancer Ther.*, 2004 3, 1365-1374.

[4] Inagaki-Ohara, K; Takamura, N; Yada, S; Alnadjim, Z; Liu, E; Yu, X; Yoshida, H; Lin, T. Radiation-induced crypt intestinal epithelial cell apoptosis in vivo involves both caspase-3-dependent and -independent pathways. *Dig. Dis. Sci.*, 2002 47, 2823-28130.

[5] Wu, B; Fujise, T; Iwakiri, R, Ootani, A, Amemori, S; Tsunada, S; Toda, S; Fujimoto, K. Venous congestion induces mucosal apoptosis via tumor necrosis factor-alpha-mediated cell death in the rat small intestine. *J. Gastroenterol.*, 2004 39, 1120-1122.

[6] Sanders, LM; Henderson, CE; Hong, MY; Barhoumi, R, Burghardt, RC; Wang, N; Spinka, CM; Carroll, RJ; Turner, ND; Chapkin, RS; Lupton, JR. An increase in reactive oxygen species by dietary fish oil coupled with the attenuation of antioxidant defenses by dietary pectin enhances rat colonocyte apoptosis. *J. Nutrition.*, 2004 134, 3233-3238.

[7] Ozen, S; Akisu, M; Baka, M; Yalaz, M; Sozmen, EY; Berdeli, A; Kultursay, N. Insulin-like growth factor attenuates apoptosis and mucosal damage in hypoxia/reoxygenation-induced intestinal injury. *Biol. Neonate.*, 2005 87, 97-98.

[8] Duncan, AM; Heddle, JA. The frequency and distribution of apoptosis induced by three non-carcinogenic agents in mouse colonic crypts. *Cancer Lett.* 1984; 23:307-311.

[9] González, NV; Badrán, AF; Barbeito, CG. Daily variation in colchicine-induced apoptosis in duodenal crypts. *Chronobiol. Int.*, 2005 22, 79-88.

[10] Inoué, S. Cell division and the mitotic spindle. *J. Cell Biol.,* 1981 91, 131-147.

[11] Salmon, ED; McKeel, M; Hays, T. Rapid rate of tubulin dissociation from microtubules in the mitotic spindle in vivo measured by blocking polymerization with colchicine. *J. Cell. Biol.,* 1984 99, 1066-1075.

[12] Abolmaali, K; Balakrishnan, A; Stearns, AT ; Rounds, J; Rhoads, DB; Ashley, SW; Tavakkolizadeh, A. Circadian variation in intestinal dihydropyrimidine dehydrogenase (DPD) expression: A potential mechanism for benefits of 5FU chrono-chemotherapy. *Surgery*, 2009 146, 269-273.

[13] Stearns, AT; Balakrishnan, A; Rhoads, DB, . Ashley, SW; Tavakkolizadeh, A. Diurnal Rhythmicity in the Transcription of Jejunal Drug Transporters. *J. Pharmacol. Sci.*, 2008 108, 144-148.

[14] Bjarnason, GA; Jordan, R. Rhythms in human gastrointestinal mucosa and skin. *Chronobiol. Int.,* 2002 19, 129-140.

[15] Barbeito, CG; Albarenque, SM; Reyna, JC; Flamini, MA; Andrés Laube, PF; Badrán, AF. Mitotic activity of the duodenal crypts enterocytes in mice transplanted with EA21a mammary carcinoma. *Cell Biol. Int.*, 2002 26, 123-125.

[16] Iwakiri, R; Gotoh, Y; Noda, T; Sugihara, H; Fujimoto, K; Fuseler, J; Aw, TY. Programmed cell death in rat intestine: effect of feeding and fasting. *Scand. J. Gastroenterol.,* 2001 36, 39-47.

[17] Barbeito, CG; González, NV; Badrán, AF. Sex and age-related temporal variations in intestinal-epithelium proliferation in the suckling mouse. *Chronobiol. Int.* 2003; 20:37-47.

[18] Reyna, JC; Barbeito, CG; Moreno, FR; Badrán, AF. Mitotic activity of duodenal crypt enterocytes in mice bearing a transplanted hepatocarcinoma. *Medicina (Buenos Aires)*, 1997 57, 708-712.

[19] Barbeito, CG; Surur, JM; Badrán AF. Mitotic activity of the pars intermedia in the female mouse: age-associated variations in proliferation rate and circadian periodicity. *Chronobiol. Int.*, 2000 17, 751-756.

[20] Ijiri, K; Potten, CS. Radiation-hypersensitivity cells in the small intestinal crypts: their relationships to clonogenic cells. *Brit. J. Cancer.,* 1986 53, 20-22.

[21] Ijiri, K; Potten, CS. Circadian rhythms in the incidence of apoptotic cells and number of clonogenic cells in the intestinal crypts after radiation using normal and reversed light conditions. *Int. J. Radiat. Biol.*, 1988 53, 717-727.

[22] Kerr, JA; Wyllie, A; Currie, A. Apoptosis: A basic biological phenomenon with wide-ranging implications in tissue kinetics. *Br. J. Cancer.*, 1972 26, 239-257.

[23] Colombo, LL; Mazzoni, EO; Meiss, RP. The time of tumor cell division and death depends on the site of growth. *Oncol. Rep.*, 2000 7, 1363-1366.

[24] Coopersmith, CM; Stromberg, PE; Dunne, WM; Davis, DG; Amiot DM; Buchman, TG; Karl, IE; Hotchkiss, RS. Inhibition of Intestinal Epithelial Apoptosis and Survival in a Murine Model of Pneumonia-Induced Sepsis. *JAMA,* 2002 287, 1716-1721.

[25] Vyas, D; Robertson, CM; Stromberg, PE; Martin, JR; Dunne, DW; Houchen, CH; Barrett, TA, Ayala, A; Mario Perl, M; Buchman, TG; Coopersmith CM. Epithelial apoptosis in mechanistically distinct methods of injury in the murine small intestine. *Histol Histopathol.*, 2007 22, 623-630.

[26] Potten, CS. Regeneration in epithelial proliferative units as exemplified by small intestinal crypts. Regeneration of vertebrate sensory reception cells. *Ciba Foundation Symposium,* 1991 160, 54-76.

[27] Suzuki, Y. Cell death, phagocytosis, and neurogenesis in mouse olfactory epithelium and vomeronasal organ after colchicine treatment. *Ann. N. Y. Acad. Sci.,* 1998 855, 252-254.

[28] Suzuki, Y; Takeda, M; Obara, N; Suzuki, N. Colchicine-induced cell death and proliferation in the olfactory ephithelium and vomeronasal organ of the mouse. *Anat. Embryol. (Berl),* 1998 198, 43-51.

[29] Takeda, M; Suzuki, Y; Obara, N; Nagai, Y. Induction of apoptosis by colchicine in taste bud and epithelial cells of the mouse circumvallate papillae. *Cell Tissue Res.,* 2000 302, 391-395.

[30] Cervinka, M; Cerman, J; Rudolph, E. Apoptosis in Hep2 cells treated with etoposide and colchicine. *Cancer Detect. Preven.,* 2004 28, 214-226.

[31] Kalvelyte, A; Imbrasaite, A; Bukauskiene, A; Verselis, VK; Bukauskas, FF. Connexins and apoptotic transformation. *Biochem. Pharmacol.*, 2003 66, 1661-1672.

[32] Jorda, EG; Verdaguer, E; Jimenez, A; Arriba, SG; Allgaier, C; Pallas, M; Camins, A. Evaluation of the neuronal apoptotic pathways involved in cytoskeletal disruption-induced apoptosis. *Biochem. Pharmacol.*, 2005 70, 470-480.

[33] Yamada, M; Kobayashi, Y; Furuoka, H; Matsui, T. Comparison of enterotoxicity between autumn crocus (*Colchicum autumnale* L.) and colchicine in the guinea pig and mouse: enterotoxicity in the guinea pig differs from that in the mouse. *J. Vet. Med. Sci.* 2000; 62:809-813.

[34] Potten, CS. The significance of spontaneous and induced apoptosis in the gastrointestinal tract of mice. *Cancer Metast. Rev.*, 1992 11, 179-195.

[35] Gibson, RJ; Bowen, JM; Cummins, AG; Keefe, DMK. Relationship between dose of methotrexate, apoptosis, p53/p21 expression and intestinal crypt proliferation in the rat. *Clin. Exp. Med.,* 2005 4, 188-195.

[36] Roberts, RA; Nebert, DW; Hickman, JA; Richburg, JH; Goldsworthy, TL. Perturbation of the mitosis/apoptosis balance: A Fundamental Mechanism in Toxicology. *Fund. Appl. Tox.*, 1997 38, 107-115.

[37] Gauthier, R; Harnois, C; Drolet, JF; Reed, JC; Vézina, A; Vachon, PH. Human intestinal epithelial cell survival: differentiation state-specific control mechanisms. *Am. J Physiol. Cell Physiol.,* 2001 280, C1540-C1554.

[38] Ijiri, K; Potten, CS. The circadian rhythm for the number and sensitivity of radiation-induced apoptosis in the crypts of mouse small intestine. *Int. J. Radiat. Biol.*, 1990 58, 165-75.

[39] Duncan, AM; Ronen, A; Blakey, DH. Diurnal variation in the response of gamma-ray-induced apoptosis in the mouse intestinal epithelium. *Cancer Lett.,* 1983 21, 163-166.

[40] Fabre, AJ. The autumn crocus: two millenniums of actuality. *Hist. Sci. Med.,* 2005 39, 143-154.

[41] Sullivan, TP; King, LE Jr; Boyd, AS. Colchicine in dermatology. *J. Am. Acad. Dermatol.,* 1998 39, 93-99.

[42] Medlej-Hashim, M; Loiselet, J; Lefranc, G; Megarbane, A. Familial Mediterranean Fever (FMF): from diagnosis to treatment. *Sante.*, 2004 14, 261-266.

[43] Arroyo, MP; Sanders, HY; Schwartz, D; Kamino, H; Strober, BE. Toxic epidermal necrolysis-like reaction secondary to colchicine overdose. *Br. J. Derm.,* 2004 150, 581-588.

[44] Brvar, M; Kozelj, G; Mozina, M; Bunc, M. Acute poisoning with autumn crocus (*Colchicum autumnale* L.). *Wien Klin Wochenschr.*, 2004 116, 205-208.

[45] Touitou, Y; Portaluppi, F; Smolensky, MH; Rensing, L. Ethical principles and standards for the conduct of human and animal biological rhythm research. *Chronobiol. Int.,* 2004 21, 161-170.

In: Vertebrate and Invertebrate Animal Approaches
Editor: José Rosa Gomes
ISBN: 978-1-63117-102-4

Chapter 4

THYROID DEREGULATION CAUSES CELLULAR DAMAGE IN PANCREATIC ISLETS AND SPLEENS IN RATS

***Srikanta Guria, Nisha Balmiki, Sudipta Chakrabarti and Madhusudan Das*[*]**

Department of Zoology, University of Calcutta, Kolkata, India

ABSTRACT

Thyroid hormones regulate a number of physiological functions in the body (Fujita, 1988). We hypothesized thyroid hormone may have an important role in regulating the function and cytomorphology of pancreatic islets and spleen in rat. We treated experimental groups of rat with methimazole 20 mg/kg and thyroxine 600 μg/kg body weight/day for 14 days intraperitoneally to induce hypo and hyperthyroid like conditions respectively. Serum T_3, T_4 hormones were measured by radioimmunoassay (RIA). Serum Glutamic pyruvate Transaminase (SGPT) or Alanine Amino Transferase (ALT) measured by Spectrophotometric method. Thyroxine induced hyperthyroid rat exhibited elevation of serum SGPT level. Hyperthyroid rat exhibited the changes in the architecture of pancreatic islets. Consequently hypothyroidism induced death of the splenocytes.

Keywords: Hypothyroid, hyperthyroid, methimazole, thyroxine

INTRODUCTION

The thyroid gland is an important endocrine gland in the body. The gland secretes thyroid hormones which control a number of homeostatic functions in the body (Fujita, 1988). The current investigation revealed that dysfunction of many other organs may be due to thyroid malfunction (Dimitriadis et al., 1991; Hoppner et al., 1989). But the mechanism of thyroidal effects on other organs in the body is still imprecise. Earlier research revealed the possible mode of regulatory function of thyroid hormone in pancreas (Lenzen et al., 1978) as well as

[*] Corresponding author: Phone: 91-9831281756, Email address: madhuzoo@yahoo.com.

the development of cellular subsets in spleen (Angelin et al., 2005; Watanabe et al., 1995). We hypothesized thyroid hormone may have an important role in maintaining the cytomorphology of pancreas that affect insulin secretion as well as glucose homeostasis and spleen which regulate immune function as well as erythropoiesis in rat. Therefore, we induced hypothyroidism with the treatment of Methimazole and hyperthyroidism with thyroxine in rats to understand this complex interaction. We measured the serum T_3 and T_4 hormones level to confirm hypo and hyperthyroid like conditions. The recent research revealed that changes in serum glutamic pyruvic transaminase (SGPT) is associated with pancreatic function. In the present investigation serum SGPT level was examined in both control and experimental animals. We also investigated the histological changes of pancreatic islets and spleen.

MATERIALS AND METHODS

The laboratory experiments were performed using rat (*Rattus rattus*). Normal young adult rat aged 8-10 weeks and weight 100-110 g were housed in polypropylene cages and were acclimatized in laboratory condition for a week with natural light and dark schedules prior to experimentation. The animals were fed standard rodent diet and water was provided *ad libitum*. They were divided into hypothyroid, hyperthyroid and their respective control groups. Group I animals were treated with Methimazole (Sigma, USA) 20 mg/kg body weight/day for 14 days. Group II animals were treated with Thyroxine (Glaxo, India) 600 μg/kg body weight/day for 14 days. Before autopsy, Rats were anesthetized with chloroform. Blood was drawn directly from the heart for serum T_3, T_4 hormones and SGPT level. Pancreas and spleen were dissected out and fixed in Bouin's fixative (Parakkal, 1961). Tissues were embedded in paraffin, sectioned (5 μm), mounted on glass slide and stained with hematoxylin-eosin for histological analysis. Pancreatic cell destruction was determined by nuclear DNA fragmentation by using DAPI. T_3 or T_4 concentrations in rat serum were determined by radioimmunoassay (RIA). Serum SGPT or alanine amino transferase (ALT) was measured colorimetrically using the method of Reitman (Reitman, 1957).

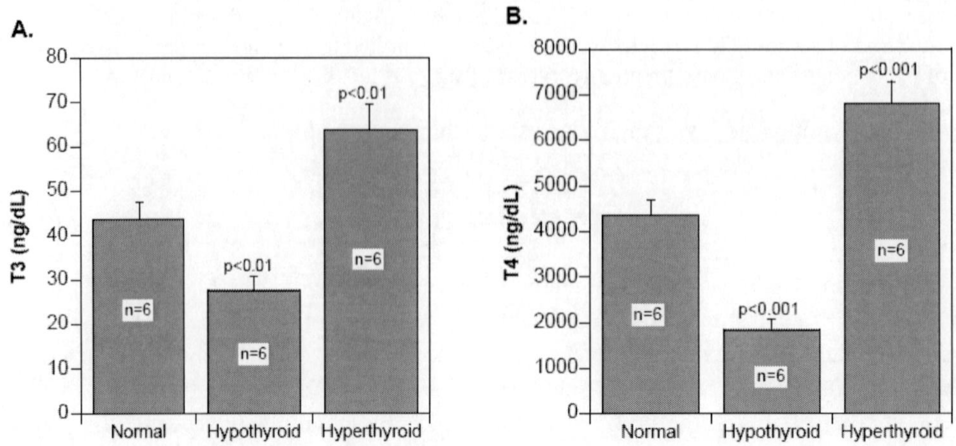

Figure 1. Plasma concentrations (ng/dL) of T3 (A) and T4 (B) in hypothyroid and hyperthyroid rats. Values are expressed as Mean ± SEM.

Results

Serum T_3 and T_4 Level

The results showed significant changes in T_3, T_4 hormones (Figure 1A & B). The serum T_3 and T_4 levels were significantly reduced in hypothyroid rat whereas in contrast the levels significantly elevated in hyperthyroid rat as expected and confirming that they were indeed in a hypothyroid and hyperthyroid state.

Serum SGPT Level

Hyperthyroid rats showed dramatic increments in SGPT level (by ~9.8-fold) as compared to a modest (by ~0.5 fold) increase in hypothyroid rats (Figure 2).

Histology of Pancreas

Hyperthyroid rats exhibited marked changes in the general cytomorphology of islets of Langerhans as evidenced by the enlargement and disruption of islets and disorientation of the cells. A significant number of islets cells were found to be picnotic. The islets cells lost their individual boundary. The number of acinar and islets cells were reduced in hyperthyroid rat (Figure 3A & B). No discernible change in pancreas histology was detected in hypothyroid rats (picture not shown).

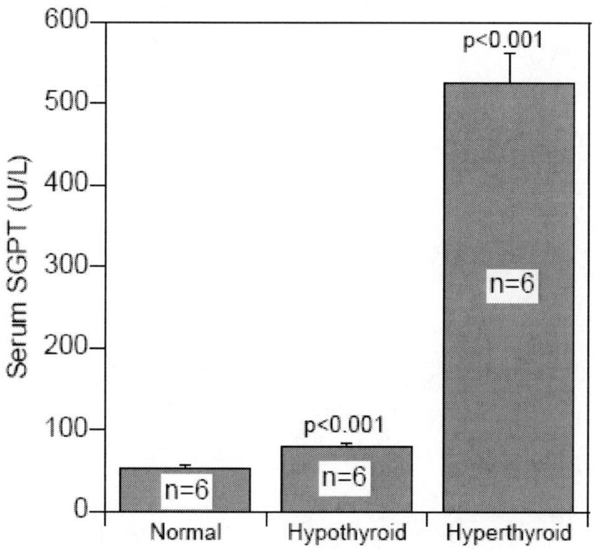

Figure 2. Serum glutamic pyruvic transaminase (SGPT) level (U/L) in normal, hypothyroid and hyperthyroid rats. Values are expressed as mean ± SE from 6 rats. *P* values: *<0.001; **<0.01.

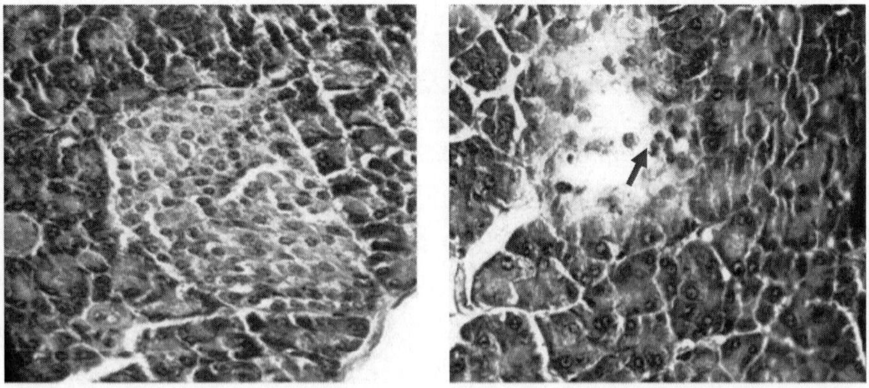

3(A) 3(B)

Figure 3 (A, B). (A) Normal rat pancreatic islets (x 400), (B) Hyperthyroid rat pancreatic islets (x 400) as indicated by the arrow.

DAPI Staining of Pancreas

Hyperthyroid pancreas showed different stages of nuclear fragmentation (Figure 4A & B). DAPI stained all the nuclei blue.

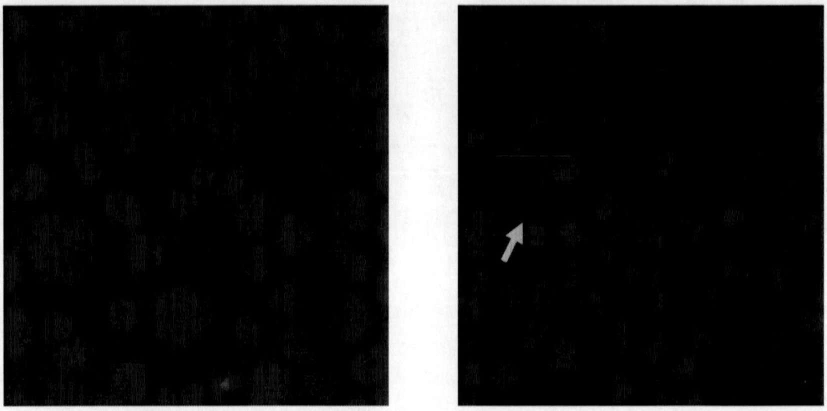

4(A) 4(B)

Figure 4 (A, B). Fluorescent microscopic analysis of (A) normal pancreatic cells nuclei and (B) hyperthyroid pancreatic cells nuclei stained with DAPI.

Histology of Spleen

Significant changes were observed in the histology of spleen in the hypothyroid group when compared with the control group under light microscope (Figure 5A & B). Hypothyroid rat exhibited cellular damage in the spleen. A significant percentage of splenocytes become picnotic and reduced in number. In the hypothyroid spleen, the number and size of macrophages were increased. No discernible change in spleen histology was detected in hyperthyroid rats (picture not shown).

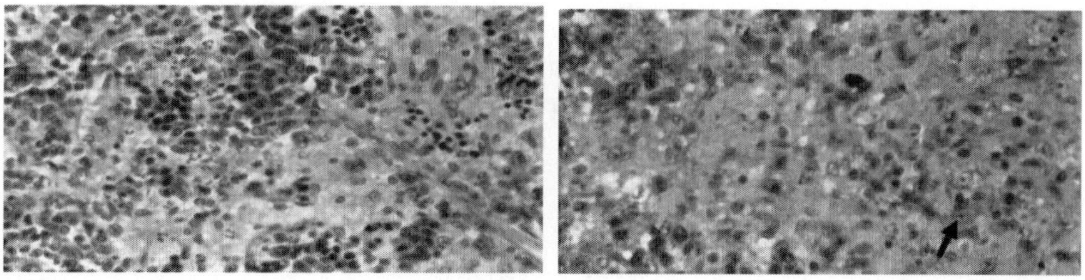

5(A) 5(B)

Figure 5 (A, B). (A) Normal rat spleen (x 400), (B) Hypothyroid rat spleen (x 400) as indicated by the arrow.

DISCUSSION

In the present investigation, we used methimazole to induce hypothyroidism and thyroxine to induce hyperthyroidism, as reflected in the serum T_3, T_4 levels (Pantos et al., 2005; Grofte et al, 1997). We found changes in the histology of pancreatic islets after induction of hyperthyroidism. The size, shape and morphology of cells were significantly changed. The histology of hyperthyroid pancreas exhibited the enlargement and disruption of islets and disorientation of cells. Significant number of cells were found to be picnotic. The number of acinar cells and islets cells were reduced in hyperthyroid rat. The reason for the reduction of the islets cells number of the pancreas after thyroxine treatment is apparently an increased rate of islets cell apoptosis (Jorns et al., 2002). In parallel, thyroxine treatment increases the rate of apoptosis in rat pancreatic ductal cells which are considered to contribute to the pool of stem cells from which insulinproducing beta cells originate (Jorns et al., 2002). An increased rate of beta cell death due to apoptosis causes a decrease of insulin content and glucose-induced insulin secretion from the pancreas in hyperthyroidism. The DAPI stained hyperthyroid pancreas showed different stages of nuclear fragmentation indicating apoptosis. The resulting reduction of islets cells in the pancreas can provide an explanation for the decrease of glucose tolerance in hyperthyroidism. (Jorns et al., 2002). The condition may affect glucose homeostasis (Chakrabarti et al., 2007). This altered condition may affect both insulin secretion and degradation in humans (Dimitriadis et al., 1985; Lenzen et al., 1976).

Serum glutamic pyruvate transaminase (SGPT) or alanine amino transferase (ALT) is found mainly in the liver, kidneys, heart, muscles, and pancreas. However, when these tissues are damaged or diseased, it releases ALT into the bloodstream, causing levels of the enzyme to rise (Ni et al., 2008). Serum SGPT level was measured to test whether it changes with the changes of pancreatic tissues. In our experiment thyroxine induced hyperthyroid rat exhibited dramatic elevation of serum SGPT level. Profound alteration in pancreas histology in hyperthyroid rats is a reflection of changes in SGPT level in blood. Increased SGPT level also indicates acute pancreatitis (Ni et al., 2008).

We studied the histological changes of spleen upon experimentally induced hypo and hyperthyroid conditions in rat. Results showed hypothyroidism altered the general morphology of spleen cells. Methimazole induced hypothyroidism showed cellular damage in the spleen and the resulting reduction of cells is due to the apoptosis of splenocytes (Edgar et al., 2009). In the hypothyroid spleen, the number and size of macrophages were increased.

The increase in the number and size of macrophages observed in the spleen of hypothyroid rat may reveal hyperactivation of reticuloendothelial cells (Sokkar et al., 2000). Hypothyroidism may affect erythropoiesis in spleen environment in the rat (Cristina et al., 2005).

The result indicates thyroid hormone may play an important regulatory function in mammalian spleen.

REFERENCES

Bonner-Weir S. Life and death of the pancreatic beta cells. *Trends Endocrinol Metab.*, 2000. 11, 375–378.

Cristina Angelin-Duclos; Chantal Domenget; Andrea Kolbus; Hartmut Beug; Pierre Jurdic and Jacques Samaru. Thyroid hormone T_3 acting through the thyroid hormone receptor is necessary for implementation of erythropoiesis in the neonatal spleen environment in the mouse. *Development.*, 2005. 132, 925-934.

Chakrabarti, S; Guria, S; Samanta, I.; Das, M. Thyroid dysfunction modulates glucoregulatory mechanism in rat. *Indian J Exp Biol.*, 2007. 45(6), 549-553.

Cortizo, AM; Chazenbalk, GD; de Gagliardino, EE; Garcia, ME; Pisarev, MA; Gagliardino, JJ; Thyroid hormone binding and deiodination by pancreatic islets: relationship with the in vitro effect upon insulin secretion. *Acta Endocrinol.*, 1987. 116, 66–72.

Dimitriadis, G; Baker, B; Marsh, H; Mandarino, L; Rizza, R; Bergman, R; Haymond, M; Gerich, J. Effect of thyroid hormone excess on action, secretion, and metabolism of insulin in humans. *Am J Physiol.*, 1985. 248, E593-E601.

Edgar Cano-Europa; Vanessa Blas-Valdivia; Margarita Franco-Colin; CarlosAngel Gallardo-Casas and Rocio Ortiz. Methimazole-induced hypothyroidism causes cellular damage in the spleen, heart, liver, lung and kidney. *Acta Histochem,* 2009.

Fujita H. Functional morphology of thyroid. *Int Rev Cytol.*, 1988. 113, 145-150.

Grofte, T; Wolthers, T; Jensen, DS; Moller, N; Jorgensen, JO; Orskov, H; Vilstrup, H. Hepatic amino nitrogen conversion and organ N-contents in hypothyroidism with thyroxine replacement, and in hyperthyroid rats. *J Hepatol.*, 1997. 26, 409-416.

Hoppner, W. and Seitz, HJ. Effect of thyroid hormones on glucokinase gene transcription in rat liver. *J Biol Chem.,* 1989. 264(34), 20643-20647.

Jörns, A; Tiedge, M;. Lenzen, S. Thyroxine induces pancreatic beta-cell apoptosis in rats. *Diabetologia.*, 2002. 45, 851–855.

Lenzen, S; Joost, HG; Hasselblatt, A. Thyroid function and insulin secretion from the perfused pancreas in the rat. *Endocrinology.*, 1976. 99, 125–129.

Lenzen, S; Klöppel, G. Insulin secretion and the morphological and metabolic characteristics of pancreatic islets of hyperthyroid ob/ob mice. *Endocrinology.*, 1978. 103, 1546– 1555.

Lenzen, S; Joost, HG; Hasselblatt, A. The inhibition of insulin secretion from the perfused rat pancreas after thyroxine treatment. *Diabetologia.*, 2000. 12, 495–500.

Mostaghni, K; Badiei, K; Khodakaram-Tafti, A ; Maafi, A. B. Pathological and biochemical studies of experimental hypothyroidism in sheep. *Vet. Arhiv.*, 2008. 78, 209-216.

Ni, Y; Wu, JS; Fang, PP; Wu, XL; Sun, XC; Jia, GB; Zhang, RZ. Mechanism of liver injury in severe acute pancreatitis rats and role of melatonin. *Zhonghua Yi Xue Za Zhi.*, 2008. 88(40), 2867-71.

Pantos, C; Mourouzis, I; Malliopoulou, V; Paizis, I; Tzeis, S; Moraitis, P; Sfakianoudis, K; Varonos, DD; Cokkinos, DV. Dronedarone administration prevents body weight gain and increases tolerance of the heart to ischemic stress: a possible involvement of thyroid hormone receptor alpha1. *Thyroid.*, 2005. 15, 16-23.

Parakkal, PF. Mordanting fixation as a means of facilitating the staining of pancreatic cells of mouse. *Stain Technol.*, 1961. 36, 33-34.

Reitman, S and Frankel, S. A colorimetric method for the determination of serum glutamic oxaloacetic and glutamic pyruvic transaminases. *Am J Clin Patho l.*, 1957. 28, 56.

Sokkar, M; Soror, A. H; Ahmed Y. F; Ezzo, O. H; Hamouda, M. A. Pathology and biochemical studies on experimental hypothyroidism in growing lambs. *J. Vet. Med. B.*, 2000. 639-652.

Watanabe, K; Iwatani, Y; Hidaka, Y; Watanabe, M. Long-term effects of thyroid hormone on lymphocyte subsets in spleens and thymuses of mice. *Endocr J.*, 1995. 42(5), 661-8.

In: Vertebrate and Invertebrate Animal Approaches
Editor: José Rosa Gomes
ISBN: 978-1-63117-102-4

Chapter 5

HYPOTHYROIDISM CAUSES CELLULAR DAMAGE IN RETINAL TISSUE AND BICEPS IN RATS

Manalee Guha[2], Preeta Sen[2], Sudakshina Ghosh[2], Soumee Ghosh[2], Anisha Chaudhuri[2], Pritha Gupta[2], Somtirtha Das[2], Sweta Mukherjee[2], Reni Ghosh[2], Rini Roychowdhury[2], Srikanta Guria[1] and Madhusudan Das[2*#]

[1]Post Graduate Department of Zoology, Parasitology and Medical Entomology Laboratory, Darjeeling Govt. College, Darjeeling, West Bengal, India

[2]Department of Zoology, University of Calcutta, Kolkata, West Bengal, India

ABSTRACT

Thyroid hormones regulate a number of physiological functions in the body (Fujita, 1988). We hypothesized thyroid hormone may have an important role in regulating the function and cytomorphology of retina of eye and bicep muscle in rat. We treated experimental group of rat with methimazole intraperitoneally to induce hypothyroid like conditions. Methimazole induced hypothyroid rat exhibited drastic changes of retinal tissue. Hypothyroid rat exhibited the changes in the architecture of rat biceps. Therefore, hypothyroidism may be an important cause for visual problem as well as thyroid myopathy.

Keywords: Hypothyroid, methimazole, thyroxine, biceps, retina

INTRODUCTION

Thyroid hormones control a number of homeostatic functions in the body (Fujita, 1988). The current investigation revealed that dysfunction of many other organs may be due to

* Corresponding author: Phone: 91-9831281756. Email: madhuzoo@yahoo.com.
# All are equal contributors.

thyroid malfunction (Dimitriadis et al., 1991; Hoppner et al., 1989). But the mechanism of thyroidal effects on other organs in the body is still imprecise. We hypothesized thyroid hormone may have an important role in maintaining the cytomorphology of retina of eye and bicep muscle. Although it has been recognized for many years that symptoms and signs indicating disordered function of the voluntary musculature may be prominent in patients with myxoedema, the muscular manifestations of hypothyroidism have received comparatively less attention. Kocher (1892) reported that patients with sporadic cretinism or myxoedema often showed bulky musculature, diminished muscular power, and sluggish movement with or without "tetany". Hoffmann (1897) later described features suggestive of myotonia in a patient who had undergone thyroidectomy several times; he pointed out that in this case there was a striking delay in muscular relaxation, reminiscent of that which occurs in myotonia congenita (Thomsen's disease). In fact, however, the abnormality of muscle function which results from hypothyroidism must be distinguished from true myotonia, as the overall slowness of muscular activity affects muscular contraction as well as relaxation and the typical electrical after-discharge of myotonia is not seen. Hence the term "pseudomyotonia" is a more appropriate description for the myxoedematous phenomenon. Thyroxine also act on eye tissue through receptor in orbital fibroblasts and in normal orbital-adipose cells (Heufelder et al., 1993). Therefore, we induced hypothyroidism with the treatment of methimazole in rats to understand this complex interaction. We also investigated the histological changes of retinal tissue and biceps muscle.

MATERIALS AND METHODS

The laboratory experiments were performed using rat (*Rattus rattus*). Normal young adult rat aged 8-10 weeks and weight 100-110 g were housed in polypropylene cages and were acclimatized in laboratory condition for a week with natural light and dark schedules prior to experimentation. The animals were fed standard rodent diet and water was provided *ad libitum*. They were divided into hypothyroid and their respective control groups. Group I animals were treated with Methimazole (Sigma, USA) 20 mg/kg body weight/day for 14 days. Before autopsy, rats were anesthetized with chloroform. Blood was drawn directly from the heart for serum T_3, T_4 hormones. Eye tissues and biceps were dissected out and fixed in Bouin's fixative (Parakkal, 1961). Tissues were embedded in paraffin, sectioned (5 μm), mounted on glass slide and stained with hematoxylin-eosin for histological analysis. Retinal tissues were also stained with Masson's Trichrome for collagen study. T3 or T4 concentrations in rat serum were determined by radioimmunoassay (RIA).

RESULTS

Serum T_3 and T_4 Level

The results showed significant changes in T_3, T_4 hormones (Figure 1A and B). The serum T_3 and T_4 levels were significantly reduced in hypothyroid rat confirming that they were indeed in a hypothyroid state.

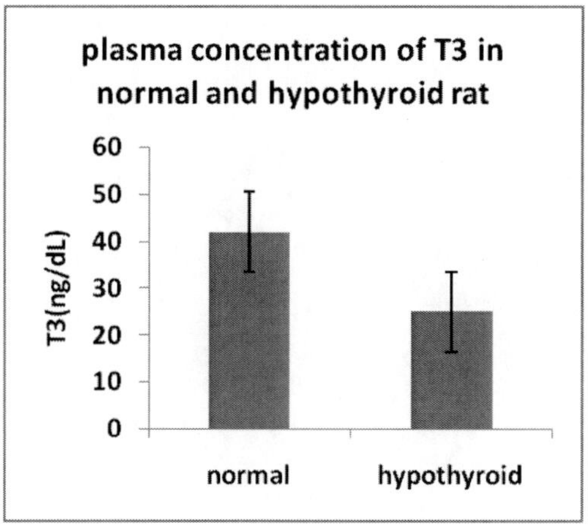

1(A)

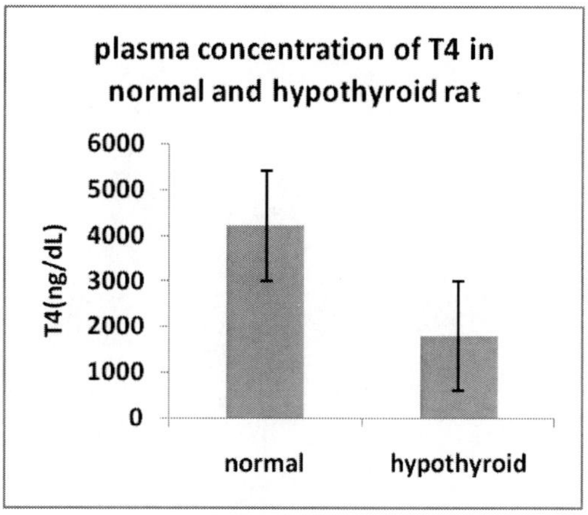

1(B)

Figure 1. Plasma concentrations (ng/dL) of T3 (A) and T4 (B) in hypothyroid and control rats. Values are expressed as Mean ± SEM.

HISTOLOGY OF RETINA

Hypothyroid rats exhibited marked changes in the general cytomorphology of retina. A significant number of cells were found to be reduced in outer and inner nuclear layer in hypothyroid group. The outer and inner nuclear layer lost their individual boundary (Figure 2A and B). The amount of collagen fibres were reduced in retinal tissue in hypothyroid rat (Figure 3A and B).

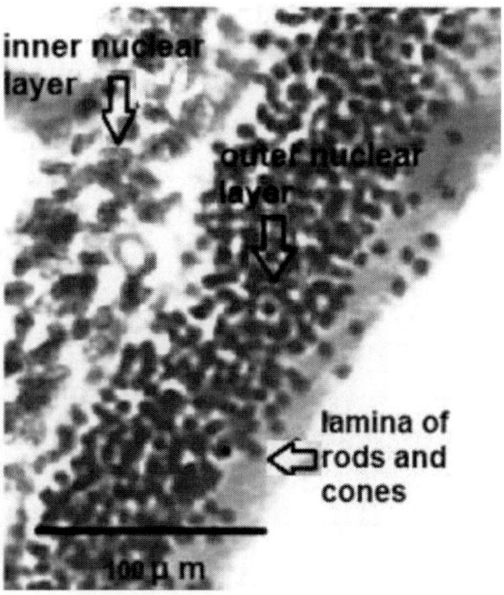

2(A)

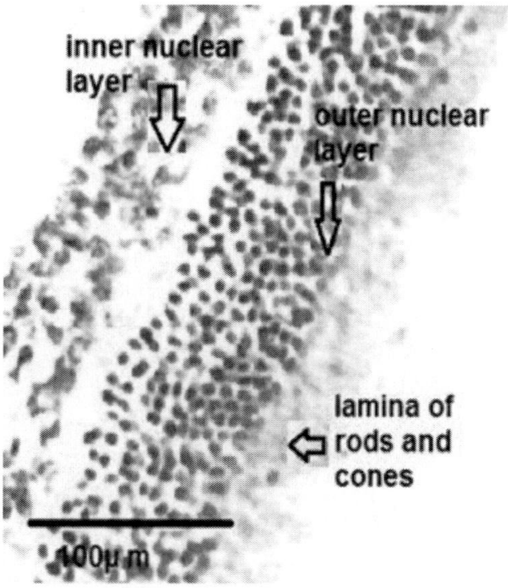

2(B)

Figure 2. (A) Normal rat retina(x 400) (B) Hypothyroid rat retina(x 400).

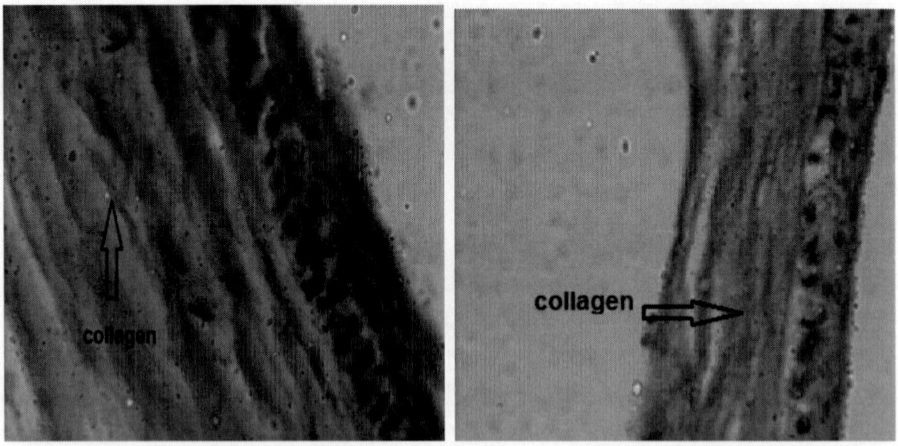

3(A) 3(B)

Figure 3. (A) Collagen in normal rat retina(x 400) (B) Collagen in hypothyroid rat retina(x 400).

HISTOLOGY OF BICEPS

Important changes were observed in the histology of biceps in the hypothyroid group when compared with the control group under light microscope (Figure 4A and B). Hypothyroid rat exhibited degenerated muscle fibres. Normal biceps muscle showed relatively uniform fiber diameter, peripherally located nuclei, with no fiber degeneration (4A). By contrast, the hypothyroid sections showed variation in fiber diameter, distinctive perifasicular fiber atrophy (4B). The existence of grouped atrophy and pyknotic nuclear clump in muscular fibres suggest muscular degeneration (4C).

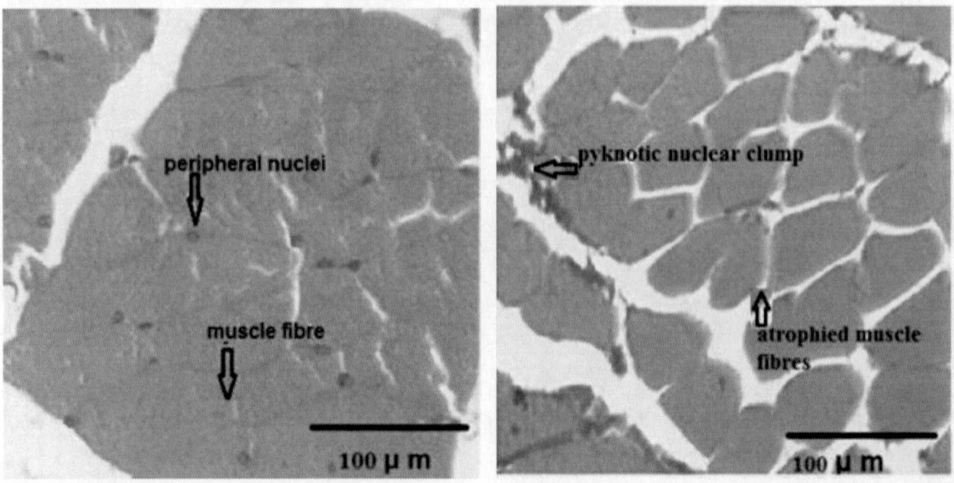

4(A) 4(B)

Figure 4. (A) Control rat bicep(x 400) 4 (B) hypo rat bicep(x 400) (variation in fiber diameter).

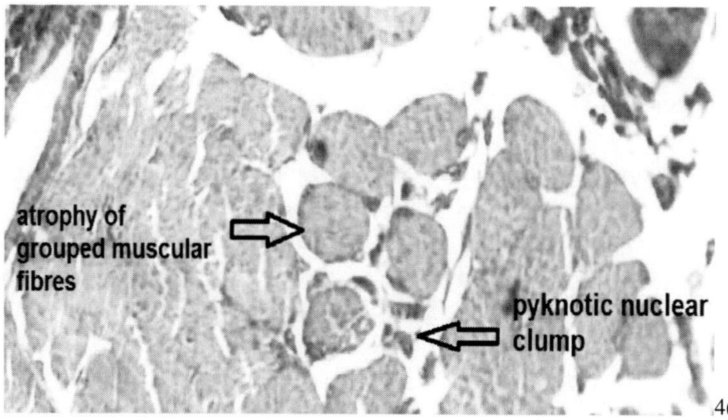

Figure 4. (C) Hypothyroid rat biceps(x 400) (existence of grouped atrophy and pyknotic nuclear clump in muscular fibres).

DISCUSSION

In the present investigation, we used methimazole to induce hypothyroidism as reflected in the serum T3, T4 levels (Pantos et al., 2005; Grofte et al, 1997). It has been previously established that the clinical symptoms of thyroid eye disease (TED) is associated with hyperthyroidism (Prummel et al., 1990). In the case of thyroid eye disease, the damage is directed to fatty tissue behind the eye. Swelling of the damaged tissues behind the eyes can cause the eyes to become red.

It may also cause the eyes to be pushed forward ('starey eyes', 'proptosis'). Occasionally, the swelling behind the eyes is bad enough to press on the nerve from the eyes to the brain affecting vision (Kendler et al., 1993). Graves' disease is the only kind of hyperthyroidism that has inflammation of the eyes, swelling of the tissues around the eyes, and bulging of the eyes (called Graves' ophthalmopathy). It is almost always associated with Graves' disease (GD) but may rarely be seen in Hashimoto's thyroiditis, primary hypothyroidism, or thyroid cancer. But in our result we found changes in the histology of retina after induction of hypothyroidism.

Number of cells became reduced in outer and inner nuclear layer in hypothyroid group. The outer and inner nuclear layer lost their individual boundary. The amount of collagen fibres were reduced in retinal tissue in hypothyroid rat.

Our result also showed the changes of cytomorphology in biceps of hypothyroid rats. Hypothyroid sections showed variation in fiber diameter, distinctive fiber atrophy. The existence of grouped atrophy and pyknotic nuclear clump in muscular fibres suggest muscular degeneration. Hypothyroid myopathy has so far been reported in long standing cases of hypothyroidism (Astrom et al., 1961; Wilson et al., 1959). It presents with severe muscle aches and cramps, stiffness and spasms (McKeran et al., 1975, 1979). Histologically, myofibrillar degeneration is present. The atrophic form may resemble the more familiar muscular dystrophies. Therefore, hypothyroidism may be an important cause for visual problem as well as thyroid myopathy.

REFERENCES

Astrom, K.E; Kugelberg, E. and Muller, R. Hypothyroid myopathy. *Arch. Neurol.*, 1961. 5, 472- 482.

Debre, R. and Semelaigne, G. Syndrome of diffuse muscular hypertrophy in infants causing athletic appearance. *Am. J. Dis. Child.,* 1935. 50, 1351-1361.

Dimitriadis, G; Baker, B; Marsh, H; Mandarino, L; Rizza, R; Bergman, R; Haymond, M; Gerich, J. Effect of thyroid hormone excess on action, secretion, and metabolism of insulin in humans. *Am. J. Physiol.*, 1985. 248, 593-601.

Fujita, H. Functional morphology of thyroid. *Int Rev Cytol.*, 1988. 113, 145-150.

Grofte, T; Wolthers, T; Jensen, DS ; Moller, N; Jorgensen, JO ; Orskov, H; Vilstrup, H. Hepatic amino nitrogen conversion and organ N-contents in hypothyroidism, with thyroxine replacement, and in hyperthyroid rats. *J. Hepatol.*, 1997. 26, 409.

Heufelder, AE; Dutton, CM; Sarkar, G; Donovan, KA; Bahn, RS. Detection of TSH receptor RNA in cultured fibroblasts from patients with Graves' ophthalmopathy and pretibial dermopathy. *Thyroid.,* 1993. 3, 297-300.

Hoffmann, J. *Dtsch. Z. Nervenheilk.*, 1897. 9, 278.

Hoppner, W. and Seitz, HJ. Effect of thyroid hormones on glucokinase gene transcription in rat liver. *J Biol Chem.,* 1989. 264(34), 20643-20647.

Kendler, DL; Lippa, J; Rootman, J. The initial clinical characteristics of Graves' orbitopathy vary with age and sex. *Arch. Ophlhalmol.*, 1993. 111, 197-201.

Kocher, T. *Dtsch. Z. Chir.*, 1892. 34, 566.

McKeran, R.O; Slavin, G; Andrews, T.M; Ward, P. and Mair, W.G.P. Muscle fibre type changes in hypothyroid myopathy. *J. Clin. Pathol.*, 1975. 28, 659-63.

McKeran, R.O; Ward, P; Slavin, G. and Paul, E.A. Central nuclear counts in muscle fibres before and during treatment in hypothyroid myopathy. *J. Clin. Pathol.,* 1979. 32, 229-233.

Norris, F.H. and Panner, B.J. Hypothyroid myopathy. *Arch. Neurol..*, 1966. 14, 574- 589.

Pantos, C; Mourouzis, I; Malliopoulou, V; Paizis, I; Tzeis, S; Moraitis, P; Sfakianoudis, K; Varonos, DD; Cokkinos, DV. Dronedarone administration prevents body weight gain and increases tolerance of the heart to ischemic stress: a possible involvement of thyroid hormone receptor alpha1. *Thyroid.*, 2005. 15, 16-23.

Parakkal, PF. Mordanting fixation as a means of facilitating the staining of pancreatic cells of mouse. *Stain Technol.,* 1961. 36, 33-4.

Prummel, MF; Wiersinga, WM; Mourits, MP; Koornneef, L; Berghout, A; Van der Gaag, R. Effect of abnormal thyroid function on the severity of Graves' ophthalmopathy. *Arch. Intern Med.,* 1990. 150, 1098-101.

Wilson, J. and Walton, J.N. Some muscular manifestations of hypothyroidism. *J. Neurol. Neurosurg. Psychiat.,* 1959. 22, 320-324.

In: Vertebrate and Invertebrate Animal Approaches
Editor: José Rosa Gomes ISBN: 978-1-63117-102-4

Chapter 6

METHIMAZOLE-INDUCED HYPOTHYROIDISM MODULATES CYTOMORPHOLOGY OF SKIN EPIDERMIS AND COLLAGEN SYNTHESIS IN RATS

Srikanta Guria, Nisha Balmiki, Sudipta Chakrabarti and Madhusudan Das*
Department of Zoology, University of Calcutta, Kolkata, India

ABSTRACT

Thyroid dysfunction may have an important role in skin in mammal. We aimed to understand the role of thyroid hormones in skin development in rat. Methimazole induced hypothyroid rat exhibited significant histological changes in epidermal skin tissue. The amount of hair follicles, mesenchymal fibroblast cells and collagen was reduced in hypothyroid group. Therefore, thyroid dysfunction may be an important cause for epidermal dysfunction and skin related diseases in mammal.

Keywords: Thyroid, methimazole, thyroxine, epidermis, dermis, hair follicle, collagen, fibroblast

INTRODUCTION

The thyroid gland is specialized organ for endocrine function in the body. The gland secretes thyroid hormones which promote normal growth and development and regulate a number of homeostatic functions in the body (Fujita, 1988). In recent times, thyroid diseases are being diagnosed very common and its relation with growth of epidermis and dermis has been the topic of intense investigation. Several observations provide some clues as to the possible mode of the regulatory action of thyroid hormone (TH) in the skin indicating that in hypothyroidism there is a slow metabolic rate due to alteration of glandular function.

* Corresponding author: Phone: 91-9831281756. Email: madhuzoo@yahoo.com.

Hypothyroidism may be an important cause of dryness of skin, poor wound healing and loss of hair all over scalp (Joshua et al., 2003).

Our study investigated the histological changes of skin epidermis in rat under experimentally induced hypo- and hyperthyroid condition. The observation from the study may help to understand how thyroid dysfunction is correlated with the skin function.

MATERIALS AND METHODS

The laboratory experiments were performed using rat (*Rattus rattus*) as a mammalian model. Normal young adult rat aged 8-10 weeks and weight 100-110 g were housed in polypropylene cages and were acclimatized in laboratory condition for a week with natural light and dark schedules prior to experimentation. The animals were fed standard rodent diet and water was provided *ad libitum*.

They were divided into hypothyroid, hyperthyroid and their respective control groups. Group I animals were treated with Methimazole 20 mg/kg body weight/day for 14 days. Group II animals were treated with thyroxine 600 µg/kg body weight/day for 14 days. Before autopsy, Rats were anesthetized with chloroform. Blood was drawn directly from heart for serum T3, T4 hormones. Dorsal skin was dissected out and fixed in Bouin's fixative (Parakkal, 1961). Then tissues were processed through graded alcohols, paraffin embedded, sectioned (5µ), mounted on glass slide and stained with hematoxylin-eosin for histological analysis.

Tissues were also stained with Masson's Trichrome for collagen study. Rat skin cells viability was determined by Trypan Blue staining. T3 or T4 concentrations in rat serum were determined by radioimmunoassay (RIA).

RESULT

Serum T3, T4 Level

The results showed significant changes in T_3, T_4 hormones (Figure 1A and B). The serum T_3 and T_4 levels were significantly reduced in hypothyroid rat whereas in contrast the levels significantly elevated in hyperthyroid rat as expected and confirming that they were indeed in a hypothyroid and hyperthyroid state

Histology of Skin Epidermis

Significant changes were observed in the cytomorphology of skin in the hypothyroid group when compared with the control group under light microscopy (Figure 2A and B). General morphology of skin epidermis was altered in hypothyroid rat. Hypothyroidism showed thin, folded epidermis and reduced hair follicles. The number of mesenchymal fibroblast cells were reduced significantly in hypothyroid rat (Figure 3A and B).

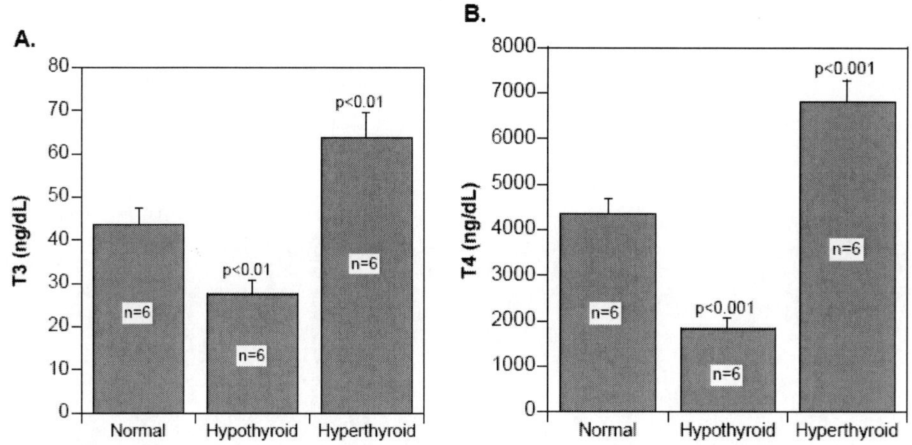

Figure 1. Plasma concentrations (ng/dL) of T3 (A) and T4 (B) in hypothyroid and hyperthyroid rats. Values are expressed as Mean ± SEM.

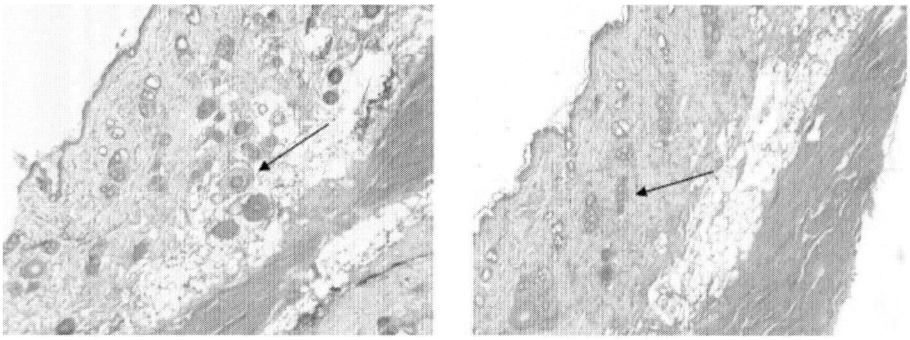

2(A) 2(B)

Figure 2. (A, B): (A) Normal rat skin epidermis (x 400), (B) Hypothyroid rat skin epidermis (x 400). Hair follicles are indicated by the arrow.

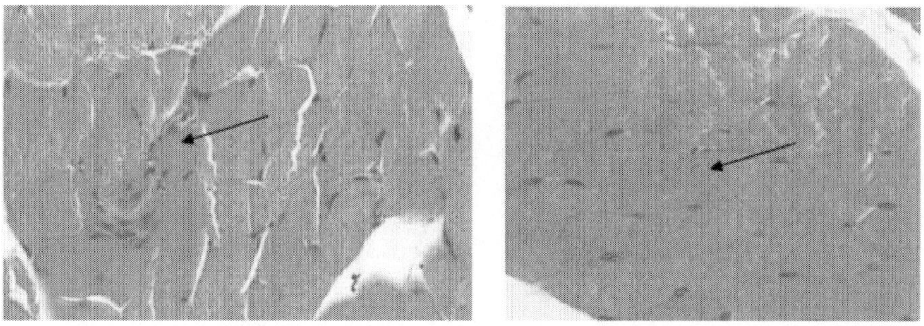

3 (A) 3 (B)

Figure 3. (A, B): (A) Normal rat skin fibroblast cells (x 400), (B) Hypothyroid rat skin fibroblast cells (x 400) as indicated by the arrow.

Trypan Blue Staining of Skin Cells

Significant number of hypothyroid rat skin mesenchymal cells showed trypan blue positive response. Dead cells of hypothyroid rat skin were blue in colour whereas the viable cells of controls were white (Figure 4A and B).

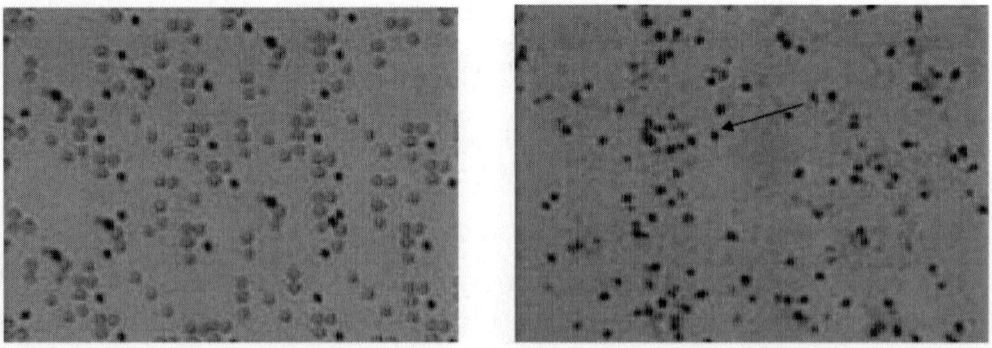

4 (A) 4 (B)

Figure 4. (A, B): (A) Normal rat viable skin cells, (B) Hypothyroid rat skin cells. Dead cells are indicated by the arrow.

Masson's Trichrome Staining of Collagen

Hyporthyroid epidermis showed reduced amount of collagen (Figure 5 A and B). Masson trichrome stained all the collagen fibres blue whereas nuclei were black and cytoplasm was red.

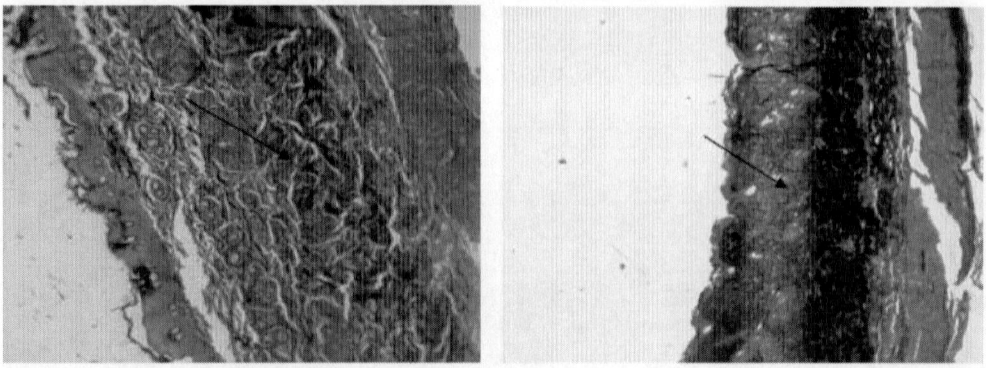

5 (A) 5 (B)

Figure 5. (A, B): (A) Normal rat skin (x 400), (B) Hypothyroid rat skin (x 400).Collagen fibres are indicated by the arrow.

DISCUSSION

In the present investigation, we used methimazole to induce hypothyroidism and thyroxine to induce hyperthyroidism, as reflected in the serum T3, T4 levels (Pantos et al.,

2005; Grofte et al, 1997). The present study clearly indicates that hypothyroidism adversely affects skin function. Thyroid hormones are essential for normal function of keratinocytes and fibroblasts. Hair growth particularly depends on thyroid hormone. Their effect may be either direct or indirect (Joshua, 2003). We observed that the size, shape and number of hair follicles were significantly reduced in hypothyroid rats. Previous research had revealed that thyroid hormone exerts its effect on skin via thyroid hormone receptor (TR) (Joshua, 2001). Collagen structures are essential for wound tensile strength. Without adequate tensile strength of collagen structures, dehiscence and evisceration can occur (Konstantinos, 2006). We found that total collagen of the skin was lower in the hypothyroid rats than in the controls. The results of the present study suggest that the rate of collagen synthesis is decreased in hypothyroidism. In addition hypothyroidism is accompanied by decreased rate of catabolism of collagen (Kari, 1967). We found hypothyroidism significantly affects the skin cells. As the fibroblast cells are responsible for the synthesis of collagen, the fewer number of fibroblast cells in hypothyroid rat may be associated with decreased rate of collagen biosynthesis (Smith, 2002). Therefore, thyroid dysfunction may be an important cause for various skin diseases as well as delayed wound healing.

References

Fujita, H. Functional morphology of thyroid. *Int Rev Cytol.*, 1988. 113, 145-150.

Grofte, T; Wolthers, T; Jensen, DS ; Moller, N; Jorgensen, JO ; Orskov, H; Vilstrup, H. Hepatic amino nitrogen conversion and organ N-contents in hypothyroidism, with thyroxine replacement, and in hyperthyroid rats. *J. Hepatol.*, 1997. 26, 409.

Joshua, D; Tara, M; Lisa, M; Ray, S; Tai, C; Holick, MF. Thyroid hormone action on skin: Diverging effects of topical versus intraperitoneal administration. *Thyroid.,* 2003. 13, 159-165

Joshua, D; Tara, M; Lisa, M; Ray, S; Tai, C; Holick, MF. Topical triiodothyronine stimulates epidermal proliferation, dermal thickening, and hair growth in mice and rats. *Thyroid.*, 2001. 11(8), 717-724.

Kari, I K; Ossi, L; Jouko, H. Metabolism of Collagen in experimental hyperthyroidism and hypothyroidism in the rat. *Endocrinology. 1967. 80(6), 1051- 1061.*

Konstantinos, A E; Georgios, C Z. A concomitant review of the effects of diabetes mellitus and hypothyroidism in wound healing. *World J Gastroenterol.*, 2006. 12(17), 2721-2729.

Pantos, C; Mourouzis, I; Malliopoulou, V; Paizis, I; Tzeis, S; Moraitis, P; Sfakianoudis, K; Varonos, DD; Cokkinos, DV. Dronedarone administration prevents body weight gain and increases tolerance of the heart to ischemic stress: a possible involvement of thyroid hormone receptor alpha1. *Thyroid.*, 2005. 15, 16-23.

Parakkal PF. Mordanting fixation as a means of facilitating the staining of pancreatic cells of mouse. *Stain Technol.,* 1961. 36, 33-4.

Parviz, Y; Bo, C; Aarne, O; Juha, R; Jan F. Thyroid hormone analogue, triiodothyroacetic Acid, corrects corticosteroid-downregulated collagen synthesis *Thyroid.*, 2004. 14, 345-353.

Parviz, Y; Bo, C; Xing, Y; Xiao, HZ; Thomas H; Jan, F. Action of topical thyroid hormone analogues on glucocorticoid-induced skin atrophy in Mice. *Thyroid.,* 2006. 3, 273-280.

Smith; Terry, J. Fibroblast biology in thyroid diseases. *Thyroid.,* 2002. 9, 393-400.

Vahlquist, H; Carlsson, B. In efficacy of topical thyroid hormone analogue TriAc in plaque psoriasis: results of a double-blind placebo-controlled trial A., *British Journal of Dermatology.,* 2004. 151, 489.

In: Vertebrate and Invertebrate Animal Approaches
Editor: José Rosa Gomes
ISBN: 978-1-63117-102-4

Chapter 7

HYPOTHYROIDISM REDUCES PHAGOCYTIC ACTIVITY OF RAT MACROPHAGES

Srikanta Guria[1] and Madhusudan Das[2*]

[1]Post Graduate Department of Zoology, Parasitology and Medical Entomology Laboratory, Darjeeling Govt. College, Darjeeling, West Bengal, India
[2]Department of Zoology, University of Calcutta, Kolkata, West Bengal, India

ABSTRACT

Thyroid dysfunction may have an important role in immune response in mammal. We aimed to understand the role of thyroid hormones in peritoneal macrophages activity in rat. Methimazole induced hypothyroid rat exhibited significant morphological changes in peritoneal macrophages.

The phagocytic activity of peritoneal macrophages was reduced in hypothyroid group. Consequently hypothyroidism induced apoptosis of the macrophages. Therefore, thyroid dysfunction may be an important cause for immune dysfunction in mammal.

Keywords: Thyroid, methimazole, thyroxine, macrophage, phagocytosis, hypothyroidism

INTRODUCTION

Thyroid hormones promote normal growth and development and regulate a number of homeostatic functions in the body (Fujita, 1988). In recent times, thyroid diseases are being diagnosed very common and its relation with immune modulation has been the topic of intense investigation.

Hypothyroidism may be an important cause of altered macrophage function in mammal (Liu et al., 1993; Schoenfeld et al., 1995).

---

* Corresponding author: Phone: 91-9831281756. Email: madhuzoo@yahoo.com.

Our study investigated the changes of peritoneal macrophage phagocytic activity in rat under experimentally induced hypo and hyperthyroid condition.

The observation from the study may help to understand how thyroid dysfunction is correlated with immune disorder.

MATERIALS AND METHODS

Animals and Housing

The laboratory experiments were performed using rat (*Rattus rattus*) as a mammalian model. Normal young adult rat aged 8-10 weeks and weight 100-110 g were housed in polypropylene cages and were acclimatized in laboratory condition for a week with natural light and dark schedules prior to experimentation. The animals were fed standard rodent diet and water was provided *ad libitum*.

Treatment of Animals

Rats were divided into hypothyroid, hyperthyroid and their respective control groups. Group I animals were treated with methimazole 20 mg/kg body weight/day for 14 days. Group II animals were treated with thyroxine 600 μg/kg body weight/day for 14 days.

Macrophage Isolation or Extraction

Activated charcoal particles suspended in normal saline (0·9% NaCl) or cold 0·1 M phosphate buffer saline (PBS, pH 7·2) was injected into rat peritoneum and the aspirate was taken for macrophage study (after Iyengar et al., 1985).

Fixation and Staining of Macrophages

Peritoneal fluid was placed and smeared directly on sterilized glass slides and incubated at 37°C in a humid chamber for 3 hours. After incubation the nonadherent cells were removed by washing three times with 0·9% NaCl. The adherent macrophages were fixed by methanol. After fixation macrophages were stained by Giemsa and observed under light microscope (after Iyengar et al., 1985). Different stages of phagocytosis of charcoal particles by peritoneal macrophages on glass slides were determined in both control and treated rats. Macrophage morphology was examined by hematoxylin-eosin staining.

Flurescence Microscopic Study of Macrophages

Macrophage destruction was determined by nuclear DNA fragmentation by using DAPI under fluorescence microscope.

Counting of Macrophages

We used peritoneal aspirate for cell counting by hemacytometer.

Thyroid Hormones (T3 and T4) Assay

Rats were anesthetized with chloroform. Blood was drawn directly from heart for serum T3, T4 hormones. T3 or T4 concentrations in rat serum were determined by radioimmunoassay (RIA).

Statistical Analysis

All the data were presented as mean ±SEM. The data were analyzed by *t*-test. Results with $P < 0.05$, $P<0.01$, $P<0.001$ were considered statistically significant.

RESULT

Serum T3, T4 Level

The results showed significant changes in T_3, T_4 hormones (Figure 1A and B). The serum T_3 and T_4 levels were significantly reduced in hypothyroid rat whereas in contrast the levels significantly elevated in hyperthyroid rat as expected and confirming that they were indeed in a hypothyroid and hyperthyroid state.

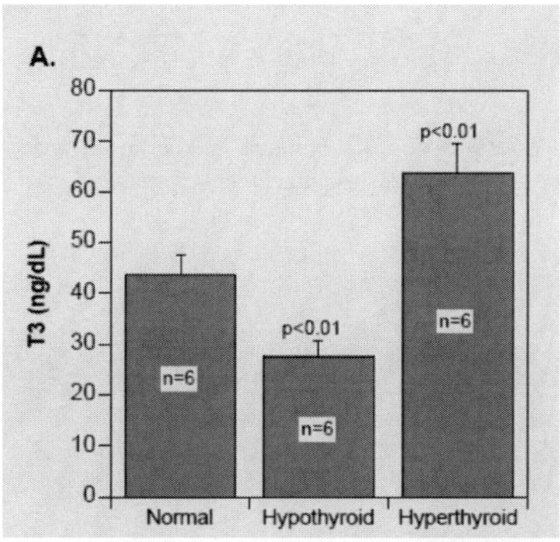

Figure 1. Plasma concentrations (ng/dL) of T3 (A) and T4 (B) in hypothyroid and hyperthyroid rats. Values are expressed as Mean ± SE.

Giemsa Staining of Macrophages

Significant changes were observed in the cytomorphology of peritoneal macrophages in the hypothyroid group when compared with the control group under light microscopy (Figure 2A and B). Hypothyroid rat exhibited cellular damage. A significant percentage of peritoneal macrophages become picnotic in hypothyroidism. Normal macrophages showed different stages of phagocytosis. But the necrotic macrophages in hypothyroid group were not able to phagocytose the charcoal particles. No discernible change in peritoneal macrophages was detected in hyperthyroid rats (picture not shown).

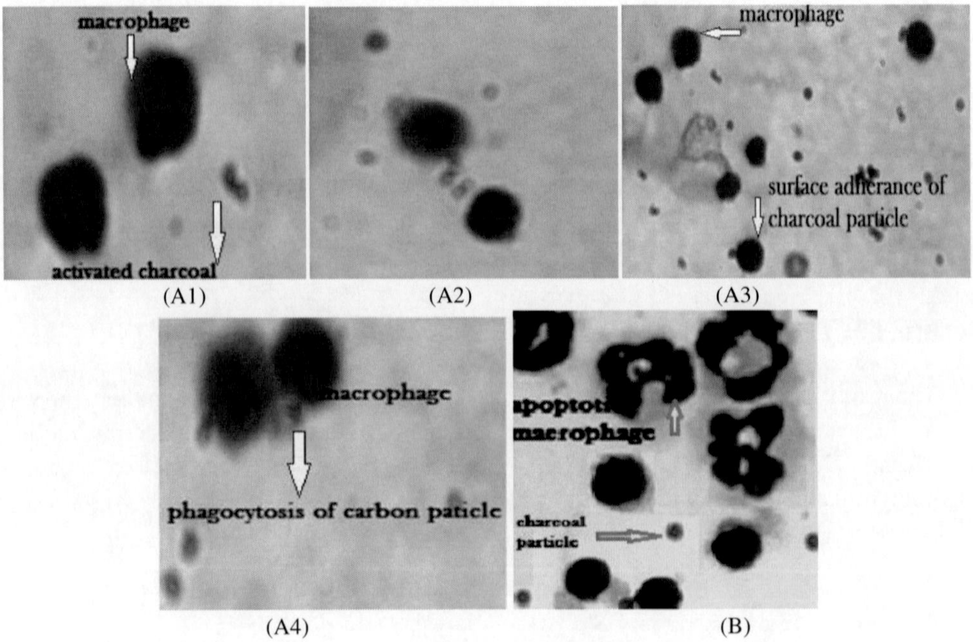

Figure 2. (A) Different stages of phagocytosis in normal rat peritoneal macrophage, A1-A2: movement of macrophage towards charcoal particles, A3: attachment of charcoal particle on macrophage surface, A4: phagocytosis of charcoal by macrophages. (B) Apoptotic peritoneal macrophages in hypothyroid rat.

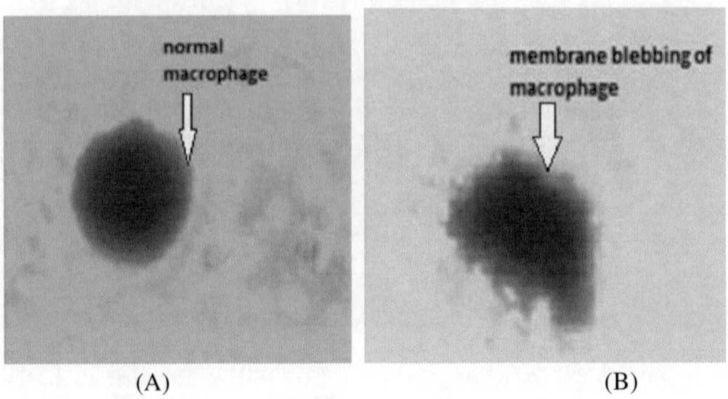

Figure 3. (A, B): A) Normal rat peritoneal macrophage (x 400), (B) Hypothyroid rat peritoneal macrophage (x 400).

Eosin-Haematoxylin Staining of Macrophages

General morphology of macrophage was altered in hypothyroid rat. Significant number of hypothyroid rat peritoneal macrophages showed membrane blebbing (Figure 3A and B).

DAPI Staining of Macrophages

Hypothyroid macrophages showed different stages of nuclear fragmentation (Figure 4A and B). DAPI stained all the nuclei blue.

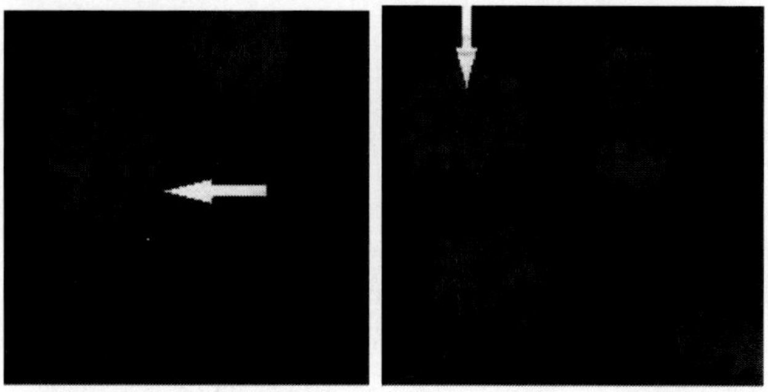

(A) (B)

Figure 4. (A, B): Fluorescent microscopic analysis of (A) normal macrophage nuclei and (B) hypothyroid peritoneal macrophages nuclei stained with DAPI.

Macrophage Counting

Significant number of macrophages was reduced in hypothyroid group. (13.0 ± 1.7 vs. 1.9 ± 1.3 macrophages /mm^2, $P < 0.05$; Figure 5)

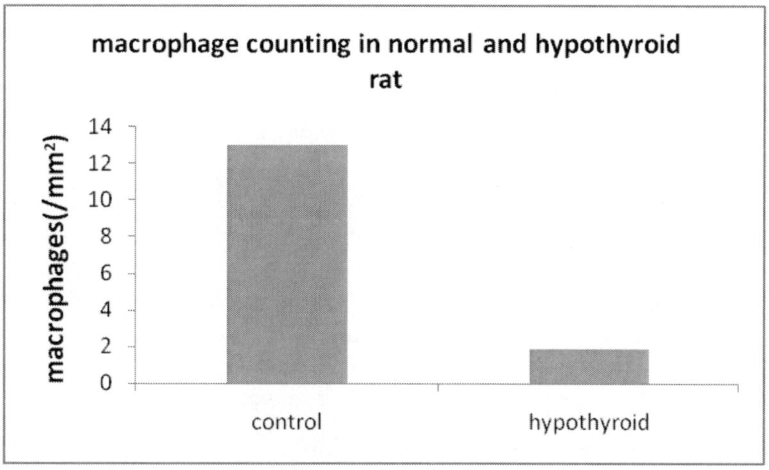

Figure 5. Macrophage counting in normal and hypothyroid group. Values are expressed as Mean ± SE.

Discussion

In the present investigation, we used methimazole to induce hypothyroidism and thyroxine to induce hyperthyroidism, as reflected in the serum T3, T4 levels (Pantos et al., 2005; Grofte et al., 1997). The present study clearly indicates that hypothyroidism adversely affects peritoneal macrophages function. Thyroid hormones are essential for normal function of macrophages. Their effect may be either direct or indirect (Liu et al., 1993; Rosa et al., 1995). In hypothyroidism a significant number of peritoneal macrophages were found to be picnotic. The number of peritoneal macrophages reduced in hypothyroid rat. The reason for the reduction of cell number after methemazole treatment is apparently due to an increased rate of peritoneal cell apoptosis (Liu et al., 1993)

An increased rate of macrophage death due to apoptosis causes a decrease of immune function in hypothyroidism (Liu et al., 1993; Schoenfeld et al., *1995*). The DAPI stained hypothyroid peritoneal macrophages showed different stages of nuclear fragmentation indicating apoptosis. The cellular death was also confirmed by membrane blebbing. The previous histological studies revealed that methimazole induced hypothyroidism caused cell damage. This damage is associated with an increase of oxidative stress by producing ROS. In methimazole induced hypothyroidism catalase activity is reduced and this allows H2O2caused damage (Cano-Europa E et al., 2010). The phagocytic behavior of macrophages was altered in hypothyroid group. The resulting reduction and death of peritoneal macrophages in hypothyroid rat can provide an explanation for the decrease of immune responses and phagocytosis in hypothyroidism (Liu et al., 1993; Schoenfeld et al., *1995*).

References

Cano-Europa E, et al. Methimazole-induced hypothyroidism causes alteration of the REDOX environment, oxidative stress, and hepatic damage; events not caused by hypothyroidism itself. *Annals of hepatology*., 2010. 9 (1), 80-88.

Chatterjee, S and Chandel, AS. Immunomodulatory role of thyroid hormones: in vivo effect of thyroid hormones on the blastogenic response to lymphoid tissue. *Acta Endocrinologica.,* 1983. 103, 95–100.

Dorshkind, K and Horseman, ND. The roles of prolactin, growth hormone, insulin-like growth factor-I, and thyroid hormones in lymphocyte development and function: insights from genetic models of hormone and hormone receptor deficiency. *Endocrine Reviews.,* 2000. 21, 292–312.

Dorshkind, K and Horseman, ND. Anterior pituitary hormones, stress and immune system homeostasis. *Bioessays.,* 2001. 23, 288–294.

Fabris, N; Mocchegiani, E; and Provinciali, M. Pituitary-thyroid axis and immune system: a reciprocal neuroendocrine-immune interaction. *Hormone Research.,* 1995. 43, 29–38.

Fujita, H. Functional morphology of thyroid. *Int Rev Cytol*., 1988. 113, 145-150.

Grofte, T; Wolthers, T; Jensen, DS ; Moller, N; Jorgensen, JO ; Orskov, H; Vilstrup, H. Hepatic amino nitrogen conversion and organ N-contents in hypothyroidism, with thyroxine replacement, and in hyperthyroid rats. *J Hepatol*., 1997. 26, 409.

Hallengren, B; Forsgren, A and Melander, A. Effects of anti-thyroid drugs on lymphocyte function in vitro. *Journal of Clinical Endocrinology and Metabolism.,* 1980. 51, 298–301.

Iyengar, U. R. and Vakil, U. K. Enumeration and structural assessment of peritoneal macrophages during progressive protein deficiency in rats. *J. Biosci*., 1985. 7, 15–26.

Liu, WK; Tsui, KW; Wong, CC. Repressed activity of peritoneal macrophages in methimazole induced hypothyroid mice. *Virchows Arch B Cell Pathol Incl Mol Pathol.,* 1993. 63(2), 131-6.

Liu, WK; Wong, CC; Mak, N K. Localization of thyroid hormone in subpopulations of rat alveolar macrophages. *The Histochemical Journal.*, 1989. 21 (2), 115-121.

Liu, WK and Ng, TB. Effect of methimazole induced hypothyroidism on alveolar macrophages. *Virchows Arch B Cell Pathol Incl Mol Pathol.,* 1991. 60, 21-25.

Liu, WK. Expression of tumor necrosis factor and c-fos genes in peritoneal macrophages of hypothyroid mice. *Inflammation., 1993. 17(2), 217-25.*

Pantos, C; Mourouzis, I; Malliopoulou, V; Paizis, I; Tzeis, S; Moraitis, P; Sfakianoudis, K; Varonos, DD; Cokkinos, DV. Dronedarone administration prevents body weight gain and increases tolerance of the heart to ischemic stress: a possible involvement of thyroid hormone receptor alpha1. *Thyroid.*, 2005. 15, 16-23.

Pawlikowski, M; Stepien, H and Komorowski, J. Hypothalamic-pituitary-thyroid axis and the immune system. *Neuroimmunomodulation.,* 1994. 1, 149–152.

Rosa, LF; Safi, DA; Curi, R. Effect of hypo and hyperthyroidism on the function and metabolism of macrophages in rats. *Cell Biochem Funct.,* 1995. 13(2), 141-7.\Schoenfeld, PS; Myers, JW; Myers, L and LaRocque, JC. Suppression of cell-mediated immunity in hypothyroidism. *Southern Medical Journal.,* 1995. 88, 347–349.

Volpe, R. The immunomodulatory effects of anti-thyroid drugs are mediated via actions on thyroid cells, affecting thyrocyte-immunocyte signalling: a review. *Current Pharmaceutical Design.,* 2001. 7, 451– 460.

In: Vertebrate and Invertebrate Animal Approaches
Editor: José Rosa Gomes
ISBN: 978-1-63117-102-4

Chapter 8

Phagocytosis of Yeast Particles by Peritoneal Macrophage in Normal and Hyperthyroid Rats

Srikanta Guria[1] and Madhusudan Das[2*]

[1]Department of Zoology, Parasitology and Medical Entomology Laboratory, Darjeeling Govt. College, Darjeeling, West Bengal, India

[2]Department of Zoology, University of Calcutta, Kolkata, West Bengal, India

Abstract

Thyroid dysfunction may have an important role in immune response in mammal. We aimed to understand the role of thyroid hormones in peritoneal macrophage phagocytic activity in rat by challenging them with yeast particles. Eltroxin induced hyperthyroid rat exhibited significant changes in phagocytic activity in peritoneal macrophages. Therefore, thyroid dysfunction may be an important cause for immune dysfunction in mammal.

Keywords: Thyroid, thyroxine, macrophage, yeast, hyperthyroidism, phagocytosis, phagocytic index

Introduction

The thyroid gland is specialized organ for endocrine function in the body. The gland secretes thyroid hormones which promote normal growth and development and regulate a number of homeostatic functions in the body (Fujita, 1988).

In recent times, thyroid diseases are being diagnosed very common and its relation with immune modulation has been the topic of intense investigation. Hyperthyroidism may be an

* Corresponding author: Phone: 91-9831281756. Email: madhuzoo@yahoo.com.

important cause of altered macrophage function in mammal (Liu et al., 1989; Schoenfeld et al., 1995).

Our study investigated the changes of peritoneal macrophage phagocytic activity in rat under experimentally induced hyperthyroid condition. The observation from the study may help to understand how thyroid dysfunction is correlated with immune disorder.

MATERIALS AND METHODS

Animals and Housing

The laboratory experiments were performed using rat (*Rattus rattus*) as a mammalian model. Normal young adult rat aged 8-10 weeks and weight 100-110 g were housed in polypropylene cages and were acclimatized in laboratory condition for a week with natural light and dark schedules prior to experimentation. The animals were fed standard rodent diet and water was provided *ad libitum*.

Treatment of Animals

Rats were divided into hyperthyroid and their respective control groups. Group I animals were treated with thyroxine (eltroxin) 600 μg/kg body weight/day for 14 days.

Staining of Yeast Particles

Powdered yeast was used for phagocytic studies.Yeast particles were stained with eosin and suspended in normal vertebrate saline or PBS.

Macrophage Isolation or Extraction

Normal saline (0·9% NaCl) or cold 0·1 M phosphate buffer saline (PBS, pH 7·2) was injected into rat peritoneum and the aspirate was taken for macrophage study (after Iyengar et al., 1985).

Fixation and Staining of Macrophages

Peritoneal fluid was placed and smeared directly on sterilized glass slides and incubated at 37˚C. The phagocytic efficiency of control and hyperthyroid macrophages was examined by challenging them with eosin stained yeast suspension *in vitro* over slide and incubated at 37˚C in a humid chamber for 3 hours. After incubation the nonadherent cells were removed by washing three times with 0·9% NaCl. The adherent macrophages were fixed by methanol. After fixation macrophages were stained by Giemsa and observed under light microscope (after Iyengar et al., 1985).

Study of Phagocytosis

Different stages of phagocytosis of yeast particles by peritoneal macrophages on glass slides were determined in both control and treated rats. The phagocytic index was calculated as per the following formula.

$$\text{Phagocytic index} = \frac{\text{Number of macrophages with phagocytosed yeast cells}}{\text{Total number of macrophages counted}} \times 100$$

Counting of Macrophages

We used peritoneal aspirate for cell counting by hemocytometer.

Assay of T3, T4 Level

Blood was drawn directly from heart for serum T3, T4 hormones. T3 or T4 concentrations in rat serum were determined by radioimmunoassay (RIA).

RESULT

Serum T3, T4 Level

The results showed significant changes in T_3, T_4 hormones (Figure 1A and B). The serum T_3 and T_4 levels were significantly elevated in hyperthyroid rat as expected and confirming that they were indeed in a hyperthyroid state.

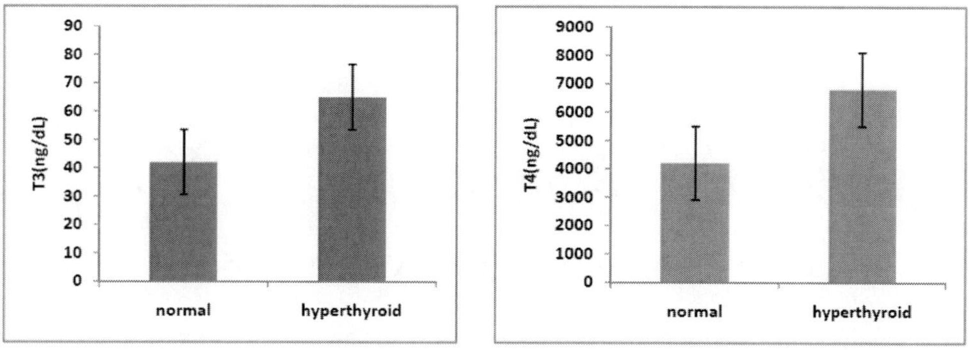

A B

Figure 1. Plasma concentrations (ng/dL) of T3 (A) and T4 (B) in control and hyperthyroid rats. Values are expressed as Mean ± SEM.

GIEMSA STAINING OF MACROPHAGES

Significant changes were observed in the cytomorphology of peritoneal macrophages in the hyperthyroid group when compared with the control group under light microscopy (Figure 2A and B). Macrophages in control rat showed different stages of phagocytosis of yeast particles like chemotaxis of macrophage, surface adherence of yeast, engulfment of yeast. General morphology of macrophage was altered in hyperthyroid rat. Hyperthyroid macrophages showed membrane blebbing. A significant percentage of peritoneal macrophages become picnotic and reduced in number in eltroxin treated group (Figure 2A and B).

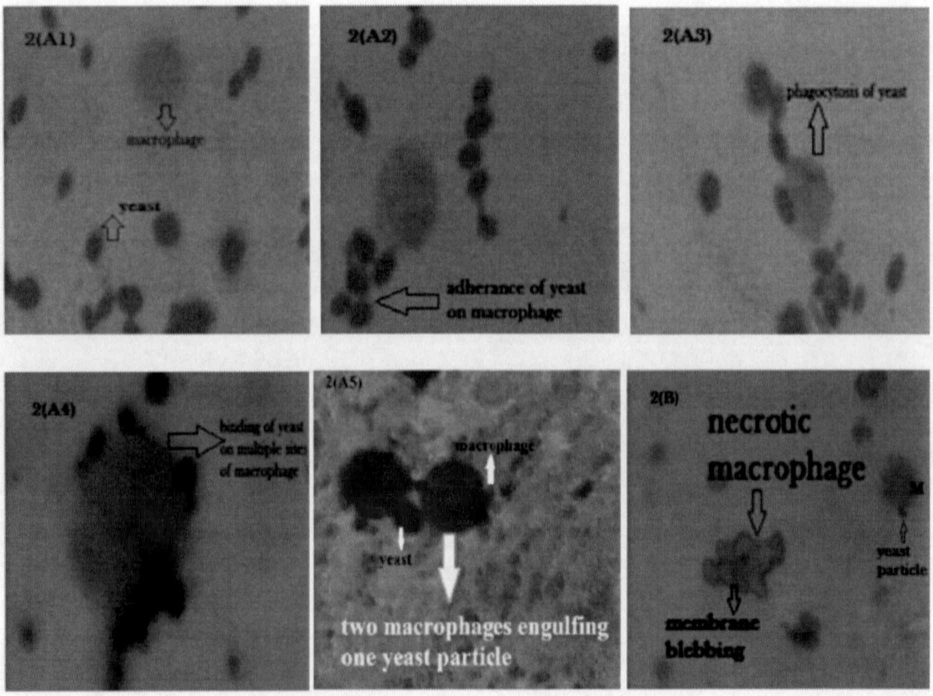

Figure 2. (A, B): (A) Normal rat peritoneal macrophage (x 400), 2(A1) Chemotaxis of macrophage towards yeast particles, 2(A2) Surface adherence of yeast on m, 2(A3) Engulfment of yeast by m, 2(A4) Binding of yeasts on multiple sites of m, 2(A5) Two macrophages engulfing one yeast particle and (B) Hyperthyroid rat peritoneal macrophage (x 400). m = macrophage.

COUNTING OF MACROPHAGES CONTAINING PHAGOCYTOSED YEAST

A significant number of peritoneal macrophages containing yeasts reduced in number in eltroxin treated hyperthyroid group.

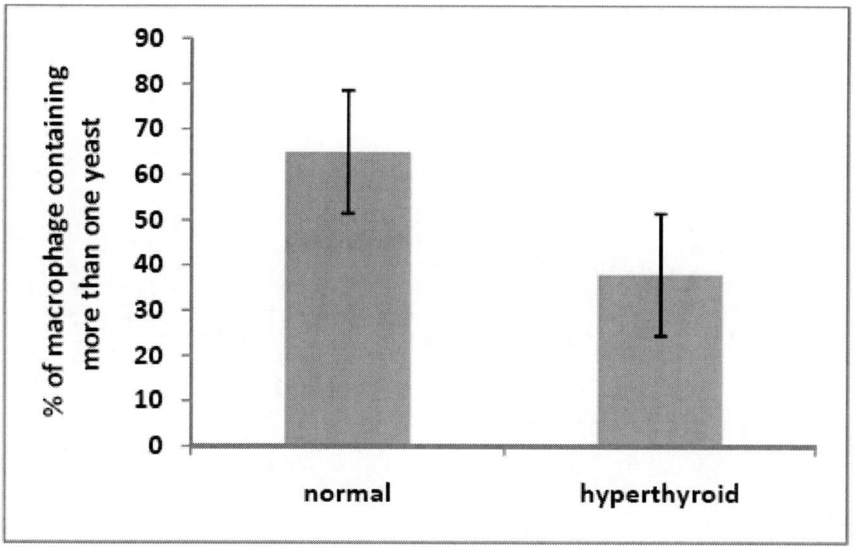

Figure 3. Percentage of macrophages containing more than one yeast in control and hyperthyroid rat. Values are expressed as Mean ± SEM.

CALCULATION OF PHAGOCYTIC INDEX

Mean phagocytic index was significantly reduced in hyperthyroid group.

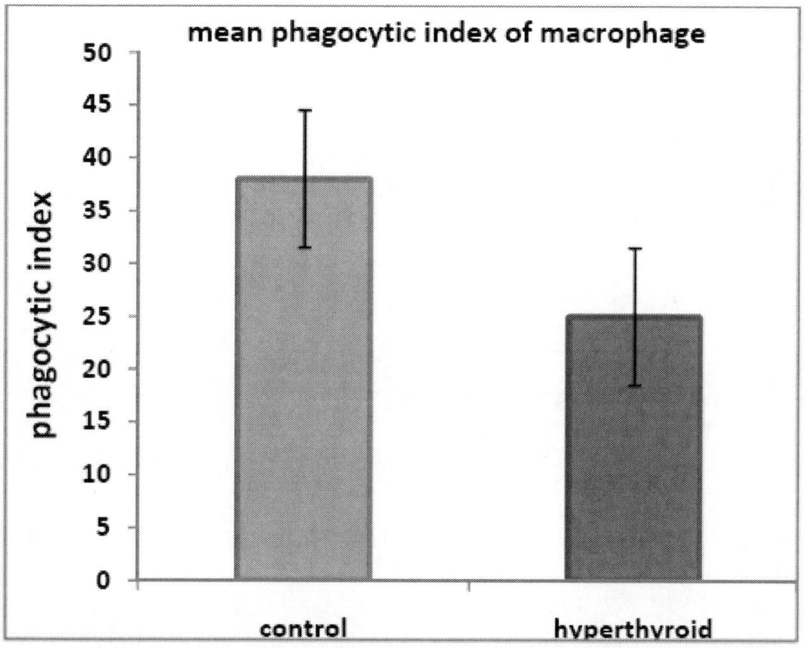

Figure 4. Mean phagocytic index in normal and hyperthyroid group. Values are expressed as Mean ± SEM.

MACROPHAGE COUNTING

Significant number of macrophages was reduced in hyperthyroid group.

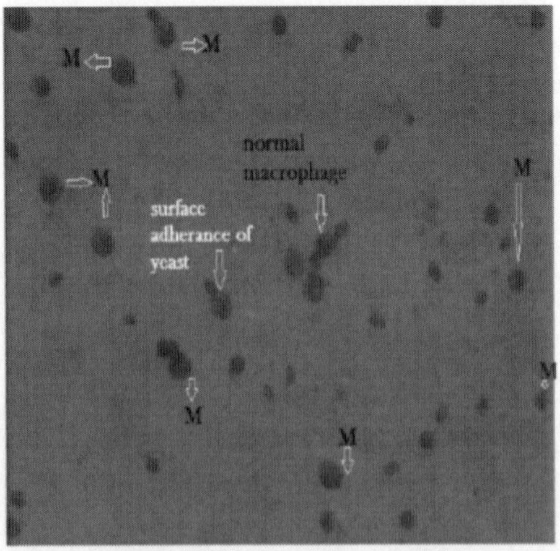

Figure (A) Field with normal M

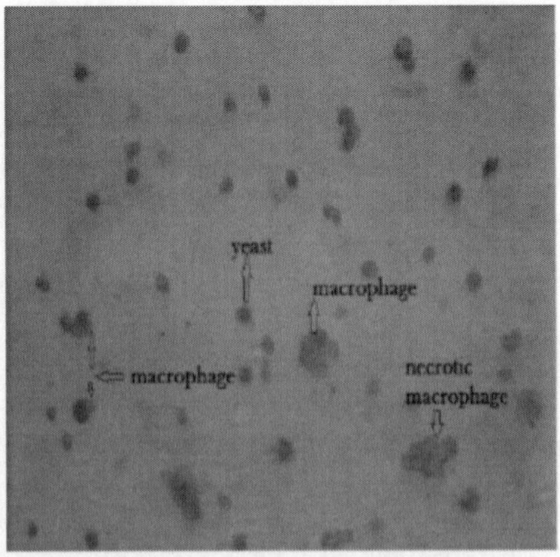

Figure (B) Field with hyperthyroid M

Figure 5. Macrophage counting in two different fields.

DISCUSSION

In the present investigation, we used thyroxine (eltroxin) to induce hyperthyroidism, as reflected in the serum T3, T4 levels (Pantos et al., 2005; Grofte et al., 1997). The present study clearly indicates that hyperthyroidism adversely affects peritoneal macrophages function. Thyroid hormones are essential for normal function of macrophages. Their effect may be either direct or indirect (Rosa et al., 1995).Significant number of peritoneal macrophages were found to be picnotic. The number of peritoneal macrophages reduced in hyperthyroid rat. The reason for the reduction of cell number after in hyperthyroidism is apparently due to an increased rate of peritoneal cell apoptosis (Liu et al., 1989). A significant number of peritoneal macrophages containing yeasts reduced in number in hyperthyroid group. The phagocytic efficiency of control and hyperthyroid macrophages was examined by calculating phagocytic index. Mean phagocytic index was significantly reduced in hyperthyroid group. So hyperthyroidism altered rate of phagocytosis of yeast by macrophages (Diamond, 1988; Filler et al., 2006).

An increased rate of macrophage death due to apoptosis causes a decrease of immune function in hyperthyroidism (Liu et al., 1989; Schoenfeld et al., 1995). The resulting reduction of phagocytosis in macrophages in hyperthyroid rat can provide an explanation for the decrease of immune responses in hyperthyroidism.

REFERENCES

Abia-Bassey, LN; Utsalo, SJ. Yeast associated with human infections in south-eastern Nigeria. *Mycoses.*, 2006. 49, 510-515.

Aderem, A. and Underhill, DM. Mechanisms of phagocytosis in macrophages. *Ann. Reû. Immunol.*, 1999. 17, 593–623.

Chatterjee, S and Chandel, AS. Immunomodulatory role of thyroid hormones: in vivo effect of thyroid hormones on the blastogenic response to lymphoid tissue. *Acta Endocrinologica.,* 1983. 103, 95–100.

Diamond, RD. Fungal surfaces: effects of interactions with phagocytic cells. *Rev. Infect. Dis.* 1988. 10, 428–431.

Fabris, N; Mocchegiani, E and Provinciali, M. Pituitary-thyroid axis and immune system: a reciprocal neuroendocrine-immune interaction. *Hormone Research.,* 1995. 43, 29–38.

Filler, SG; Sheppard, DC. Fungal invasion of normally non-phagocytic host cells. *PLoS Pathog.,* 2006. 2, 129.

Fujita, H. Functional morphology of thyroid. *Int Rev Cytol.*, 1988. 113, 145-150.

Gow, NA; Brown, AJP and Odds, F. Fungal morphogenesis and host invasion. *Curr Opin Microbiol.,* 2002. 5, 366- 371.

Grofte, T; Wolthers, T; Jensen, DS ; Moller, N; Jorgensen, JO ; Orskov, H; Vilstrup, H. Hepatic amino nitrogen conversion and organ N-contents in hypothyroidism, with thyroxine replacement, and in hyperthyroid rats. *J Hepatol.*, 1997. 26, 409.

Ibrahim-Granet, O; Philippe, B; Boleti, H; Boisvieux-Ulrich, E; Grenet, D; Stern, M; Latge, JP. Phagocytosis and intracellular fate of Aspergillus fumigatus conidia in alveolar macrophages. *Infect Immun.,* 2003. 71, 891-903.

Iyengar, U R. and Vakil, U K. Enumeration and structural assessment of peritoneal macrophages during progressive protein deficiency in rats. *J. Biosci.*, 1985. 7, 15–26.

Kaur, R; Domergue, R; Zupancic, ML; Cormack, BP. A yeast by any other name: Candida glabrata and its interaction with the host. *Curr Opin Microbiol.*, 2005. 8, 378-384.

Liu, WK; Wong, CC; Mak, N K. Localization of thyroid hormone in subpopulations of rat alveolar macrophages. *The Histochemical Journal.*, 1989. 21 (2), 115-121.

Newman, SL; Bhugra, B; Holly, A and Morris, RE. Enhanced killing of Candida albicans by human macrophages adherent to type 1 collagen matrices via induction of phagolysosomal fusion. *Infect Immun.*, 2005.73, 770-777.

Pantos, C; Mourouzis, I; Malliopoulou, V; Paizis, I; Tzeis, S; Moraitis, P; Sfakianoudis, K; Varonos, DD; Cokkinos, DV. Dronedarone administration prevents body weight gain and increases tolerance of the heart to ischemic stress: a possible involvement of thyroid hormone receptor alpha1. *Thyroid.*, 2005. 15, 16-23.

Pawlikowski, M; Stepien, H and Komorowski, J. Hypothalamic-pituitary-thyroid axis and the immune system. *Neuroimmunomodulation.,* 1994. 1, 149–152.

Rosa, LF; Safi, DA; Curi, R. Effect of hypo and hyperthyroidism on the function and metabolism of macrophages in rats. *Cell Biochem Funct.,* 1995. 13(2), 141-7.

Schoenfeld, PS; Myers, JW; Myers, L and LaRocque, JC. Suppression of cell-mediated immunity in hypothyroidism. *Southern Medical Journal.,* 1995. 88, 347–349.

Waldorf, AR; Levitz, SM and Diamond, RD. In vivo bronchoalveolar macrophage defense against Rhizopus oryzae and Aspergillus fumigatus. *J. Infect. Dis,* . 1984.150, 752–760.

In: Vertebrate and Invertebrate Animal Approaches
Editor: José Rosa Gomes
ISBN: 978-1-63117-102-4

Chapter 9

PULMONARY INVOLVEMENT OF ARSENIC POISONING AND EFFECT OF ARSENIC TOXICITY ON PERITONEAL MACROPHAGES AND SPLEEN CELLS IN RATS

Srikanta Guria*[*1], Swapna Chhetri[1], Sayantan Saha[1], Nikita Chetri[1], Gitanjali Singh[1], Piali Bala Saha[1], Bilkis Sultana Sarkar[1] and Madhusudan Das[2]

[1]Post Graduate Department of Zoology, Parasitology and Medical Entomology Laboratory, Darjeeling Govt. College,
Darjeeling, West Bengal, India
[2]Department of Zoology, University of Calcutta, Kolkata,
West Bengal, India

ABSTRACT

Chronic arsenic poisoning due to ingestion of contaminated ground-water is a major public health problem in West Bengal as well as in India. It causes multi organ damage. The present study attempts to objectively investigate the pulmonary involvement in arsenic toxicity. The cellular architecture of lung changes was also evaluated. Arsenic toxicity may have an important role in immune response in mammal. We aimed to understand the role of chronic arsenic poisoning by using sodium arsenate on peritoneal macrophages activity and spleen cells in rat. Arsenic induced toxicity exhibited significant morphological changes in peritoneal macrophages and spleen cells. Consequently sodium arsenate induced toxicity showed death of the macrophages and splenic cells. Therefore, arsenic poisoning may be an important cause for immune dysfunction in mammal.

Keywords: Sodium arsenate, macrophage, spleen cell, lung

* Corresponding author: Phone: 91-9830286956. Email: srikantarna@yahoo.com; guriasrikanta@gmail.com.

INTRODUCTION

Arsenic, as an environmental agent, is considered to be a very high priority toxic substance largely due to its carcinogenic potential in humans (Squibb, 1983). In exposed population, arsenic is associated with various tumours of the lung, skin, bladder and liver (Quatrehomme et al., 1992; Smith et al., 1998). The molecular and cellular basis of chemically induced immunotoxicity and its relation to immune suppression has been reviewed (Luster, 1987). Alteration in macrophage functions by environmental chemicals like arsenic has also been reported (Gonsebatt et al., 1992). Evidence for arsenic as the immunosuppressive component has also been demonstrated.

Previous studies have shown that arsenic may exert marked immunomodulatory effects (Bustamante et al., 1997). However, the mechanism of their influence on the immune system is unclear. The objective of the present study was to examine the functions of rat macrophages and splenic cells from sodium arsenate exposure to determine the cause of their modulation in function (Lantz et al., 1995). Increased mortality from lung cancer associated with chronic arsenic poisoning has been reported (Hopenhayn-Rich et al., 1998). A recent epidemiological survey in West Bengal found clinical evidence of respiratory effects due to chronic arsenic poisoning (Guha Mazumder et al., 2000). We have attempted to evaluate the nature of lung changes compared them with arsenic-unaffected controls.

MATERIALS AND METHODS

Animals and Housing

The laboratory experiments were performed using rat (*Rattus rattus*) as a mammalian model. Normal young adult rats aged 8-10 weeks and weight 100-110 g were housed in polypropylene cages and were acclimatized in laboratory condition for a week with natural light and dark schedules prior to experimentation. The animals were fed standard rodent diet and water was provided *ad libitum*.

Arsenic Treatment of Animals

Rats were divided into experimental and their respective control groups. Group I animals were treated with sodium arsenate (NaH_2AsO_4). 1 mg of sodium arsenate was dissolved in 1 ml. of sterile 0.9% saline, and 10 µl was injected i.p. to each mouse for two weeks.

Macrophage and Spleen Cell Isolation or Extraction

Normal saline (0·9% NaCl) or cold 0·1 M phosphate buffer saline (PBS, pH 7·2) was injected into rat peritoneum and the aspirate was taken for macrophage study (after Iyengar et al., 1985). Spleen was removed using the forceps and mashed through the cell strainer into the petridish containing 0·1 M phosphate buffer saline (PBS, pH 7·2). Cell suspension was subjected for centrifugation at 800xg for 3 minutes. Supernatant was discarded and pellet was resuspended in PBS. Cell suspension was taken for study.

Fixation and Staining of Macrophages and Splenic Cell

Peritoneal fluid was placed and smeared directly on sterilized glass slides and incubated at 37°C in a humid chamber for 3 hours. After incubation the non adherent cells were removed by washing three times with 0·9% NaCl. The adherent macrophages were fixed by methanol. After fixation macrophages were stained by Giemsa and observed under light microscope (after Iyengar et al., 1985). Spleen cell suspension was placed and smeared directly on sterilized glass slides and were fixed by methanol and were stained by Giemsa.

Counting of Macrophages and Spleen Cell

We used peritoneal aspirate and spleen cell suspension for cell counting by haemocytometer.

Trypan Blue Dye Exclusion Test

After the incubation cells were treated with 50 μl of 0.25 % trypan blue dye solution for 5 minutes, the percentage of macrophages and spleen cells with cytoplasm stained blue can be obtained by counting ten optical fields under a light microscope at magnification 1, 000. Cells that have taken up the dye are dead, since the dye is normally excluded by the membranes which maintain their semi permeability intact and therefore, the percentage of blue-stained cells represents a mortality index.

$$\text{Mortality index} = \frac{\text{Number of cells with blue stained cytoplasm}}{\text{Total number of cells}} \times 100$$

Histological Analysis

Before autopsy, rats were anesthetized with chloroform. Lung tissues were dissected out and fixed in Bouin's fixative (Parakkal, 1961). Tissues were embedded in paraffin, sectioned (5 μm), mounted on glass slide and stained with hematoxylin-eosin for histological analysis.

RESULTS

Giemsa Staining of Macrophages

Significant changes were observed in the cytomorphology of peritoneal macrophages in the arsenic treated group when compared with the control group under light microscopy

(Figure 1A and B). General morphology of macrophage was altered in arsenic treated rat. Arsenic toxicity showed cellular damage. A significant percentage of peritoneal macrophages become picnotic and reduced in number (Figure 1A and B).

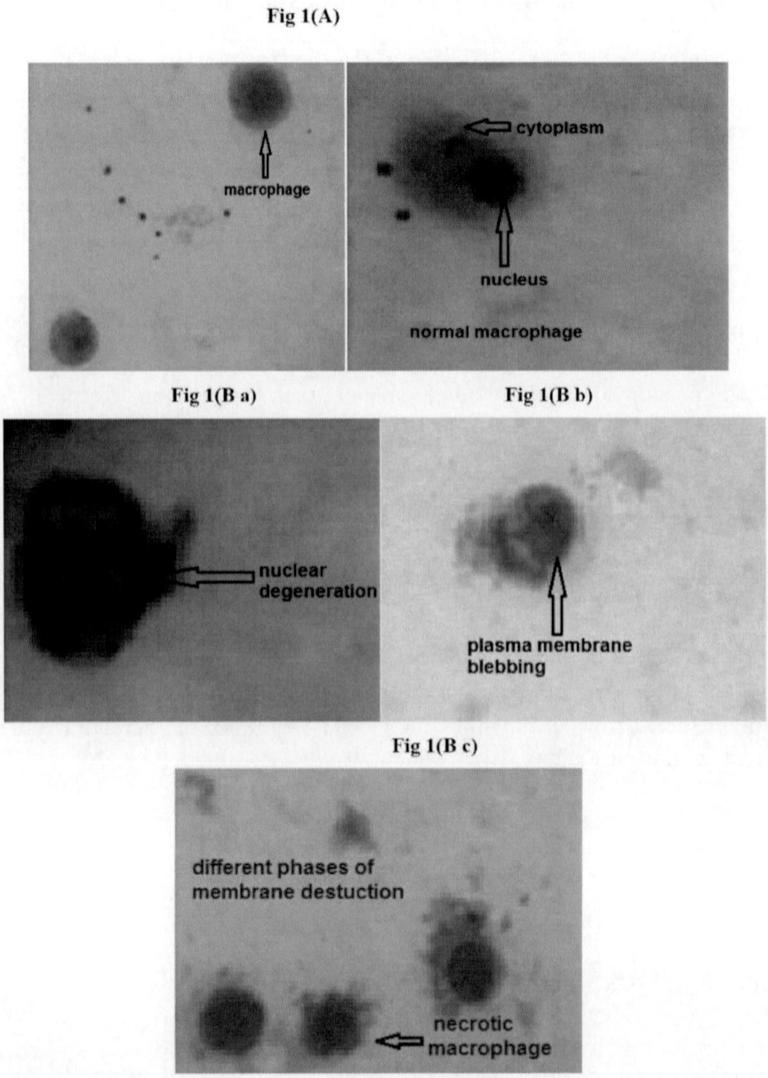

Figure 1. (A, B): (A) Normal rat peritoneal macrophage (x 400), (B) (a, b, c) Morphological variations of macrophages under sodium arsenate exposure (x 400). Necrotic macrophages are indicated by the arrow.

Giemsa Staining of Spleen Cells

General morphology of spleen cells was altered in arsenic treated rat. Arsenic treated rat exhibited cellular damage. Arsenic treated cells showed membrane blebbing and fragmentation (Figure 2A and B).

Fig 2(A)

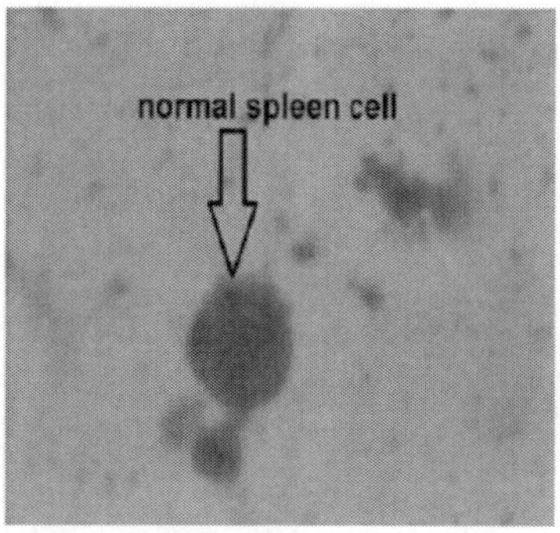

Fig 2(B a)

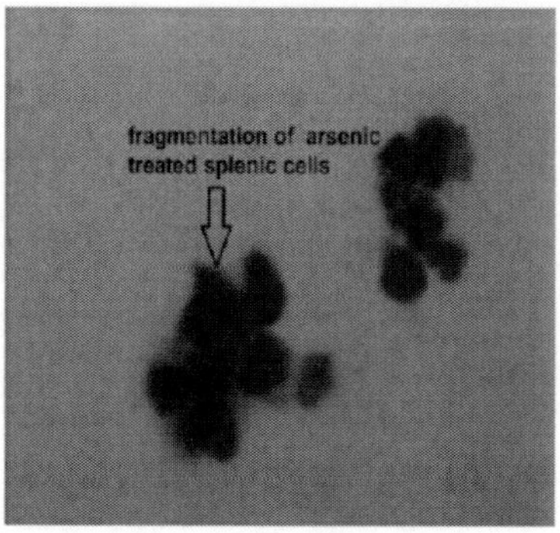

Fig 2(B b)

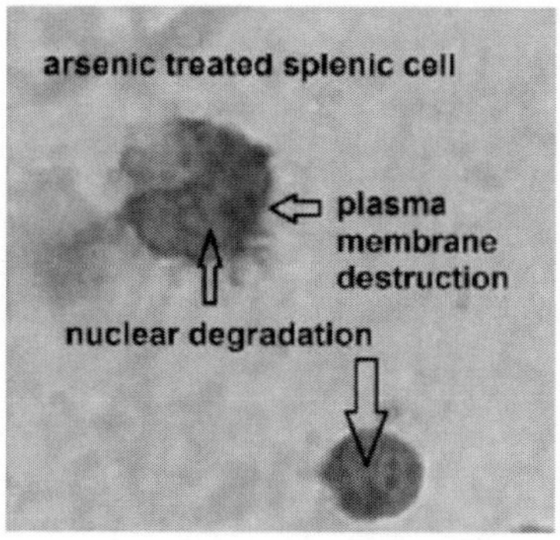

Figure 2. (A, B): (A) Normal rat spleen cells (x 400), (B: a, b) Morphological variations of spleen cells under sodium arsenate exposure (x 400). Necrotic cells showed membrane blebbing and fragmentation.

Trypan Blue Staining of Peritoneal Macrophages

Significant number of arsenic treated rat peritoneal macrophages showed Trypan blue positive response. Dead macrophages of rat were blue in colour whereas the viable cells of controls were white (Figure 3A and B).

Trypan Blue Staining of Spleen Cells

Significant number of arsenic treated spleen cells showed Trypan blue positive response. Dead cells were blue in colour whereas the viable cells of controls were white (Figure 4A and B).

Macrophage Counting

Significant number of macrophages was reduced in arsenic treated group. (12.0 ± 1.5 vs. 5.9 ± 1.3 macrophages /mm^2, $P < 0.05$; Figure 5A)

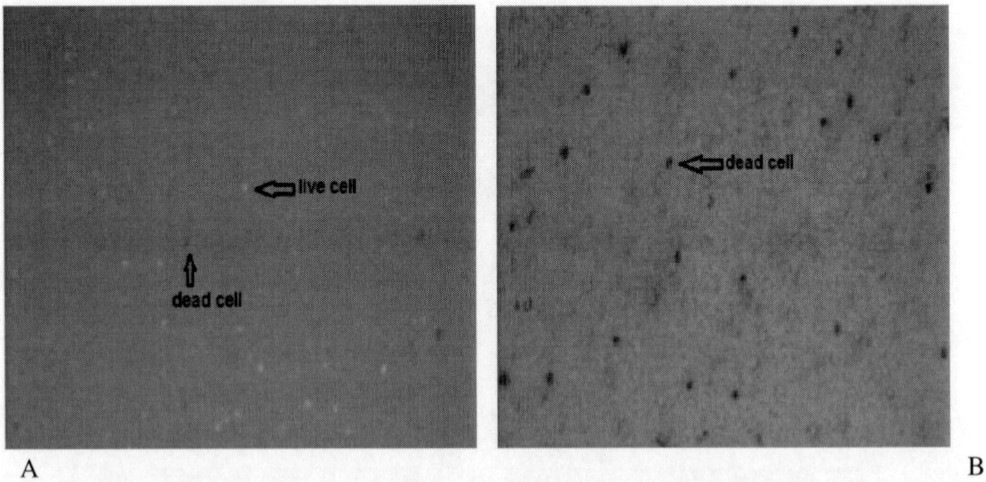

Figure 3. (A, B): (A) Normal rat viable macrophages, (B) Arsenic treated rat dead macrophages. Dead cells are indicated by the arrow.

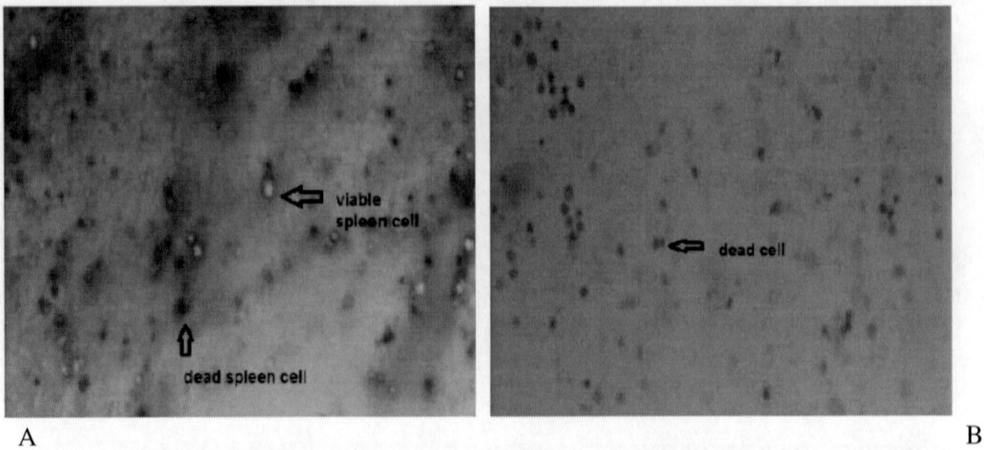

Figure 4. (A, B): (A) Normal rat viable spleen cells, (B) Arsenic treated rat dead splenic cells. Dead cells are indicated by the arrow.

Splenic Cell Counting

Significant number of spleen cells was reduced in arsenic treated group. (18.0 ± 1.5 vs. 12.9 ± 1.3 macrophages /mm^2, $P < 0.05$; Figure 5B)

Calculation of Mortality Index

Mean mortality index was significantly increased in arsenic treated group (Figure 6).

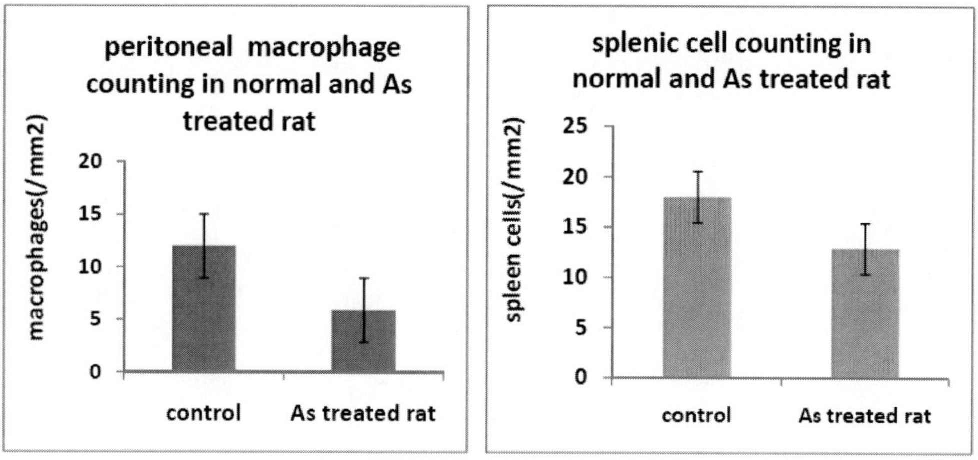

A B

Figure 5. Macrophage and spleen cell counting in normal and As treated group. Values are expressed as Mean ± SE.

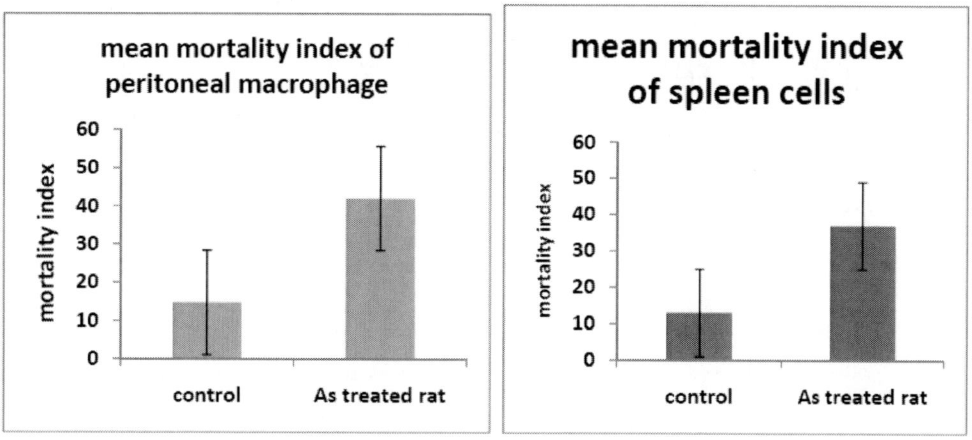

A B

Figure 6. Mean mortality index in normal and arsenic treated group. Values are expressed as Mean ± SEM.

Histology of Lung

The section of control rat lung showed the normal lung architecture with thin interalveolar septa. Notice the clear alveoli (a) and alveolar sacs (as) (Figure 7). A higher magnification showed normal alveoli lined by flat type I pneumocytes and cuboidal type II pneumocytcs (Figure 9A). A photomicrograph of section of the lung of arsenic treated rat showing marked thickening of the interalveolar septa (Figure 8A).

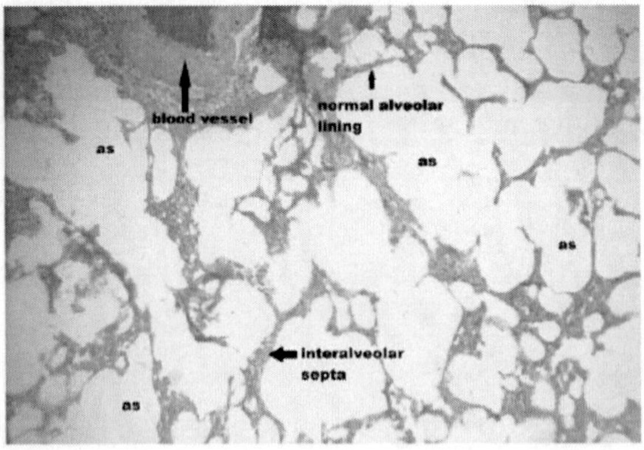

Figure 7. Normal lung architecture.

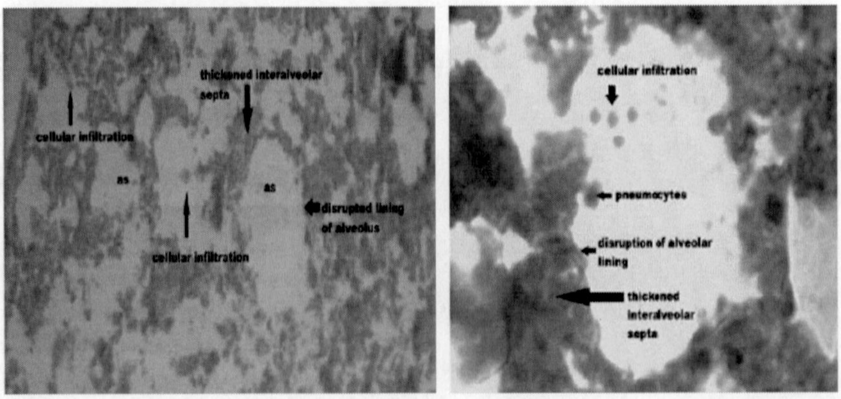

A B

Figure 8(A). Arsenic treated lung architecture. 8(B). cellular infiltration.

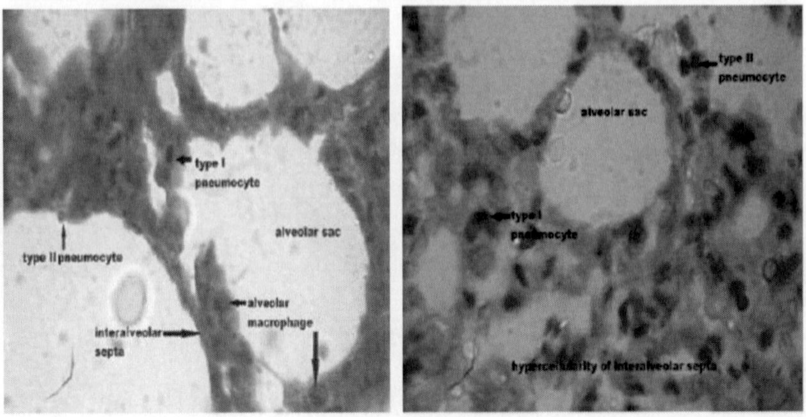

A B

Figure 9(A). Control lung pneumocytes. 9(B). Arsenic treated lung pneumocytes.

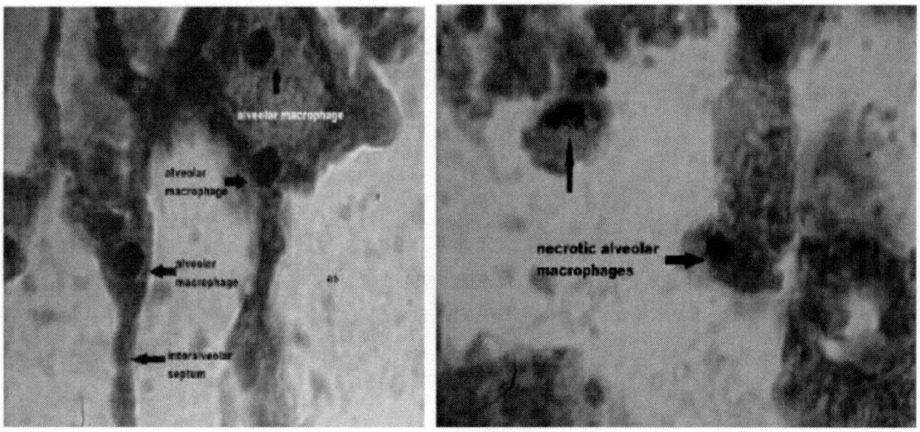

A B

Figure10(A). Control alveolar macrophages. 10(B). Arsenic treated alveolar macrophages.

Mononuclear cells appear heavily infiltrating the alveolar septa (Figure 8B). Sodium arsenate treated lung pneumocytes showed apparent increase in both size and numbers. Some areas of the affected lung tissue revealed marked hypercellularity (Figure 9B). This cellular hyperplasia included pneumocytes, capillary endothelial cells and fibroblasts. The proliferated pneumocytes were recognized by little deformed surface. Hyperplastic pneumocytes were identified by the irregularly shaped (Figure 9B).

Macrophages (M) are detected in the alveolar septa or in the alveolar space (Figure 10A). Arsenic treated rat exhibited cellular damage in alveolar macrophages. Treated macrophages become picnotic and showed membrane blebbing (Figure 10B).

DISCUSSION

The present study clearly indicates that arsenic toxicity adversely affects peritoneal macrophages and spleen cell function. Significant number of peritoneal macrophages and spleen cells were found to be picnotic and reduced in number. The cellular death was also confirmed by Trypan blue staining. It was calculated by mortality index. An increased rate of macrophage death causes a decrease of immune function in arsenic poisoning (Gardner, 1984). The dead cells took blue colour. The resulting reduction and death of peritoneal macrophages and spleen cells in rat can provide an explanation for the decrease of immune responses in arsenic toxicity (Bustamante et al., 1997; Sikorski et al., 1991).

The histological section of the lung of arsenic treated rat showed marked thickening of the interalveolar septa. Mononuclear cells appear heavily infiltrating the alveolar septa. Arsenic treated lung pneumocytes showed apparent increase in both size and numbers. Some areas of the affected lung tissue revealed marked hypercellularity. Arsenic treated rat exhibited cellular damage in alveolar macrophages. It may affect the defensive mechanism of lung (Lantz et al., 1995). Treated macrophages become picnotic and showed membrane blebbing. Pneumocytes of both types (I and II) were degenerated and had surface blebbings. Cellular hyperplasia causes thickening of the alveolar walls or even occasional collapse of the alveolar lumina. The observed pneumocyte hyperplasia is considered as an early response of the alveolar wall to injury (Smith et al., 1998). Extravascular localization of leukocytes points to acute vascular injury which may be an important component of injury caused by most pulmonary toxicants. The results demonstrate the usefulness of investigations on immune system and macrophages function as parameters in detecting the effects of arsenic toxicity.

REFERENCES

Bustamante, J; Dock, L; Vahter, M; Fowler, B; Orrenius, S. The semiconducter elements arsenic and indicum induce apoptosis in rat thymocytes. *Toxicol.,* 1997. 118, 129-36.

Gardner, DE. Alteration in macrophage function by environmental chemicals. *Environ Health Perspect.*, 1984. 55, 343-58.

Gonsebatt, ME; Vega, L; Herrera, LA; Montero, R; Rojas, S; Cebrian, ME; Ostrisky, WP. Inorganic arsenic effects on human lymphocyte stimulation and proliferation. *Mutation Res.*, 1992. 283, 91-5.

Guha Mazumder, DN; Haque, R; Ghosh, N; De, BK; Santra, A; Chakraborti, D; Smith, AH. Arsenic in drinking water and the prevalence of respiratory effects in West Bengal, India. *Int J Epidemiol.*, 2000. 29, 1047-52.

Hopenhayn-Rich, C; Biggs, ML; Smith, AH. Lung and kidney cancer mortality associated with arsenic in drinking water in Cordoba, Argentina. *Int J Epidemiol.,* 1998. 27, 561-9.

Iyengar, U. R. and Vakil, U. K. Enumeration and structural assessment of peritoneal macrophages during progressive protein deficiency in rats. *J. Biosci.*, 1985. 7, 15–26.

Lantz, RC; Parliman, G; Chen, GJ; Barber, D; Winski, S; Carter, DE. Effect of arsenic exposure on alveolar macrophage function. II. Effect of slightly soluble forms of As (III) and As (V). *Environ Res.*, 1995. 68, 59-67.

Luster, MI; Blank, JA. Molecular and cellular basis of chemically induced immunotoxicity. *Ann Rev harmacol Toxicol.,* 1987. 27, 23-49.

Parakkal, PF. Mordanting fixation as a means of facilitating the staining of pancreatic cells of mouse. *Stain Technol.*, 1961. 36, 33-34.

Quatrehomme, G; Ricq, O; Lapalus, P; Jacomet, Y; Ollier, A. Acute arsenic intoxication: forensic and toxicologic aspects (an observation). *J Forensic Sci.*, 1992. 37, 1163-71.

Sikorski, EE; Burns, LA; Stern, ML; Munson, AE. Splenic cell targets in gallium arsenide induced suppression of primary antibody response. *Toxicol Appl Pharmacol.,* 1991. 110, 129-54.

Smith, AH; Goycolea, M; Haque, R; Biggs, ML. Marked increase in bladder and lung cancer mortality in a region of Northern Chile due to arsenic in drinking water. *Am J Epidemiol.,* 1998. 147, 660-9.

Squibb, KS; Fowlar, BA. The toxicity of arsenic and its compounds. In: Fowlar BA editor. Biological and environmental effects of arsenic. New York. *Elsevier.*, 1983. 233 - 69.

In: Vertebrate and Invertebrate Animal Approaches
Editor: José Rosa Gomes
ISBN: 978-1-63117-102-4

Chapter 10

STUDY OF CYTOMORPHOLOGY OF PANCREATIC ISLETS AND PERITONEAL MACROPHAGE IN ALLOXAN-INDUCED DIABETIC RATS: A MECHANISTIC INSIGHT

Srikanta Guria*[*1], Swapna Chhetri[1], Sayantan Saha[1], Gitanjali Singh[1], Piali Bala Saha[1], Nikita Chetri[1], Bilkis Sultana Sarkar[1] and Madhusudan Das[2]

[1]Post Graduate Department of Zoology, Parasitology and Medical Entomology Laboratory, Darjeeling Govt. College, Darjeeling, West Bengal, India
[2]Department of Zoology, University of Calcutta, Kolkata, West Bengal, India

ABSTRACT

Diabetes mellitus has been considered as one of the major health concerns all around the world today. Alloxan-induced diabetes is one of the widely used model to induce Type I diabetes mellitus in the experimental animals. Alloxan has been found to be selectively toxic to pancreatic beta cells as it causes rapid destruction of beta cells. We aimed to understand the effect of alloxan induced diabetes on cytomorphology of peritoneal macrophages and islets cells in rat. Alloxan treated rat exhibited significant morphological changes in peritoneal macrophages, consequently it showed death of the pancreatic islets cells. An increased rate of macrophage death causes a decrease of antimicrobial activity of macrophages as well as immune function in alloxan diabetes. But further experiments are in progress to solve this paradox.

Keywords: Alloxan, diabetes, islets cells, macrophage

[*] Corresponding author: Phone: 91-9830286956. srikantarna@yahoo.com, guriasrikanta@gmail.com.

INTRODUCTION

Diabetes mellitus is a metabolic disorder constituting a major health concern today whose prevalence has increased worldwide over the past few decades (Kruger et al., 2012). One of the most potent methods to induce experimental diabetes mellitus is chemical induction by alloxan (Etuk et al., 2010; Mythili et al., 2004). Alloxan is a cyclic urea derivative causes selective necrosis of the β- cells of pancreatic islets through the generation of cytotoxic reactive oxygen species (ROS) (Lenzen et al., 2008). Alteration in macrophage functions in diabetes has also been reported (Mohsenin et al., 1990; Gaulton et al., 1985). However, the mechanism of influence of alloxan on the immune system is unclear. The objective of the present study was to examine the cytomorphology of rat peritoneal macrophages. Our experiments showed that macrophages of alloxan diabetic rat showed an increased death. However, it is difficult to correlate with reported increased susceptibility of diabetic subjects to infections in which macrophages play a leading defensive role.

MATERIALS AND METHODS

Animals and Housing

The laboratory experiments were performed using rat (*Rattus rattus*) as a mammalian model. Normal rats aged 7-10 weeks and weight 80-100 g were housed in polypropylene cages and were acclimatized in laboratory condition for two weeks with natural light and dark schedules prior to experimentation. The animals were fed standard rodent diet and water was provided *ad libitum*.

Alloxan Treatment of Animals

Rats were divided into experimental and their respective control groups. Group I animals were treated with intra peritoneal injection of alloxan with a concentration of 10% solution in 0.9% NaCl, at the dose of 100 mg/kg of body weight (Naglaa et al., 2010).

Glucose Tolerance Test (GTT)

For glucose tolerance test (GTT) blood was collected first from the tail veins of control and alloxan treated rats after 18 hr of fasting followed by challenge with glucose (25 mg glucose/100 g body weight) and at the following time point after glucose infusion: 1.5 hr, 2.5 hr and 24 hr. Blood glucose was measured using a blood glucose monitoring system (glucometer).

Macrophage Isolation

Normal saline (0·9% NaCl) was injected into rat peritoneum and the aspirate was taken for macrophage study (after Iyengar et al., 1985).

Fixation and Staining of Macrophages

Peritoneal fluid was placed and smeared directly on sterilized glass slides and incubated at 37°C in a humid chamber for 3 hours. After incubation the non adherent cells were removed by washing two times with 0·1 M phosphate buffer saline (PBS, pH 7·2). The adherent macrophages were fixed by methanol. After fixation macrophages were stained by Giemsa and observed under light microscope (after Iyengar et al., 1985).

Trypan Blue Dye Exclusion Test

After the incubation cells were treated with 50 μl of 0.25 % trypan blue dye solution for 5 minutes, the percentage of macrophages with cytoplasm stained blue can be obtained by counting ten optical fields under a light microscope at magnification 1, 000. Cells that have taken up the dye are dead, since the dye is normally excluded by the membranes which maintain their semi permeability intact and therefore, the percentage of blue-stained cells represents a mortality index.

$$\text{Mortality index} = \frac{\text{Number of cells wih blue stained cytoplasm}}{\text{Total number of cells}} \times 100$$

Histological Analysis

Before autopsy, rats were anesthetized with chloroform. Pancreatic tissues were dissected out and fixed in Bouin's fixative (Parakkal, 1961). Tissues were embedded in paraffin, sectioned (5 μm), mounted on glass slide and stained with hematoxylin-eosin for histological analysis.

Statistical Analysis

All the data were presented as mean ±SEM. The data were analyzed by *t*-test. Results with $P < 0.05$, $P < 0.01$, $P < 0.001$ were considered statistically significant.

Results

Blood Glucose Level

In the control rats the blood glucose level returned to the normal level after 24 hr of glucose feeding. Like control rats, in diabetic rats glucose level increased after 1.5 hr of glucose challenge but the elevated glucose didn't return to control level even after 24 hr of glucose challenge (Figure 1).

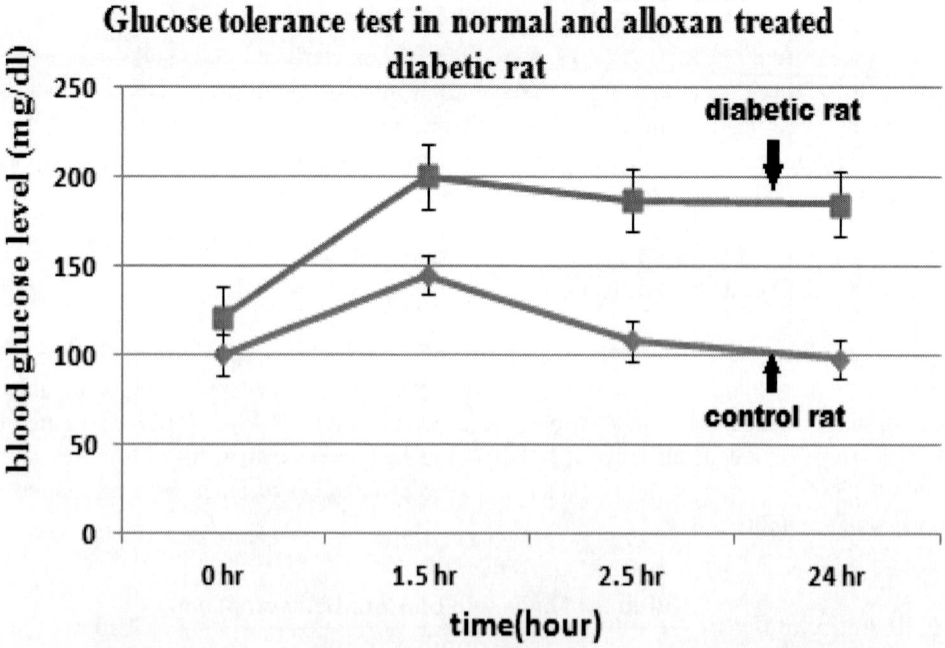

Figure 1. Blood glucose level in response to glucose tolerance test in control and alloxan treated diabetic rats. Values are expressed as mean ± SE. P values: <0.001. P-value < 0.05 is considered to be statistically significant.

Histology of Pancreas

Control group showed a normal lobular architecture of the pancreas. The islets appeared lightly stained than the surrounding acinar cells, with intact Interlobular connective tissue and interlobular duct (Figure 2A). Each islet consisted of lightly stained polygonal cells arranged in cords separated by a network of blood capillaries. The pancreatic sections of the diabetic group showed marked morphological changes. Blood vessels were seen congested and dilated. Some islets cells showed pyknosis. A significant number of islets cells were found to be reduced in number. The architecture of connective tissue septa in diabetic group was found to be altered (Figure 2B).

Giemsa Staining of Macrophages

Significant changes were observed in the cytomorphology of peritoneal macrophages in the alloxan treated group when compared with the control group under light microscopy. Peritoneal macrophages of control rat showed different morphological features like pseudopodia, long filopodia like structure, food vacuole and phagocytic cavity (Figure 3 A, B, C). General morphology of macrophages was altered in alloxan treated rat. Alloxan treated diabetic rat showed cellular damage. A significant percentage of peritoneal macrophages become picnotic. Alloxan treated macrophages showed membrane blebbing and nuclear degeneration (Figure 4B; Figure 5B; Figure 6B). Higher magnification of diabetic rat peritoneal macrophages showed different phases of cellular death like formation of membrane blebs, rupture of plasma membrane and release of cytoplasmic contents (Figure 5C).

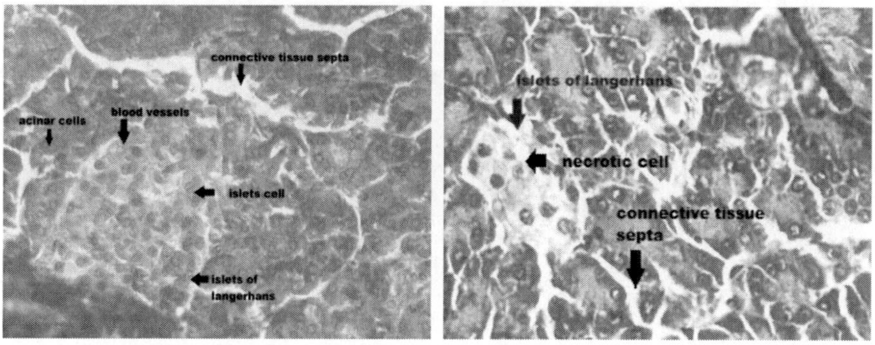

A B

Figure 2. (A, B): (A) Normal rat pancreatic islets (x 400), (B) diabetic rat pancreatic islets (x 400).

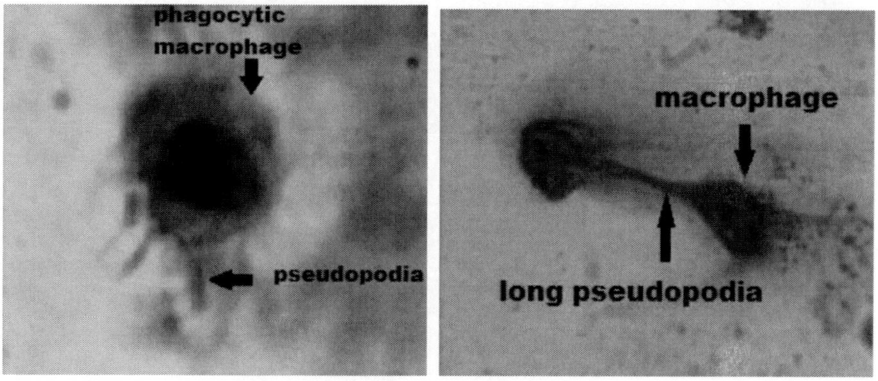

A B

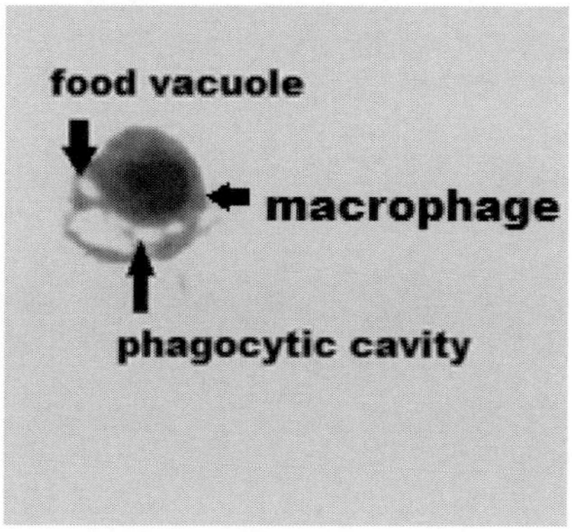

C

Figure 3. (A, B, C): Normal rat peritoneal macrophages (x 400), Pseudopodia of phagocytic macrophage (A), finger like cytoplasmic projection of macrophages (B), food vacuole, phagocytic cavity of normal peritoneal macrophages (C).

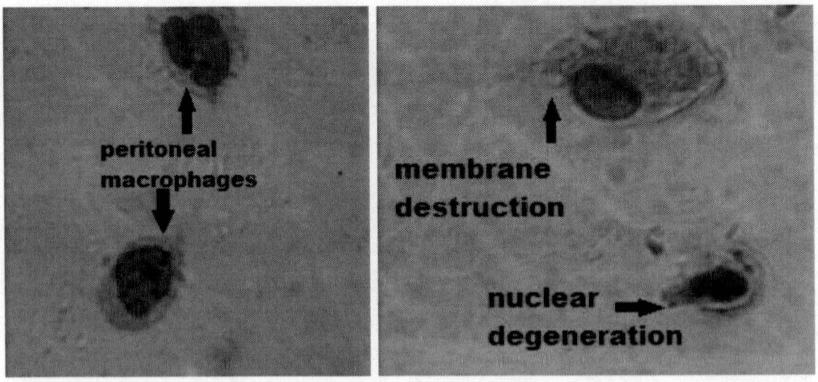

A B

Figure 4. (A, B): Control rat peritoneal macrophages with normal morphology (x 400) (A), alloxan treated macrophages with membrane destruction and nuclear degeneration (x 400) (B).

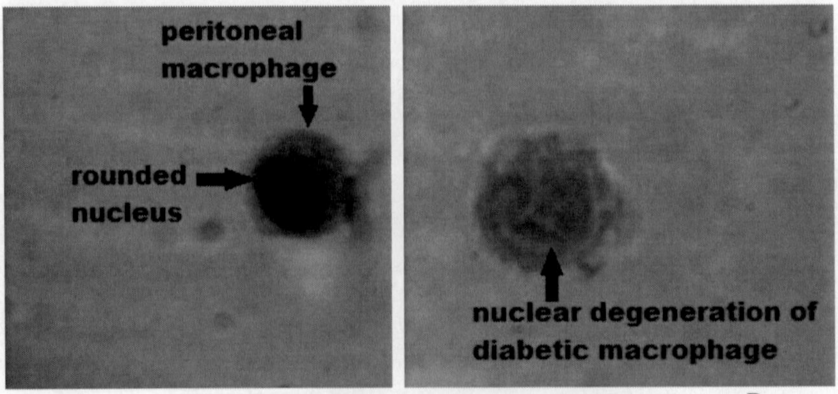

A B

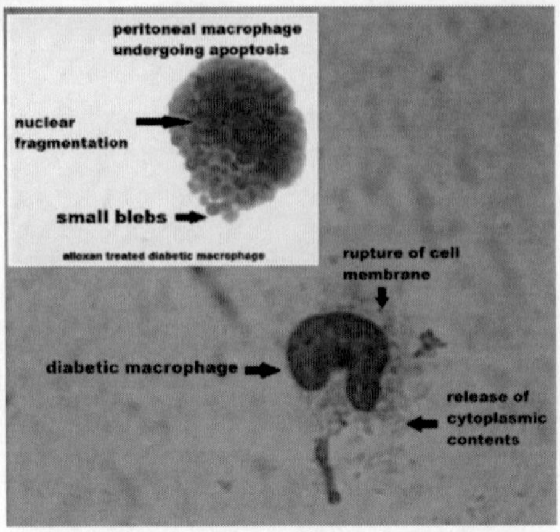

C

Figure 5. (A, B, C): Control rat peritoneal macrophages with rounded nucleus (x 400) (A), alloxan treated macrophages with nuclear degeneration (x 400) (B), different phases of cellular death found in alloxan treated macrophages i.e. macrophage apoptosis or necrosis (x 1000) (C).

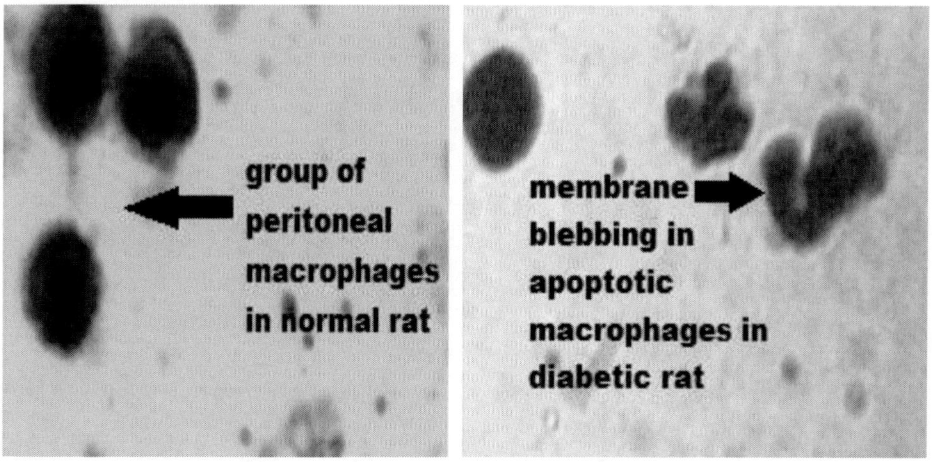

A B

Figure 6. (A, B): Control rat peritoneal macrophages with proper morphology (x 400) (A), alloxan treated apoptotic macrophages with membrane blebbing (x 400) (B).

Trypan Blue Staining of Peritoneal Macrophages

Significant number of alloxan treated rat peritoneal macrophages showed Trypan blue positive response.

Dead macrophages of rat were blue in colour whereas the viable cells of controls were white (Figure 7A and B).

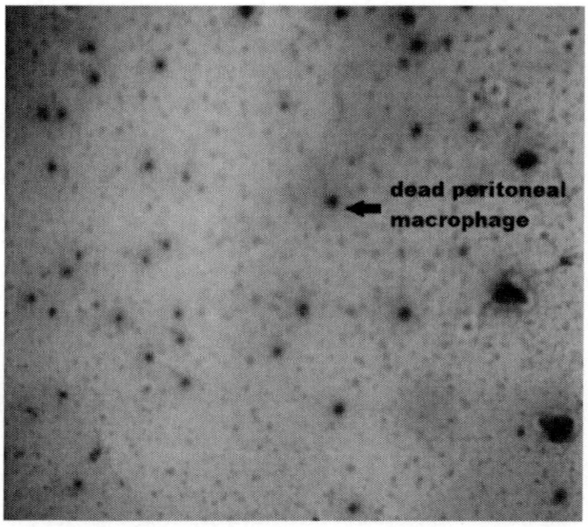

A B

Figure 7. (A, B): Normal rat viable macrophages (A), alloxan treated diabetic rat dead macrophages (B).

Calculation of Mortality Index

Mean mortality index was significantly increased in diabetic group (Figure 8).

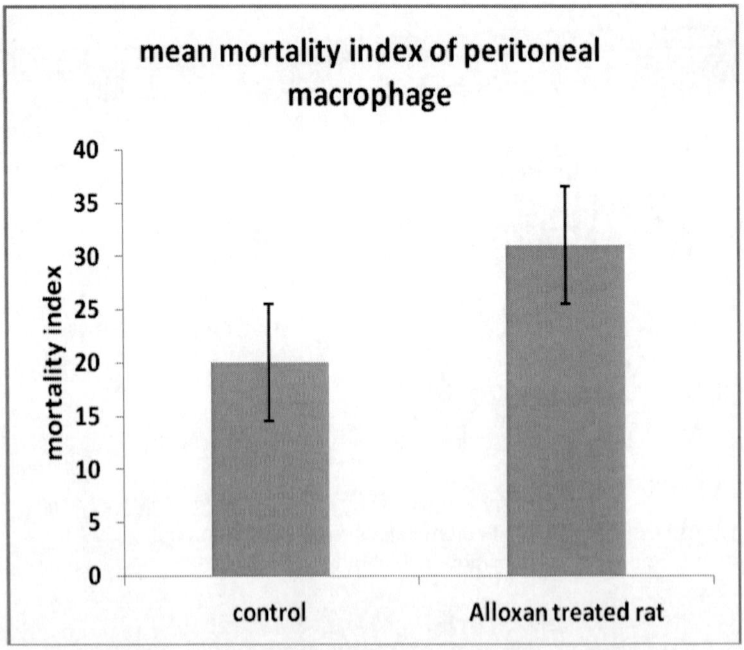

P-Value < 0.05 is considered to be statistically significant.

Figure 8. Mean mortality index in normal and alloxan treated group. Values are expressed as Mean ± SEM.

Discussion

The present study examined the effects of alloxan treated diabetes on pancreatic islets as well as peritoneal macrophages. The results showed that alloxan rat developed atrophy of pancreatic islets and pyknosis of islets cells. In alloxan treated diabetic rat increments of blood glucose levels were observed after GTT and the hyperglycemia persisted even 24 h after glucose load. It is believed that insulin response to a glucose load is relatively decreased in diabetes (Lenzen et al., 2008). It is possible that alloxan causes deleterious effects on β-cell function thereby impairing insulin secretion.

It has been shown in humans as well as in animals that in diabetic subjects *in vitro* and *in vivo* cell-mediated and humoral immune responses are significantly reduced (Larkin et al., 1985) and the degree of impairment is directly related to the blood glucose level. To explain the increased incidence of bacterial and mycotic infections in poorly controlled diabetic patients it has been argued that activities of professional phagocytes (granulocytes, macrophages) are also impaired by hypoinsulinaemia (Mowat et al., 1971; Ptak, et al., 1983; Tierney et al., 1995). It is well known that macrophages possess specific insulin binding sites on their surfaces and the saturation of these receptors by insulin is critical for some important immune functions of these cells (Bar et al., 1977; Abrass et al., 1991). It has been demonstrated that insulin deficit in diabetes affects production of monokines, tumour necrosis factor-alpha (TNF-α)) and active NO and reactive oxygen radicals from macrophage (Ptak, et al., 1981). Their diminished secretion could possibly affect the overall antimicrobial activity of diabetic macrophages (Tierney et al., 1995; Gaulton et al., 1985). Our present study clearly indicates that alloxan treated diabetes adversely affects peritoneal macrophages. Significant numbers of peritoneal macrophages were found to be picnotic. Alloxan treated macrophages showed membrane blebbing and nuclear degeneration. Diabetic rat peritoneal macrophages showed different phases of cellular death like formation of membrane blebs, rupture of plasma membrane and release of cytoplasmic contents. It indicated the cells undergoing apoptosis or necrosis. Peritoneal macrophage death was also confirmed by Trypan blue staining. It was calculated by mortality index which was significantly increased in diabetic group.

In conclusion, our results show that alloxan diabetes influence pancreatic islets morphology concomitant with cytomorphology of peritoneal macrophages. Considering that macrophage is an important component of the defence system, suppression of macrophage viability may explain, at least in part, the increased susceptibility of diabetic patients to infection. The present findings may provide a novel approach to elucidate the mechanisms of macrophage-related disorders in diabetes. But further experiments are in progress to solve this paradox.

REFERENCES

Abrass, C.K. Fc receptor-mediated phagocytosis: abnormalities associated with diabetes mellitus. *Clin. Immunol. Immunopathol.,* 1991. 58, 1–17.

Bar, R.S; Kahn, C.R; Koren, H.S. Insulin inhibition of antibody-dependent cytotoxicity and insulin receptors in macrophages. *Nature.*, 1977. 265, 632–4.

Etuk, E.U. Animals models for studying diabetes mellitus. *Agric. Biol. J. N. Am.*, 2010. 1, 130-4.

Gaulton, G.N; Schwartz, J.L; Eardley, D.D. Assessment of the diabetogenic drugs alloxan and streptozotocin as models for the study of immune defects in diabetic mice. *Diabetologia.*, 1985. 28, 769–75.

Iyengar, U. R. and Vakil, U. K. Enumeration and structural assessment of peritoneal macrophages during progressive protein deficiency in rats. *J. Biosci.*, 1985. 7, 15–26.

Kruger, D.F; Lorenzi, G.M; Dokken, B.B; Sadler, C.E; Mann, K; Valentine, V. Managing diabetes with integrated teams: maximizing your efforts with limited time. *Postgrad Med J.,* 2012. 124, 64-76.

Larkin, J.G; Ireland, J. Diabetes mellitus and infection. *Postgrad. Med. J.*, 1985. 61, 233–7.

Lenzen, S. The mechanisms of alloxan and streptozotocin induced diabetes. *Diabetologia.*, 2008. 51, 216- 26.

Mohsenin, H; Latifpour, J. Respiratory burst in alveolar macrophages of diabetic rats. *J. Appl. Physiol.*, 1990. 68, 2384–90.

Mowat, A.G; Baum, J. Chemotaxis of polymorphonuclear leukocytes from patients with diabetes mellitus. *N. Engl. J. Med.*, 1971. 284, 621–4.

Mythili., M.D; Vyas, R; Akila, G; Gunasekaran, S. Effect of streptozotocin on the ultrastructure of rat pancreatic islets. *Microsc. Res. Tech.*, 2004. 63, 274-81.

Naglaa, Z.H; Hesham, A.M ; Hosny Abdel Fadil and Abdel Motal, S.M. Impact of Metformin on Immunity and Male Fertility in Rabbits with Alloxan- Induced Diabetes. *Journal of American Science.*, 2010. 6(11), 417-26.

Parakkal, PF. Mordanting fixation as a means of facilitating the staining of pancreatic cells of mouse. *Stain. Technol.*, 1961. 36, 33-34.

Ptak, W; Rewicka, M; Bielecka, J. Macrophage function in alloxan diabetic mice: expression and activity of Fc receptors. *J. Clin. Lab. Immunol.*, 1981. 5, 121–4.

Ptak, W; Rewick, M et al. Alleviation of IgE-mediated immune reactions in hypoinsulinaemic and hyperglycaemic mice. *Clin. Exp. Immunol.*, 1983. 52, 54–60.

Tierney, M.R; Baker, A.S. Infections of the head and neck in diabetes mellitus. *Infect. Dis. Clin. North Am.*, 1995. 9, 195–216.

In: Vertebrate and Invertebrate Animal Approaches
Editor: José Rosa Gomes
ISBN: 978-1-63117-102-4

Chapter 11

ANTIOXIDANT ENZYME ACTIVITY AND LIPID PEROXIDATION IN LIVER OF WISTAR RATS EXPOSED TO ARABIAN INCENSE

Majed S. Alokail[1*], Mohammad A. Ismael[1] and Saud A. Alarifi[2]
[1]Department of Biochemistry, College of Science #5,
King Saud University, Saudi Arabia
[2]Department of Zoology, College of Science #5,
King Saud University, Saudi Arabia

ABSTRACT

Incense is an oleoresin that oozes from incisions in the trunks and leaves of trees of the genus Boswellia (B. carterii and B. papyrifera, native of Arabia, Africa and India). Incense composed with aromatic plant substances which is available in numerous forms, such as stick, cone and coil. In Arabian cultures the incense is brunt traditionally for several domestic purposes. The aim of this study was to investigate the indoor air pollution effects of Arabian incense smoke in antioxidant enzymes and lipid peroxidation in liver of male Wistar rats. Sixteen Wistar rats were exposed daily to incense smoke over a period of 15 weeks (4 grams daily). There was significant decrease in the levels of liver alkaline phosphatase (ALP), alanine aminotransferase (ALT) and aspartate aminotransferase (AST). Incense smoke also decreased levels of GSH and the activities of superoxide dismutase (SOD), catalase (CAT), and glutathione peroxidase (GPx) by 34.6%, 29%, 17.1%, and 17.4%, respectively. However, incense smoke increased lipid peroxidation by 61%. In conclusion, our findings strongly showed hepatotoxic potential causes of incense smoke and its harmful effects on health.

Keywords: Arabian incense, Wistar rats, Antioxidants enzymes, Lipids peroxidation

[*] Corresponding Author: E-mail: malokail@ksu.edu.sa.

INTRODUCTION

Arabian incense is an oleoresin oozing from incisions in the trunks and leaves of trees of the gnus *Boswellia* with a mix of some oriental oils and perfumes (Basto an Aenha, 1991). Arabian incense burning is a daily practice in the Arabian countries. According to the local customs, the incense is burnt and the individuals get in a close contact with the resultant smoke. The health effects of indoor pollution due to incense smoke have not geting much attention until recent years.

In addition, incense burning produces CO, CO_2, NO_2, SO_2, and many volatile organic compounds, such as benzene, toluene, and xylenes, as well as aldehydes and polycyclic aromatic hydrocarbons (Lee and Wang, 2004, Lin et al, 2008). Acute symptoms of incense fumes exposures include eye irritation and nose irritation, throat irritation, headaches, nausea, dizziness, and asthma exacerbation (Abdul Wahab and Mostafa, 2007). Many studies have shown chronic effects of pollutants emitted from incense burning in a close environment on human health. Exposure to volatile organic compounds increases the risk for the development cancer, liver, kidney and central nervous system damage. (Brandt and Bernstein, 2006; Jetter et al., 2002; MBride et al., 1999; Liou et al., 1996). Intensive exposure to incense smoke has accounted for obstructive lung disease and many asthmatic cases (Lee et al., 2003; Dossing et al., 1994; Dawod and Hussain, 1995).

Alarifi et al. studied the histological effects of the Arabian incense (Bokhour) on lung tissues. They found that incense smoke causes morphological changes of alveolar pneumocytes and infiltration of neutrophils into alveolar lumena in experimental rats. Alveolar walls also revealed deposition of collagen fibrils which contributed in its thickening (Alarifi et al., 2004). Similar ultrastructural pulmonary changes have also been reported in rats exposed to Arabian incense (Alarifi et al., 2004).

Reactive oxygen species (ROS) consist of superoxide anion (O_2^-), hydrogen peroxide (H_2O_2) and hydroxyl radical ($OH^\cdot$). These highly reactive species can cause a wide spectrum of cell damage including lipid peroxidation, inactivation of enzymes, alteration of intracellular oxidation–reduction stage, and damage to DNA (Szeto et al, 2009). Excess free radicals can result from overexposure to environmental factors including smoking, ultraviolet radiation, and pollutants leading to tissue damage and hypoxia (Halliwell, 1996).

Furthermore, oxidative stress is thought to play an important role in various disease processes, and can act as initiator or promoter of carcinogenesis (Poeggeler et al., 1993). Therefore, tissues are protected by expression of stress-response genes and genes encoding antioxidant enzymes. ROS generated from leucocytes in the blood or inhaled in the form of environmental oxidant pollutants are scavenged by antioxidants and antioxidant enzymes. This fact shows us the importance of the ROS scavenging and the antioxidant defense system in maintaining normal cellular physiology, facing diseases and promoting immunity.

In our knowledge, there has been no investigation or publication on the link between incense exposure and antioxidant enzymes nor lipid peroxidation. Therefore, the present study is the first investigation undertaken to determine hepatotoxic effects of Arabian incense fumes in experimental rats. The activities of antioxidant enzymes and markers of oxidative stress were evaluated in liver tissue to explore the mechanistic link between oxidative stress and liver injury.

MATERIALS AND METHODS

Animals

The experiments were performed in male Wistar albino rats (*Rattus norvegicus*) weighing about 200-300 g. The rats were obtained from the King Saud University breeding colony. and kept for a week before using in a well-ventilated room maintained at 23°C ± 2°C, 12 h light/12 h dark cycles. The animals were fed a standard rat chow diet with free access to water and were housed under controlled environmental conditions (light, temperature, feeding time, etc.). The animals were synchronized to a light-dark cycle (at 23°C ± 2°C, 12 h light/12 h dark cycles) starting at least 2 weeks before the beginning of the experiments.

Experimental animals were allotted into two groups designated group A and B of 16 rats each. Animals of group B were exposed to the Arabian incense over a period of 15 weeks in the exposure chamber. The total amount of the incense used was 420 grams at the rate of 4 grams per day. Exposure time was 30-40 min/day. Animals in group A served as unexposed controls. At the end of the experimentation period, all experimental and control rats were anesthetized and killed by cervical dislocation. Liver tissues were removed and fragments were fixed by immersion in 10% neutral buffered formalin and the rest was stored at -70°C until required.

Preparation of Liver Homogenate

Liver samples were homogenized in 20 fold volume of 50 mM potassium phosphate buffer (pH 7.5) by using a homogenizer (Ultra-Turrax T 25, IKA; Werke 24, 000 r.p.m.j. Germany). Homogenates were centrifuged at 10000 g for 30 min at 4°C and the supernatants were saved at -20°C until use.

Liver AST, ALT and AP

Aspartate aminotransferase (AST), alanine aminotransferase (ALT) and Alkaline phophatase (AP) activity in liver homogenates were determined using UDI colorimetric assay kit (United Diagnostics Industry- Dammam KSA) activities were determined using UDI kits (United Diagnostics Industry- Dammam KSA) according to the instructions provided.

Lactate dehydrogenase (LDH) activity levels were determined by a commercial assay (Boehringer Mannheim, Germany). The enzymatic activity was quantified by measuring the decrease in NADH at 340 nm.

Total protein concentration of the liver homogenate supernatant was measured using the method of Bradford (1976), with bovine serum albumin (BSA) as a standard.

Reduced Glutathione (GSH)

Reduced glutathione (GSH) was determined by the method of Ellman (1959). Supernatant of liver homogenate (0.5 ml) was added to 0.5 ml 5% trichloroacetic acid solution, and centrifuged at 10000 g immediately for 5 min to precipitate the proteins. To the supernatant, 1 ml of 0.1 M sodium phosphate buffer (pH 8) and 0.5 ml of DTNB (39.6 mg in

100 ml of 1% sodium citrate solution to give a concentration of 1 mM) were added. The absorbance of the yellow color developed was measured at 412 nm. GSH concentrations were calculated from a standard curve constructed with known concentrations of reduced GSH and are reported as μg GSH/mg protein.

Estimation of Lipid Peroxidation

Hepatic lipid peroxidation was quantified by the assessment of thiobarbituric reactive substance (TBARS) level as described by Ohkawa et al., (1979). Liver tissue was homogenized in 9 volumes of 50 mM Tris-HCl buffer (pH 7.4) containing 180 mM KCl, 10 mM EDTA, and 0.02% butylated hydroxytoluene. To 0.2 ml of the tissue homogenate, 0.2 ml of 8.1% sodium dodecyl sulfate, 1.5 ml of 20% acetic acid, 1.5 ml of 0.9% thiobarbituric acid, and 0.6 ml of distilled water were added and vortexed. The reaction mixture was placed in a water bath at 95°C for 1 hr. After cooling on ice, 1.0 ml of distilled water and 5.0 ml of butanol/pyridine mixture (15:1, v/v) were added and vortexed. After centrifugation at 10, 000*g* for 10 min, the resulting colour reaction in the lower phase was measured at 532 nm. The TBARS concentration was calculated using 1, 1, 5, 5-tetraethoxypropane as a standard. TBARS levels were expressed as nmol MDA (malondialdehyde)/mg protein

Superoxide Dismutase (SOD)

SOD activity was assayed spectrophotometrically according to the method of Misra and Fridovich, (1972). This method is based on the capacity of SOD in inhibiting autoxidation of epinephrine to epinechrome. Liver tissue extract (10 μl) was added to 970 μl 50 mM carbonate buffer (pH 10.2) containing 0.1 mM EDTA. Twenty μl of 30 mM epinephrine (in 0.05% acetic acid) was added to the mixture. Autooxidation was measured at 480 nm for 4 min. SOD activity was expressed as U/mg protein, where one unit of enzyme is defined as the amount of enzyme required to inhibits the rate of epinephrine autoxidation by 50% at 25°C

Catalase (CAT)

CAT activity was assayed colorimetrically as described by Luck (1971), which involves monitoring the disappearance of H_2O_2. The reaction mixture contained 50 mM potassium phosphate buffer (pH 7.0), hydrogen peroxide (H_2O_2) at a concentration whereby it shows an optical absorbance of 0.5, and 10 μl of liver homogenate. The decrease in optical density was measured at 240 nm for 1 min. The molar extinction coefficient of 43.6 M cm^{-1} was used to determine CAT activity. One unit of activity is equal to the moles of H_2O_2 degraded/min/mg protein.

Glutathione peroxidase (GPX)

GPx activity was measured according to the method described by Rotruck et al. (1973). The reaction mixture contained 0.2 ml of 0.4 M Tris–HCl buffer (pH 7.0), 0.1 ml of 10 mM sodium azide, 0.2 ml of liver homogenate, 0.2 ml glutathione and 0.1 ml of 0.2 mM hydrogen peroxide. The contents were incubated at 37°C for 10 min. The reaction was stopped by adding 0.4 ml of 10% TCA, and centrifuged. Supernatant was assayed for glutathione content by the method of Ellman (1959). GPX activity was expressed as μg GSH utilized/min/mg protein.

Statistics

The data are expressed as mean ± standard error of the mean (SE) of three independent experiments. Differences between control and treated groups according to the parameters were tested by Student *t* test and were considered to be significant when $P < 0.05$.

RESULTS

The present study was designed to evaluate the toxic effects of incense smoke in liver of male experimental rats. Table 1 displays the activities of hepatic enzymes that serve as a hallmark of liver damage provoked by incense smoke. The activities of serum AST, ALT, ALP and LDH significantly *(P<0.05)* decreased compared to unexposed controls. Liver GSH content was significantly higher in group A controls compared with group B ($P < 0.001$). GSH levels were about 35% lower in the incense exposed rats (21.8 ± 4.5) as compared to that in the control group (33.4 ± 6.2).

Table 1. The activities of hepatic enzymes that serve as a hallmark of liver damage provoked by incense smoke. Values are expressed as mean ± SE of three independent experiments

Enzyme parameters	Group A (control)	Group B (incense smoke)
AST	105 ± 21.2	89 ± 12.4
ALT	74.7 ± 6.7	51.4 ± 5.8
ALP	78.4 ± 7.2	42.2 ± 6.6
LDH	13.3 ± 4.3	8.9 ± 5.4

Units: AST, ALT (μmoles of pyruvate liberated/min/mg protein); LDH (moles of NADH/min/mg protein); ALP (moles phenol liberated/min/mg protein).

Table 2. Effect of incense smoke on the activities of hepatic antioxidant enzymes, levels of lipid peroxidation and glutathione (GSH). Values are expressed as mean ± SE of three independent experiments

Enzyme parameters	Group A control)	Group B incense smoke)
GSH	33.4 ± 6.2	21.8 ± 4.5
LPx	3.3 ± 2.4	5.4 ± 2.6
SOD	53.0 ± 3.7	68.4 ± 5.8
CAT	10.4 ± 1.2	8.9 ± 2
GPx	42.4 ± 4.2	59.8 ± 7.6

Units: GSH (μg/mg protein), LPx (nmol MDA/mg protein), SOD (U/mg protein), CAT (units/mg protein) and GPx (μg glutathione utilized/min/mg protein).

Lipid peroxidation is the reaction of oxidative deterioration of polyunsaturated fatty acids, which are important constituents of biological membranes. There was a significant ($P<0.001$) increase in the liver MDA levels of rats exposed to incense smoke as compared to unexposed controls (Table 2). Thus the increased lipid peroxidation products suggest the presence of increased oxidative stress. In addition, the changes in the activities of enzymic antioxidants namely SOD, CAT, and GPx in liver of control and experimental animals are shown in Table 2. Rats exposed to incense smoke showed an increased SOD activity as compared to the control ($P<0.001$). The major enzymes associated with hydrogen peroxide decomposition include CAT and GPx. Although exposed rats in group B had lower catalase activity than control rats in group A, this difference was not significant ($P= 0.071$). GPx activity of group B liver homogenates increased significantly ($p< 0.05$) when compared to controls. GPx activity increased 38 % in the liver of rats exposed to incense smoke, as compared to group A controls.

DISCUSSION

Incense smoke contains numerous compounds, many of which could be oxidants and prooxidants (Jetter et al., 2002), capable of producing free radical and enhancing the oxidative stress in vivo. Liver was selected as a target for oxidative stress evaluation because of its important metabolic functions and role in the detoxification of xenobiotics. The abnormal level of liver marker enzymes ALT, AST, ALP and LDH observed in this study are the consequence of incense smoke induced liver dysfunction and denotes the damage to the hepatic cells.

The liver, immune system is protected against reactive substances and toxic oxygen metabolites generated during the metabolism of endogenous and exogenous substance by GSH (Nakamura et al., 1997). The hepatocytes are highly specialized to synthesize GSH from its precursors or to recycle it from GSSG, as well as to utilize GSH against potential toxicants (Dröge et al, 1994). Recently, Lin et al. (2008) have reported that the gas products from burning incense include CO, CO2, NO2, SO2, and incense smoke also produces volatile organic compounds, such as benzene, toluene, and xylenes, as well as aldehydes and polycyclic aromatic hydrocarbons (PAHs). They also reported that when incense smoke inhaled, they cause respiratory system dysfunction. They suggested that incense smoke is a risk factor for elevated cord blood IgE levels and has been indicated to cause allergic contact dermatitis in human (Lin et al., 2008).

In our knowledge, the current study is the first to investigate the effect of incense smoke in antioxidant enzymes and lipid peroxidation. Several recent studies have shown that DNA is damaged induced by emissions from incense smoke by addition of CAT, in both the presence and absence of SOD, indicating that hydrogen peroxide may be involved (Szeto et al, 2009). They suggested that the incense emission-induced DNA damage is not likely to be mediated by singlet oxygen in the presence of environmental light. Moreover, Navasumrit et al. (2008) have shown that incense smoke had a significant increase in DNA damage observed as elevated levels of DNA strand breaks in temple workers. These indications have been shown to be associated with an increased risk of developing diseases such as cancer (Navasumrit et al, 2008). They previously suggested that DNA strand breaks can originate from direct modification of DNA by chemicals or their metabolites and also by interaction of ROS with DNA strands (Navasumrit et al, 2005).

This is supported by results presented here, which show that the antioxidant enzymes participate in the elimination of hydrogen peroxide and lipid hydroperoxides to prevent generation of free radical and protect the cellular constituents from oxidative damage. Increased SOD dismutates superoxide to hydrogen peroxide. CAT and GPx enzymes are important scavengers of hydrogen peroxides and GSH plays a pivotal role in the defense against free radicals, as a cofactor of GPx. To support our findings with the currently available evidence of the link between antioxidant enzymes and asthma, recent studies have suggested there is An imbalance between oxidative stress and antioxidative capacity may play an important role in the development and progression of bronchial asthma (Gumral et al, 2009). They observed that oxidative stress in the patients with bronchial asthma increased whereas the antioxidant enzymes such as SOD, CAT, GSH and MDA where decreased. Al-Rawas et al, (2009) have also investigated the influence of incense smoke in children with asthma and they concluded their study in which the incense smoke is a common trigger of wheezing among asthmatic children in Oman.

In addition, high levels of lipid peroxide were recognized, as a biomarker for lipid peroxidation. An oxidative stress condition shown as elevated level of malondialdehyde (MDA) may be an outcome of hepatic antioxidants and GSH depletion. High MDA level and enhanced GPx activity in the exposed rats indicate increased consumption of GSH in hepatic tissues. Meister (1991) reported that plasma antioxidants were depleted due to the inhibition of γ-glutamyl cysteine synthetase, the rate limiting enzyme in glutathione synthesis. Factors that lead to the depletion of GSH or influence the activity of any of these antioxidant enzymes may result in toxic responses in rates exposed to incense smoke due to hepatocellular damage.

In conclusion, the present study have shown for the first time the toxic effects of Arabian incense smoke in liver of male Wistar rats. All ALP, ALT, and AST have significantly decreased and incense smoke decreased levels of GSH and enhanced the activities of SOD, GPx and lipid peroxidation. The potential harmful effect of Arabian incense smoke is need more public health concern that requires further study and implementation of proper health protection measures for those who use it in many ceremonies and occasions.

ACKNOWLEDGMENTS

This project was supported by research grant from King Saud University College of Science Research Centre. The animal experimental protocol was approved by College of Science Research Centre Ethics Committee for animal studies.

REFERENCES

Abdul Wahab, A; Mostafa, OA. Arabian incense exposure among Qatari asmatic. Apossible risk. *Saudi Medical Journal,* 2007, 28: 476-478.

Alarifi, SA; Mubarak, M; Alokail, MS. Ultrastructural changes of pneumocytes of rat exposed to Arabian incense (Bakhour). *Saudi Medical Journal,* 2004, 25, 1689-1693.

Alarifi, SA; Mubarak, M; Alokail, MS. Ultrastructure of the pulmonary alveolar cells of rats exposed to Arabian mix incense (Ma'amoul). *Journal of Biological Sciences,* 2004, 4, 694-699.

Al-Rawas, OA; Al-Maniri, AA; Al-Riyami, BM. Home exposure to Arabian incense (bakhour) and asthma symptoms in children: a community survey in two regions in Oman. *BMC pulmonary medicine*, 2009, 9, 23-31.

Basto, As; Azenha, A. Contact dermatitis due to incense. *Contact Dermatitis*, 1991, 24, 312-313.

Bradford, MM. A rapid and sensitive method for the quantitation of icrogram quantities of protein utilizing the principles of protein-dye binding. *Analytical Biochemistry*, 1976, 72, 248–254.

Brandt, D; Bernstein, JA. Questionnaire evaluation and risk factor identification for nonallergic vasomotor rhinitis. *Annals of allergy, asthma and immunology,* 2006, 96, 526-532.

Dawod, ST; Hussain, AA. Childhood asthma in Qatar. *Annals of allergy, asthma and immunology,* 1995, 75, 560-364.

Dossing, M; Khan, J; Al-Rabiah, F. Risk factors for chronic obstructive lung disease in Saudi Arabia. *Respiratory Medcine,* 1994, 88, 519-522.

Dröge, W; Schulze-Osthoff, K; Mihm, S; Galter, D; Schenk, H; Eck, HP; Roth, S; Gmünder, H. Functions of glutathione and glutathione disulfide in immunology and immunopathology. *FASEB Journal,* 1994, 8, 1131-1138.

Ellman, GL. Tissue sulfhydryl groups. *Archives of biochemistry and biophysics,* 1959, 82, 70–77.

Gumral, N; Naziroglu, M; Ongel, K; Beydilli, ED; Ozguner, F; Sutcu, R; Caliskan, S; Akkaya, A. Antioxidant enzymes and melatonin levels in patients with bronchial asthma and chronic obstructive pulmonary disease during stable and exacerbation periods. *Cell Biochemistry Function,* 2009, 27, 276-83.

Halliwell; B. Antioxidants in human health and disease. *Annual Review of Nutrition,* 1996, 16, 33–50.

Jetter, JJ; Guo, Z; McBrain, JA; Flynn, MR. Characterization of emissions from burning incense. *The Science of the total environment.* 2002, 295, 51-67.

Lee, SC; Wang, B. Characteristics of emissions of air pollutants from burning of incense in a large environmental chamber. *Atmosphere Environment,* 2004, 38, 941-951.

Lee, YL; Lin, YC; Hsiue, TR; Hwang, BF; Guo, YL. Indoor and outdoor environmental exposures, parental atopy, and physician diagnosed asthma in Taiwanese school children. *Pediatrics,* 2003, 112, e389-e395.

Lin, TC; Krishnaswamy, G; Chi, DS. Incense smoke: clinical, structural and molecular effects on airway disease. *Clinical and Molecular Allergy,* 2008, 6, 3-12.

Liou, SH; Yang, JL; Chen, SY; Lai, FM; Respiratory symptoms and pulmonary function among wood dust exposed joss stick workers. *International archives of occupational and environmental*, 1996, 68, 154-160.

Luck, H. Catalase Methods of Enzymatic Analysis. 1971. New York: Academic Press; 1971, pp. 885–893.

McBride, SJ; Ferro, AR; Ott, WR; Switzer, P; Hildemann, LM. Investigations of the proximity effect for pollutants in the indoor environment. *Journal of exposure analysis and environmental epidemiology,* 1999, 9, 602-620.

Meister, A. Glutathione deficiency produced by inhibition of its synthesis, and its reversal; applications in research and therapy, *Pharmacology and therapeutics,* 1991, 51, 155-194.

Misra, HP; Fridovich; I. The role of superoxide anion in the autoxidation of epinephrine and a simple assay for superoxide dismutase. *Journal of Biological Chemistry,* 1972, 247, 3170–3175.

Nakamura, H; Nakamura, K; Yodoi, J. Redox regulation of cellular activation. *Annual Review of Immunology,* 1997, 15, 151-169.

Navasumrit, P; Chanvaivit, S; Intarasunanont, P; Arayasiri, M; Lauhareungpanya, N; Parnlob, V; Settachan, D; Ruchirawat, M. Environmental and occupational exposure to benzene in Thailand, *Chemico Biological Interaction,* 2005, 30, 75–83.

Navasumrit, P; Arayasiri, M; Hiang, OMT; Leechawengwongs, M; Promvijit, J; Choonvisase, S; Chantchaemsai, S; Nakngam, N; Mahidol, C; Ruchirawat, M. Potential health effects of exposure to carcinogenic compounds in incense smoke in temple workers, *Chemico Biological Interactions,* 2008, 173, 19-31.

Ohkawa, H; Ohishi, N; Yagi, K. Assay for lipid peroxidation in animal tissues by thiobarbituric acid reaction. *Analytical Biochemistry,* 1979, 44, 276–278.

Poeggeler, B; Reiter, RJ; Tan, DX; Chen, LD; Manchester, LC. Melatonin, hydroxyl radical-mediated oxidative damage, and aging: A hypothesis. *Journal of Pineal Research,* 1993, 14, 151–168.

Rotruck, JT; Pope, AL; Ganther, H; Awanson, AB; Hafeman, DG. Hoeckstra WG. Selenium biochemical role as a component of glutathione peroxidase. *Science* 1973, 179, 588-590.

Szeto, YT; Sok, Wa; Leong, K; Keong, LK; Fun, CY; Benzie, IF. Effects of incense smoke on human lymphocyte DNA. *Journal of toxicology and environmental health. Part A,* 2009, 72, 369-373.

In: Vertebrate and Invertebrate Animal Approaches
Editor: José Rosa Gomes

ISBN: 978-1-63117-102-4

Chapter 12

BETAINE REDUCES HYPERHOMOCYSTEINEMIA AND ENHANCES 5-HYDROXYINDOLEACETIC ACID IN ETHANOL-INDUCED HYPERHOMOCYSTEINEMIA IN RABBITS

***Masoud Alirezaei*[1*], *Mehdi Saeb*[1], *Katayoun Javidnia*[2], *Saeed Nazifi*[3] and *Najmeh Khalighyan*[2]**

[1]Department of Biochemistry,
School of Veterinary Medicine-Shiraz University, Shiraz, Iran
[2]Medicinal and Natural Products Chemistry Research
Center-Shiraz University of Medical Sciences, Shiraz, Iran
[3]Department of Clinical Pathology,
School of Veterinary Medicine- Shiraz University, Shiraz, Iran

ABSTRACT

A high level of total homocysteine (tHcy) is one hypothesis for the association of homocysteine with cerebrovascular diseases, neurodegenerative diseases and depression of mood. Thus, we examined whether oral betaine can act as a preventive agent in ethanol-induced hyperhomocysteinemia on the monoaminergic system. A total of 32 New Zealand White rabbits were divided into four groups (n=8); the Control group (C). The Ethanol group (E) was administered ethanol at a dosage of 4 g/kg daily. The Betaine group (B) received betaine at a dosage 1.5% (w/w) of the diet daily, and the Betaine and Ethanol group (BandE) was administered with the betaine group diet; after one hour the rabbits received ethanol at a dosage of 4 g/kg daily. Blood samples were taken in the morning of the day before beginning treatment (0.0 day) and on the 30th, 60th and 90th day of the treatment. Serum folate and vitamin B_{12} levels were determined using a radioimmunoassay, tHcy level of plasma was determined by homocysteine EIA kit, and

* Corresponding author: Department of Biochemistry, School of Veterinary Medicine, University of Shiraz, Shiraz, Iran. Dr. Masoud Alirezaei. E-mail address: Alirezaei_m54@yahoo.com. Tel.: +987112286950; Fax: +987112286940.

5- hydroxyindoleacetic acid (5-HIAA) of plasma was measured with HPLC-ECD. There was a significant negative correlation between 5-HIAA and tHcy in the E group (r=-0.473, P=0.02), and compared to the E group the concentrations of 5-HIAA in the BandE group increased considerably (p<0.05). In contrast to the E group, significantly high concentrations of 5-HIAA were observed in the B and C groups. While the serum concentrations of vitamin B_{12} showed no significant difference in the BandE group on the 90th day compared to the control group, the serum concentrations of folate on the 90th day differed. However, no significant difference was observed between tHcy and gender. Overall, oral pretreatment with betaine significantly prevented ethanol-induced hyperhomocysteinemia, subsequently increasing 5-HIAA in the plasma as well as vitamin B_{12} and folate in the serum. Thus, betaine may be recommended as a pretreatment method for depressive patients with alcoholism.

Keywords: Betaine; Hyperhomocysteinemia; 5-hydroxyindoleacetic acid; Ethanol; Vitamin B_{12}; Folate

Abbreviations

5-HIAA	5- hydroxyindoleacetic acid;
tHcy	total homocysteine;
HPLC-ECD	high performance liquid chromatography-electrochemical detector;
MTHF	5-methyl tetrahydrofolate;
BH4	tetrahydrobiopterin;
BHMT	betaine-homocysteine methyltransferase;
DHFR	dihydrofolate reductase;
HVA	homovanillic acid;
MTHFR	methyltetrahydrofolate reductase;
NMDA	N-methyl-D-aspartate;
SAMe	S-adenosyl methionine;
SAH	S-adenosyl homocysteine;
SAHH	S-adenosyl homocysteine hydrolase;
CSF	cerebrospinal fluid;
5HT	serotonin;
CBS	cystathionine beta-synthetase;
GSH	glutathione;
DMG	dimethyl glycine;
MS	methionine synthase;
MAT	methionine adenosyl transferase.

Introduction

Chronic alcoholism leads to elevated plasma homocysteine levels, as shown by clinical investigations and animal experiments (Bleich et al., 2004). Homocysteine, a metabolite of the essential amino acid methionine, can be either remethylated to methionine by enzymes that require folate or cobalamin, or catabolized by cystathionine β-synthase, a pyridoxine-dependent enzyme, to form cysteine (Supplementary file) (Kruman et al., 2000).

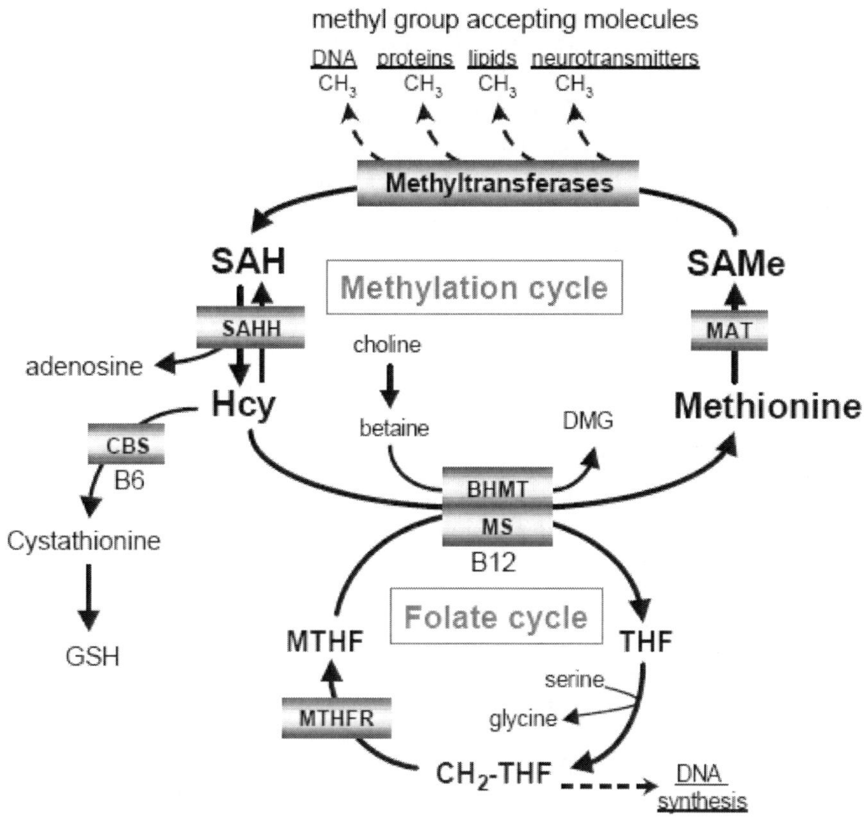

S. 1. Homocysteine metabolism. Homocysteine has three main metabolic fates: to be remethylated to methionine, to enter the cysteine biosynthetic pathway, and to be released into the extracellular medium. Hcy, homocysteine; CBS, cystathionine beta-synthetase; GSH, glutathione; DMG, dimethyl glycine; MS, methionine synthase; BHMT, betaine-homocysteine methyltransferase; MTHF, 5-methyl tetrahydrofolate; MTHFR, methyltetrahydrofolate reductase; THF, tetrahydrofolate; MAT, methionine adenosyl transferase; SAMe, S-adenosyl methionine; SAH, S-adenosyl homocysteine; SAHH, S-adenosyl homocysteine hydrolase. (Bottiglieri, 2005). T. Bottiglieri, Homocysteine and folate metabolism in depression, Progress in Neuropsychopharmacology & Biological Psychiatry 29 (2005) 1103-1112.

The formation of methionine from homocysteine can occur either via betaine or via 5-methyl-tetrahydrofolate (MTHF). Animal studies have shown that both pathways are equally important and betaine is a vital methylating agent (Craig, 2004). Liver betaine homocysteine methyl transferase (BHMT) concentrations increase when rats are fed diets supplemented with betaine or choline, showing an adaptive change in the catabolism of betaine (Finkelstein et al., 1983). It is interesting to note that the content of BHMT in rodent livers increased with both restriction and excessive dietary methionine, prompting the suggestion that this reaction could function to either conserve methionine or remove excess homocysteine (Finkelstein, 2007). Elevated total homocysteine (tHcy) level has been observed as a result of chronic alcohol consumption in rats and ethanol has altered sulfur amino acid metabolism, including a decreased conversion of methionine to S-adenosylmethionine (SAMe) and homocysteine to methionine(Cravo et al., 1996; Bottiglieri, 2005). All tissues possess the methionine cycle with methyltetrahydrofolate as the methyl donor, but the liver, kidney, pancreas, intestine and brain also contain the transsulfuration pathway (Finkelstein, 2007). Folate and homocysteine are related through the one-carbon cycle, which involves the production of S-adenosyl methionine from adenosine triphosphate and methionine. SAMe, which is uniformly distributed in the brain, serves as the major donor of the methyl groups required in the synthesis of neuronal messengers and membranes (Papakostas et al., 2005). Folate and vitamin B_{12} deficiency, hyperhomocysteinemia and the T677 allele of the methyl-enetetrahydrofolate reductase (MTHFR) gene, which cause impaired methylation reactions in the central nervous system, have been associated with depressive disorders (Kim et al., 2008). In addition, methyl folate has been proven to have an antidepressant effect and correlates with cerebrospinal fluid 5-hydroxyindoleacetic acid (5-HIAA) (Atmaca et al., 2005). Patients with severe hyperhomocysteinemia exhibit a wide range of clinical manifestations including neurological abnormalities such as mental retardation, cerebral atrophy, and depression (Kruman et al., 2000). Effective treatment of depressive disorders being advanced by the precise delineation of the neurochemical basis of the disease as well as alterations in the catecholaminergic system during depression and upon antidepressant treatment have been reported (Zangen et al., 1999). Monoaminergic abnormalities have been implicated in the pathophysiology of depression and alcoholism. For example, lower cerebrospinal fluid (CSF) 5-HIAA levels with alcoholism are associated with a higher lethality of suicide attempts in major depression (Sher et al., 2007).

It is well known that alcoholism is associated with altered CSF monoamine metabolite levels. The reduction of serotonin metabolite, 5-HIAA, has been observed in the serum samples of depressive patients (Bose et al., 2004). Taking the above into consideration, we hypothesized that the oral administration of betaine prior to ethanol can act as a methylating agent to increase the level of 5-HIAA in ethanol-induced hyperhomocysteinemia in rabbits. We also investigated how plasma tHcy varied with concentrations of vitamin B_{12} and folate.

MATERIALS AND METHODS

Materials

Alcohol (Ethanol 95%) and 1-octanesulfonic acid sodium salt were from the Merck Chemical Company (Merck, Darmstadt, Germany). Betaine (Betafin® 96%) was obtained from the Biochem Company (Lohne, Germany). 5-Hydroxyindoleacetic acid was purchased from Sigma (St Louis, MO, USA). SimulTRAC-SNB Radioassay kit vitamin B_{12} [^{57}Co]/Folate [^{125}I] was prepared by MP, Biomedical, LLC (CNI Pharmaceutical, USA) and the homocysteine kit was prepared by Axis® Homocysteine EIA (Axis-Shield AS, Germany). All other chemicals used were of analytical grade.

Animals and Experimental Design

All animal experimentation procedures were approved by the Institutional Animal Care and Use Committee of Shiraz University of Medical Sciences. A total of 32 adult New Zealand White rabbits (2.0-2.5 kg) obtained from the animal house of Shiraz University of Medical Science were housed under standard conditions of temperature (23±2 °C) and illumination (12-h light–dark cycle).

They were provided with a standard chow diet (average 50 g/kg), and water ad *libitum* in a 2-week acclimation period. The animals were divided into four equal groups (n=8); the first group (Control) received a standard chow diet plus 2ml water by using gavage daily. The second group (Ethanol) was administered ethanol at a dosage of 4 g/kg per day by using gavage plus the standard chow diet. The third group (Betaine) received the standard chow diet, plus betaine with a dosage 1.5% (w/w) of the diet soluble in water daily by using gavage, and the fourth group (Betaine and Ethanol) was administered with the betaine group diet. After one hour the rabbits received ethanol at a dosage of 4 g/kg per day by using gavage (pretreatment method) (Ji and Kaplowitz, 2003; Song et al., 2003). Each group consisted of 4 male and 4 female animals. The study period was 90 days. Weight gains and food consumption were determined at weekly intervals (data not shown).

Blood Samples and Biochemical Analyses

Blood samples were collected from the rabbits in a fasting state and were taken from the marginal ear vein in the morning of the day before treatment began (0.0 day) and on the 30th, 60th and 90th day of the treatment. 1.0 mL of whole blood was drawn into tubes of ethylenediamine tetra-acetic acid (EDTA), centrifuged, separated into plasma aliquots and the remaining whole blood placed in other tubes. The collection of serum was assessed in micro tubes.

Serum and plasma aliquots were stored at -70°C until analysis. Serum folate and vitamin B_{12} levels were determined using a radioimmunoassay; (SimulTRAC-SNB Radioassay kit vitamin B_{12}/Folate) (Ferrucci et al., 2007; Golbahar et al., 2005) and the total plasma homocysteine level was determined by Axis® homocysteine EIA kit (Golbahar et al., 2005; Karthikeyan et al., 2007).

Determination of 5-HIAA

Plasma 5-hydroxyindoleacetic acid concentrations were measured by high-performance liquid chromatography with electrochemical detection (Chi et al., 1999). In short, the HPLC system consisted of a Constametric1000 pump (Knauer, Germany), a manual Rheodyne7725 injection valve equipped with a 20-μl loop, and a 3 mm particle size (250×4.6 mm, I.D.) with a C_{18} analytical column (Knauer, Germany). End-point detection was achieved with an Introamperometric detector (EC3000, GmbH, Germany). The operating potential was 0.75 V.

The mobile phase consisted of 0.1 M KH_2PO_4 acetonitrile (84:16, v/v) and 1-octane sulphonic acid (100 mg/l) adjusted to pH 4.75 (with 0.5 M K_2HPO_4). The flow-rate was 1.0 ml/min. Peak height rather than area in the chromatography was measured normally. The concentration of 5-HIAA was calculated by the interpolation of its standard curve. Working standards for the assay were prepared using the mobile phase as the diluent and consisted of six concentration points over the range 2–32 ng/ml.

The plasma extraction procedure was a combination of a protein precipitation step via acetonitrile and centrifugation at 14500 g for 5.0 minutes at 4 °C. 20 μl of the supernatant was injected normally into the HPLC system.

Statistical Analysis

Statistical analysis was performed using a computer statistical package SPSS 11.0 for windows (SPSS, Inc., Chicago, I L., U S A). The significance of the differences between the groups was assessed with One-Way ANOVA. Tukey's test was used after One-Way ANOVA to determine statistical differences among all of the groups. The significant differences within the groups at monthly intervals were assessed with repeated measures ANOVA. The relationship between tHcy and 5-HIAA in the plasma of the E group on the 90th day was calculated by Pearson's correlation test. Independent sample t-Test was used for tHcy in both male and female rabbits from the Ethanol group. Data were expressed as mean ± SD and *p*-values of <0.05 were regarded as statistically significant.

RESULTS

5-HIAA and tHcy were compared in the treatment and control groups having significant differences between the groups only on the 90th day. Therefore, the differences between the groups on the 90th day of the treatment for 5-hydroxyindoleacetic acid and tHcy have been illustrated in Figure 1 and 2 respectively.

There was a significant negative correlation between 5-HIAA and tHcy in the E group ($r=-0.473$, $P=0.021$), and a significant increase in the concentration of 5-HIAA in the BandE group compared to the E group ($p<0.05$). Significantly high concentrations of 5-HIAA were also observed in the B and C groups, in contrast to the E group ($p<0.05$).

Figures 3 and 4 show the folate and vitamin B_{12} concentrations of both the control and the treatment groups in 1 month intervals. While vitamin B_{12} showed no significant difference in the B and E group on the 90th day compared to the control group, the serum concentrations of folate on the 90th day differed.

Significant differences were observed for folate on the 30th, 60th and 90th days of the treatment with regard to the 0.0 day for the all treatment groups (Figure 3). Also, significant differences for vitamin B_{12} on the 30th, 60th and 90th day of the treatment in the B and E groups and on the 30th and 60th day in the BandE group were observed with respect to the 0.0 day (Figure 4). The plasma total homocysteine levels in the E group are not significantly higher in male (16.32μmol/l) *versus* female rabbits (14.75μmol/l), however, no significant difference was observed between tHcy and gender.

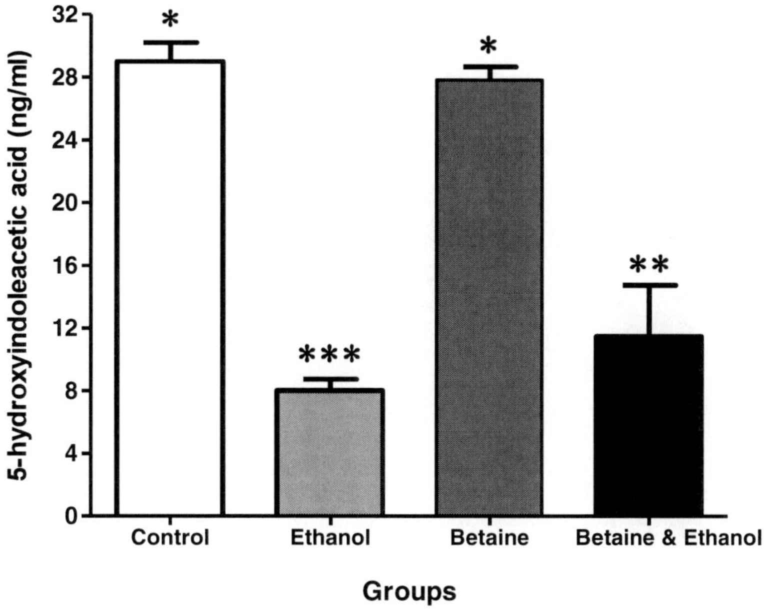

Figure 1. 5- HIAA levels of plasma between the control and treatment groups, on the 90th day of the treatment. Values represent mean ± SD of tHcy; *, **, *** indicate the statistical difference ($P < 0.05$) between the groups.

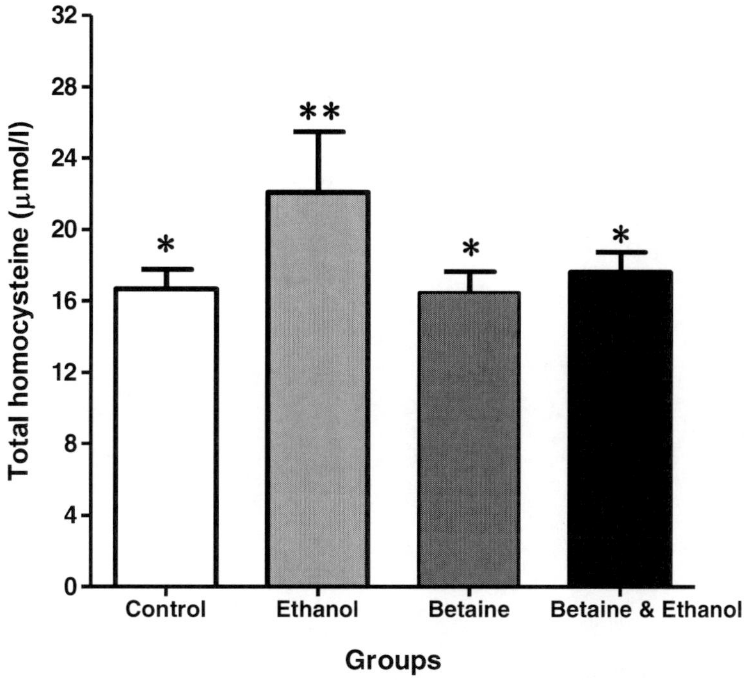

Figure 2. Total homocysteine levels of plasma between the control and treatment groups on the 90th day of the treatment. Values represent mean ± SD of tHcy; *, ** indicate the statistical difference ($P < 0.05$) between the groups.

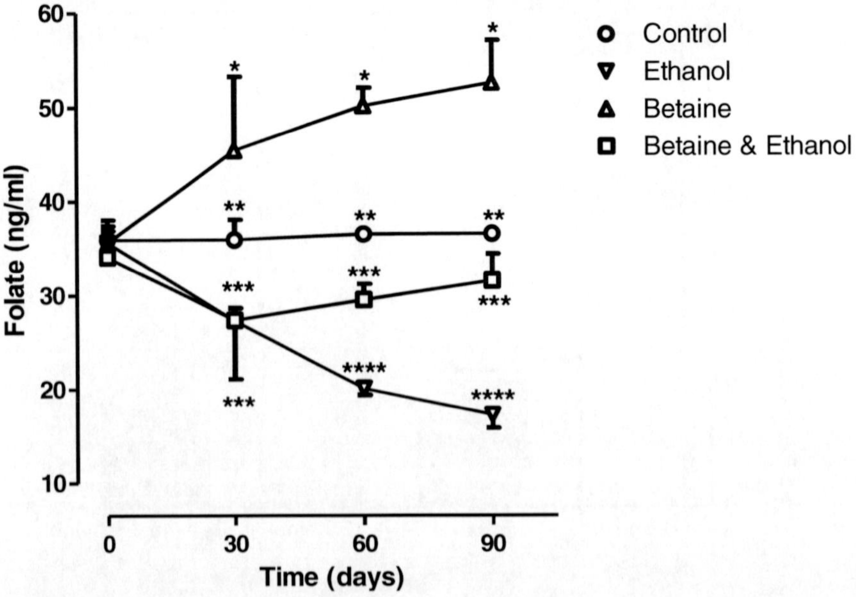

Figure 3. Effect of the administration of betaine and ethanol on the serum folate. Values represent mean ± SD of Folate; *, **, ***, **** indicate the statistical difference ($P < 0.05$) between the groups on the same days.

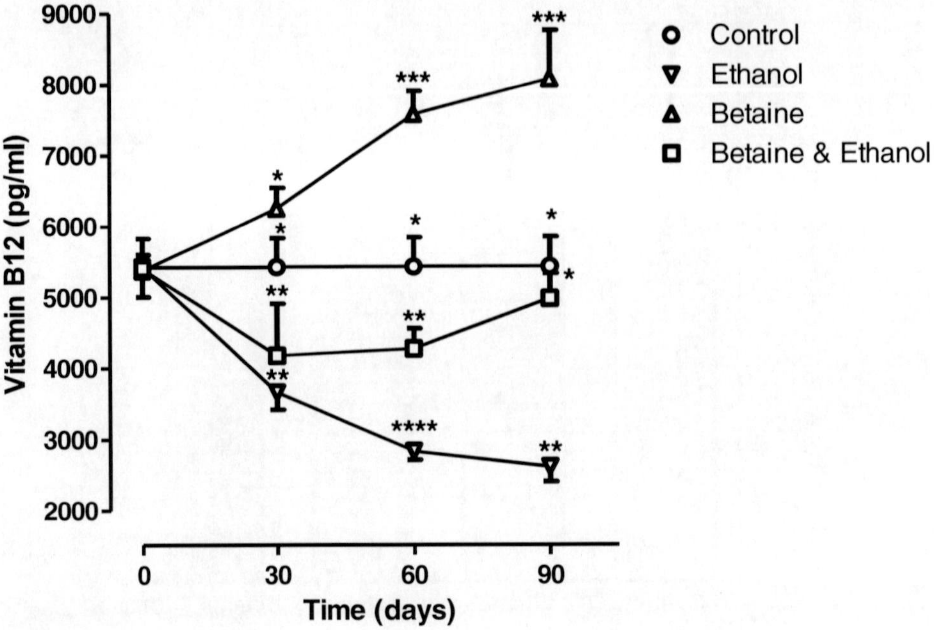

Figure 4. Effect of the administration of betaine and ethanol on the serum vitamin B_{12}.Values represent mean ± SD of vitamin B_{12}; *, **, ***, **** indicate the statistical difference ($P < 0.05$) between the groups on the same days.

DISCUSSION

Our data support the hypothesis that betaine reduces hyperhomocysteinemia (Finkelstein, 2007) and enhances 5-hydroxyindoleacetic in ethanol-induced hyperhomocysteinemia in rabbits. To the best of our knowledge, this study is the first ethanol-induced hyperhomocysteinemia investigation of associations between folate, vitamin B_{12}, total homocysteine and the serotonergic system to determine the therapeutic effects of betaine. The protective effects of betaine were confirmed by higher folate and vitamin B_{12} in serum and lower total homocysteine in plasma. The data we found provides new evidence that the tHcy of plasma decline, while 5-HIAA elevate with oral betaine. There is a negative association between the 5-HIAA (metabolite of serotonergic system) and ethanol treatment as well.

While vitamin B_{12} showed no decrease in the E group on the 90th day with regard to the 30th day of the treatment, the concentration of folate differed, and folate seems to be associated with an increased risk of hyperhomocysteinemia. Furthermore, vitamin B_{12} showed no significant difference in the Band E group on the 90th day compared to the control group, whereas the serum concentration of the folate differed significantly. Thus, the important role of folate rather than vitamin B_{12} is established and confirmed by previous studies (Bottiglieri, 2005; Kim et al., 2008; Papakostas et al., 2005). However, plasma total homocysteine levels were higher in the male *versus* the females rabbits, although not significantly. It is clear that this difference is related to sex steroids (Giltay et al., 1998).

Homocysteine is a non-essential, thiol containing and potentially cytotoxic 4-carbon α-amino acid formed during methionine metabolism through the demethylation of methionine (Nasir et al., 2007). In recent years increasingly more evidence supports the hypothesis that elevated total homocysteine is an independent risk factor for coronary vascular and neurodegenerative diseases (Bidulescu et al., 2009; Chandra et al., 2006; Folstein et al., 2007; Kim et al., 2008; Zieminska and Lazarewicz, 2006). High levels of homocysteine are associated with cerebrovascular disease, abnormality of monoamine neurotransmitters, and depression of mood. A plausible hypothesis for these associations is that high homocysteine levels cause cerebral vascular disease and neurotransmitter deficiency, causing depression of mood (Folstein et al., 2007).

Traditional explanations of the mechanism of Hcy neurotoxicity point to the key role of a disturbance in the methylation and remethylation process. SAMe accumulation in cells in hyperhomocysteinemia is a very strong competitive antagonist of many transferase (Bottiglieri et al., 2000; Zieminska and Lazarewicz, 2006).. Hcy toxicity and impaired methylation may be potential mechanisms involved in the clinical spectrum of MTHFR and cerebral folate deficiency. However folate may also be directly involved in the regulation of the neurotransmitter metabolism. Low concentrations of cerebrospinal fluid (CSF) 5-hydroxyindole acetic acid, a metabolite of serotonin (5-HT) that reflects the global central nervous system tissue levels, have been reported in folate-deficient patients with various neuropsychiatric illnesses and severe depression (Bottiglieri, 2005). The observation for vitamins, tHcy and 5-HIAA concentrations in the treatment groups of the present study supports the idea that vitamin B_{12} and folate are associated with 5-HIAA through their involvement in homocysteine remethylation. The role of folate in the central nervous system function has been established because of the essential role of folate in the one-carbon cycle that furnishes SAMe, the principal methyl donor for a broad range of reactions involving the synthesis of neuroactive substances, the formation of membrane phospholipids, and the metabolism of nucleic acids (Atmaca et al., 2005).

A possible mechanism linking folate deficiency and perturbed monoamine neurotransmitter function may involve tetrahydrobiopterin (BH4) metabolism, a co-factor required in the synthesis of monoamine neurotransmitters. Due to the structural similarities between folate and BH4, the folate enzymes MTHFR and dihydrofolate reductase (DHFR) have been postulated to be involved in theBH4 metabolism. However, in depressed patients, significant correlations between red cell folate and CSF BH4, and also between CSF monoamine metabolites have been reported (Bottiglieri, 2005; Bottiglieri et al., 2000). Low cerebrospinal fluid levels of the serotonin metabolite 5-hydroxyindoleacetic acid and the dopamine metabolite homovanillic acid (HVA) have been reported in several studies among folate-deficient patients with epilepsy, other neuropsychiatric disorders, and congenital folate-deficiency states (Atmaca et al., 2005). Plasma homocysteine levels have been found to increase with age, and it has been reported that the extent of increase is much greater in patients with Alzheimer's disease and Parkinson's disease (Chandra et al., 2006).

With regards to the fact that folate appears to influence the rate of synthesis of tetrahydrobiopterin, a cofactor in the hydroxylation of phenylalanine and tryptophan, rate-limiting steps in the biosynthesis of dopamine, norepinephrine, and serotonin, neurotransmitters are postulated to play a role in the monoamine hypothesis of affective disorders. In addition, when folate has been administered in parenteral and certain oral forms, both SAMe and methyl folate have been shown to have an antidepressant efficacy greater than the placebo and comparable to that of tricyclic antidepressants (Atmaca et al., 2005; Hofmann et al., 1996). In contrast, in some cases the causative treatment of hyperhomocysteinemia and depression to reduce the levels of Hcy in human body fluid include supplementing the diet with folic acid and/or vitamin B_{12} and vitamin B_6, which has been unsuccessful (Zieminska and Lazarewicz, 2006). Therefore, we designed the present investigation to determine the preventive effect of betaine on ethanol-induced hyperhomocysteinemia in rabbit.

The results of the present study demonstrated that betaine supplementation to alcohol-fed rabbits promotes the generation of hepatic S-adenosyl methionine due to the stimulation of methionine synthesis by the alternate BHMT pathway (Bottiglieri, 2005; Finkelstein, 2007). Betaine, a methyl-donor that continuously generates S-adenosyl methionine, is shown to lead to long-term lowering of plasma homocysteine during supplementation in the dietary intake range of 1.5% (w/w) (Ji and Kaplowitz, 2003).

It seems the intracellular content of SAMe is the likely gauge for the availability of methionine, either excess or insufficiency. Changes in the concentrations of SAMe and S-adenosyl homocysteine (SAH) affect the methionine conserving enzymes of both of the methionine cycles. SAMe inhibits the activity of methionine adenosyl transferase (MAT) and also down-regulates the expression of the MAT gene in hepatocytes. Conversely, low concentrations of SAMe allow the expression of this gene in liver cells.

SAMe is an allosteric inhibitor of MTHFR, the sole source of the methyltetrahydrofolate (MTHF) substrate for methionine resynthesis via the methionine synthase (MS) reaction (Finkelstein, 2007). Thus, SAMe, which is provided from the BHMT pathway in our study via betaine supplementation can probably inhibit MTHFR, subsequently MS, and ultimately increase vitamins (B_{12}, folate) by saving its consumption through the classical pathway involving MS (Supplementary file).

With regard to some limitations, the measurement of depression and the molecular mechanism of homocysteine neurotoxicity are not evaluated from the present investigation. One limitation of the present study is that there were no instruments or program to define the depression of rabbits. Another limitation is the insufficiently sensitive index to measure the changes in depression and the timing of clinical improvement.

In contrast, previous studies have reported that homocysteine, being an excitatory neurotoxic by the excessive accumulation of cytosolic calcium, N-methyl D-aspartate (NMDA) receptor overstimulation, generates oxidative stress and the activation of the apoptotic pathway (Chandra et al., 2006).

According to the monoamine hypothesis of affective disorders, depression is due to a deficiency of 5-HT, norepinephrine, or both of these monoamines (Atmaca et al., 2005; Folstein et al., 2007). In the present study the results of the ethanol group of rabbits showed an abnormality of tHcy and 5-HIAA that is related to the deficiency of folate and vitamin B_{12}. In animal studies, low dietary intake of choline and betaine results in aberrant DNA methylation and possible increased atherogenesis, while independent of folate, the dietary intake of choline and betaine are inversely associated with plasma homocysteine, a putative cardiovascular and neurodegenerative disease risk factor (Bidulescu et al., 2009).

In the present study it was concluded that betaine is a preventable metabolic agent and that the ingestion of lipotropic agents is part of a preventative strategy. Betaine prevented a decrease in the content of vitamin B_{12} and folate and decreased the tHcy concentration in both the Betaine and the Betaine and Ethanol groups of rabbits.

It is well known that depression may be the consequence of a developmental pathology affecting serotonergic and/or dopaminergic neuronal systems which have also been suggested for the etiopathogenesis of depression (Atmaca et al., 2005).

The present study provides evidence for the serotonergic neurotoxicity of homocysteine (Bleich et al., 2004); it also indicates that the elevated levels of this excitatory amino acid in the Ethanol group have declined in the Betaine andEthanol group of rabbits with long-term betaine therapy. Therefore, these relationships have probably resulted in the elevation of serotonin, which plays a major role in the configuration of mood and depression (Bleich et al., 2004; Bottiglieri, 2005; Kim et al., 2008).

In summary, the results of the present study may have important implications in understanding the possible role of betaine on the treatment of depression and betaine may be recommended in the pretreatment method of depressive patients with alcoholism. It appears that a limited betaine-dependent remethylation of homocysteine to methionine (BHMT pathway) also exists within mammalian brains (Finkelstein, 2007). Interestingly, the existence of a brain BHMT pathway would explain the details of our results. Nevertheless, future investigations with controlled human studies are required to confirm whether the betaine exactly increases 5-HT output or not.

ACKNOWLEDGMENT

This research was financially supported by grant 87-GR-VT-1 from the Shiraz University Research Council. The authors wish to thank Dr. R. Miri (Manager of Medicinal and Natural Products Chemistry Research Center-Shiraz University of Medical Science, Shiraz, Iran), for his kind cooperation. We are most grateful to Dr. N. Tanideh and Dr. A. Tamadon (The members of Stem cell and Transgenic Technology Research Center – Shiraz University of Medical Sciences, Shiraz, Iran) for their financial and scientific cooperation respectively.

REFERENCES

Atmaca, M.; Tezcan, E.; Kuloglu, M.; Kirtas, O.; Ustundag, B. (2005) Serum folate and homocysteine levels in patients with obsessive-compulsive disorder. *Psychiatry and Clinical Neurosciences*, 5, 616-620.

Bidulescu, A.; Chambless, LE.; Siega-Riz, AM.; Zeisel, SH.; Heiss, G. (2009) Repeatability and measurement error in the assessment of choline and betaine dietary intake: the Atherosclerosis Risk in Communities (ARIC) Study. *Nutrition Journal*, 1, 14-20.

Bleich, S.; Degner, D.; Sperling, W.; Bönsch, D.; Thürauf, N.; Kornhuber, J. (2004) Homocysteine as a neurotoxin in chronic alcoholism. *Progress in Neuropsychopharmacology and Biological Psychiatry*, 3, 453-464.

Bose, D.; Durgbanshi, A.; Capella-Peiró, ME.; Gil-Agustí, M.; Esteve-Romero, J.; Carda-Broch, S. (2004)Micellar liquid chromatography determination of some biogenic amines with electrochemical detection. *Journal of pharmaceutical and biomedical analysis*, 2, 357-363.

Bottiglieri, T. (2005) Homocysteine and folate metabolism in depression. *Progress in Neuropsychopharmacology and Biological Psychiatry*, 7, 1103-1112.

Bottiglieri, T.; Laundy, M.; Crellin, R.; Toone, BK.; Carney, MWP.; Reynolds, EH. (2000) Homocysteine, folate, methylation, and monoamine metabolism in depression. *British Medical Journal*, 2, 228-232.

Chandra, G.; Gangopadhyay, PK.; Senthil Kumar, KS.; Mohanakumar, KP. (2006) Acute intranigral homocysteine administration produces stereotypic behavioral changes and striatal dopamine depletion in Sprague–Dawley rats. *Brain research*, 1, 81-92.

Chi, JD.; Odontiadis, J.; Franklin, M. (1999) Simultaneous determination of catecholamines in rat brain tissue by high-performance liquid chromatography. *Journal of Chromatography B: Biomedical Sciences and Applications*, 2, 361-367.

Craig, SA. (2004) Betaine in human nutrition. *The American journal of clinical nutrition*, 3, 539-549.

Cravo, ML.; Gloria, LM.; Selhub, J.; Nadeau, MR.; Camilo, ME.; Resende, MP.; Cardoso, JN.; Leitao, CN.; Mira, FC (1996) Hyperhomocysteinemia in chronic alcoholism: correlation with folate, vitamin B-12, and vitamin B-6 status, vol 63, pp 220-224. *Am. Soc. Nutrition.*

Ferrucci, L.; Guralnik, JM.; Bandinelli, S.; Semba, RD.; Lauretani, F.; Corsi, A.; Ruggiero, C.; Ershler, WB.; Longo, DL. (2007) Unexplained anaemia in older persons is characterised by low erythropoietin and low levels of pro-inflammatory markers. *British journal of haematology*, 6, 849-855.

Finkelstein, JD. (2007) Metabolic regulatory properties of S-adenosylmethionine and S-adenosylhomocysteine. *Clinical Chemistry and Laboratory Medicine*, 12, 1694-1699.

Finkelstein, JD.; Martin, JJ.; Harris, BJ.; Kyle, WE. (1983) Regulation of hepatic betaine-homocysteine methyltransferase by dietary betaine. *Journal of Nutrition*, 3, 519.

Folstein, M.; Liu, T.; Peter, I.; Buel, J.; Arsenault, L.; Scott, T.; Qiu, WW. (2007) The homocysteine hypothesis of depression. *American Journal of Psychiatry*, 6, 861-867.

Giltay, EJ.; Hoogeveen, EK.; Elbers, JMH.; Gooren, LJG.; Asscheman, H.; Stehouwer, CDA. (1998) Effects of sex steroids on plasma total homocysteine levels: a study in transsexual males and females, vol 83, pp 550-553. *Endocrine Soc.*

Golbahar, J.; Aminzadeh, MA.; Hamidi, SA.; Omrani, GR. (2005) Association of red blood cell 5-methyltetrahydrfoate folate with bone mineral density in postmenopausal Iranian women. *Osteoporosis International*, 12, 1894-1898.

Hofmann, P.; Loimer, N.; Chaudhry, HR.; Pfersmann, D.; Schmid, R.; Wieselmann, G. (1996) 5-Hydroxy-indolacetic-acid (5-HIAA) serum levels in depressive patients and ECT. *Journal of psychiatric research*, 3, 209-216.

Ji, C.; Kaplowitz, N. (2003) Betaine decreases hyperhomocysteinemia, endoplasmic reticulum stress, and liver injury in alcohol-fed mice. *Gastroenterology*, 5, 1488-1499.

Karthikeyan, G.; Thachil, A.; Sharma, S.; Kalaivani, M.; Ramakrishnan, L. (2007) Elevated high sensitivity CRP levels in patients with mitral stenosis and left atrial thrombus. *International Journal of Cardiology*, 3, 252-254.

Kim, JM.; Stewart, R.; Kim, SW.; Yang, SJ.; Shin, IS.; Yoon, JS. (2008) Predictive value of folate, vitamin B12 and homocysteine levels in late-life depression. *The British Journal of Psychiatry*, 4, 268-274.

Kruman, II.; Culmsee, C.; Chan, SL.; Kruman, Y.; Guo, Z.; Penix, LR.; Mattson, MP. (2000) Homocysteine elicits a DNA damage response in neurons that promotes apoptosis and hypersensitivity to excitotoxicity. *Journal of Neuroscience*, 18, 6920-6926.

Nasir, K.; Tsai, M.; Rosen, BD.; Fernandes, V.; Bluemke, DA.; Folsom, AR.; Lima, JAC. (2007) Elevated homocysteine is associated with reduced regional left ventricular function: the Multi-Ethnic Study of Atherosclerosis. *Circulation*, 2, 180-187.

Papakostas, GI.; Petersen, T.; Lebowitz, BD.; Mischoulon, D.; Ryan, JL.; Nierenberg, AA.; Bottiglieri, T.;Alpert, JE.; Rosenbaum, JF.; Fava, M. (2005) The relationship between serum folate, vitamin B12, and homocysteine levels in major depressive disorder and the timing of improvement with fluoxetine. *The International Journal of Neuropsychopharmacology*, 04, 523-528.

Sher, L.; Oquendo, MA.; Grunebaum, MF.; Burke, AK.; Huang, Y.; Mann, JJ. (2007) CSF monoamine metabolites and lethality of suicide attempts in depressed patients with alcohol dependence. *European Neuropsychopharmacology*, 1, 12-15.

Song, Z.; Zhou, Z.; Chen, T.; Hill, D.; Kang, J.; Barve, S.; McClain, C. (2003) S-adenosylmethionine (SAMe) protects against acute alcohol induced hepatotoxicity in mice. *The Journal of Nutritional Biochemistry*, 10, 591-597.

Zangen, A.; Overstreet, DH.; Yadid, G. (1999) Increased catecholamine levels in specific brain regions of a rat model of depression: normalization by chronic antidepressant treatment. *Brain research*, 2, 243-250.

Zieminska, E.; Lazarewicz, JW. (2006) Excitotoxic neuronal injury in chronic homocysteine neurotoxicity studied in vitro: the role of NMDA and group I metabotropic glutamate receptors. *Acta Neurobiol Exp*, 301-309.

Reviewed by Sang Yoon
Professor JS Yoon, Department of Psychiatry and Depression
Clinical Research Centre, Chonnam National University Medical School, Kwang
Republic of Korea., Email: jsyoon@chonnam.ac.kr

In: Vertebrate and Invertebrate Animal Approaches
Editor: José Rosa Gomes

ISBN: 978-1-63117-102-4

Chapter 13

VOLATILE SUBSTANCES OF MANDIBULAR GLAND SECRETION OF A STINGLESS BEE: *SCAPTOTRIGONA POSTICA* LATREILLE

***Luciana Fioretti Gracioli -Vitti*[*1], *Carminda da Cruz-Landim*[1] *and Fábio Camargo Abdalla*[2]**

[1]Departamento de Biologia, Instituto de Biociências de Rio Claro, UNESP-Univ. Estadual Paulista. Rio Claro, São Paulo, Brasil

[2]Universidade Federal de São Carlos – UFSCar - Sorocaba, São Paulo, Brasil

ABSTRACT

The composition of secretion of mandibular gland of *Scaptotrigona postica* was determined by gas chromatography and mass espectrometry and the composition in workers in different life phases, virgin queens and sexually mature males was compared. The secretion in all individual classes studied consists of a rich mixture of aliphatic hydrocarbons, alcohols, acetones and aldehydes. Although compounds were shared among the individuals studied, each class presented a particular secretion profile. The secretion of the forager worker contained the most varied composition, and that of virgin queen the least varied composition. Mandibular glands of forager workers and mature males produced secretions with similar profiles. The composition of the secretion is discussed in relation to individual's behavior.

Keywords: Stingless bee; *Scaptotrigona postica*; mandibular gland secretion; chromatography; spectrometry

INTRODUCTION

The colonies of eusocial bees function as coherent units in which the individuals perform different functions in an auto-organized way. The key factor for these species success is the

[*] Corresponding author: lufgvitti@hotmail.com.

high sophisticated capacity for social regulation throughout communication among the colony members (Wilson and Sober 1989; Michener 2000).

Bees, as all insects, produce chemicals species-specific molecules that play roles in communication, defense, mating and also release physiologic changes on the target individuals. In this way, some chemical molecules are high conserved tools for a plethora of functions in insects, and of great importance for eusociability organization (Abdalla 2006).

Highly eusocial bees, such as Meliponines, use sophisticated mechanisms for recognition of nest-mate's sex, caste and physiological state, as well as for communication of food resources location (Von Frisch 1967; Kerr 1969; Wilson 1971; Blum 1992; Nieh 2004) in which semiochemicals known as pheromones are predominant.

Pheromones are important as carriers of information in bees. These compounds are produced in exocrine glands from epidermal origin (Cruz-Landim 1994, 2002; Cruz-Landim et al., 2005, 2006). One of the most complex systems of pheromonal communication is found in *Apis mellifera,* where the mandibular glands secretion is responsible for establish the social integration between the female castes. These glands produce caste specific compounds that, according to the task displayed by the caste play role as sexual attractant, alarm, defense or recognition of nest mates (Butler 1954; Nedel 1960; Mota and Cruz-Landim 1988; Winston 1989; Winston et al., 1992; Plettner et al., 1996; Engels et al., 1997). Honeybee queen pheromone produced in the mandibular gland can also act modulating workers division of labor by controlling the juvenile hormone production (Pankiw et al., 1998). In workers the mandibular glands produce alarm pheromones that elicit defense behavior but, in queen less colonies, the glands of workers might releases queen specific chemicals simulating a "false queen identity" (Crewe and Velthius 1980), showing that these glands are able to respond to environmental conditions.

The function of the mandibular glands in stingless bees is not so specific, as in *A. mellifera*, playing diverse functions according the species considered. To the secretion of the mandibular glands from workers of *Scaptotrigona postica* and *S. xantothrica* was attributed the function of marking trails to signalize for foragers, the path between the colony and food resources (Lindauer and Kerr 1958, 1960; Kerr et al., 1963; Kerr 1969, 1994). However, Schorkopf et al. (2009) demonstrated through bioassays that in *Trigona spinipes* and *Scaptotrigona aff. depilis* the mandibular gland products are used to communicate intra- and inter-specific behaviors of defense and aggression. The caustic secretion of the mandibular gland of *Oxytrigona tataira* workers is used for defense (Kerr and Costa-Cruz 1961) and in *Lestrimelitta limao*, a cleptobiotic species, the secretion act as repellent that allows the stealing of other species colonies. The use of the mandibular gland secretion for defensive/aggressive behavior by workers seems to be common among bees, such as *A. mellifera* and some species of *Melipona* (Smith and Roubik 1983; Cruz-López et al., 2005).

Several papers about the bee's mandibular gland morphology are available, some of which compare the degree of development and cell contents between the sexes, castes and ages (Butler 1954; Callow et al., 1959; Gary 1962; Pain and Ruttner 1963; Butler 1967; Strang 1970; Crewe 1976; Crewe and Velthuis 1980; Costa-Leonardo 1980, 1981; Mota and Cruz-Landim 1988; Salles and Gracioli 2002; Gracioli-Vitti and Abdalla 2006). The morphological studies show changes in secretion production according the tasks displayed by the castes in the colony but few is known about the secretion chemical composition in Meliponine workers and much less in queens and males. Gracioli-Vitti et al. (2004) using gas chromatography (GC) and mass spectrometry (MS) found caste and sex differences in the chemical composition of *Melipona bicolor* mandibular gland extracts suggesting different functions to the gland according to the gender and physiological status of workers, as demonstrated for *A. mellifera* queens and workers and is in accordance with the morphology variations seen among the individuals and their ages.

The authors Gracioli-Vitti and Abdalla (2006) shown that the ultrastructure of the mandibular gland cells of *S. postica* changes according to the ethary polyethism of the workers, the reproductive condition of the queens (virgin or physogastric) and also between females and males. Therefore, the present investigation compares the composition of mandibular gland of workers, queens and males of *S. postica* in order to verify a possible correspondence between the cell's gland morphological changes described the chemical composition of the secretion and bee's functions in colony.

MATERIAL AND METHODS

Gas Chromatography/Mass Spectrometry

The bees were captured from colonies kept in the apiary at the Departamento de Biologia da Universidade Estadual Paulista, Campus de Rio Claro (UNESP), SP. Brazil. Samples of newly-emerged, nurse and forager workers were captured from the same colony. Virgin queens were captured from several colonies in the same apiary and sexually-mature males from male groupments outside the colonys. Cold was used for anesthetize the specimens to prevent anesthetics stress, which could promote emptying of the gland reservoir. Forty glands were dissected from each worker phase and twelve from each mature males and virgin queens.

The glands were dissected in distilled water and transferred to dichloromethane to extract the secretion compounds. The extracts were analyzed by GC-MS-EI, in a Hewlett Packard-6890 gas chromatograph, with a system of capillary columns (30 m x 0.25 mm x 0.25 μm, HP-5MS, Crossbond 5%-phenyl-95%-dimethilpolysiloxane) directly coupled to a Hewlett Packard-5973 mass detector. The samples were heated to 240° C, at a rate of 3° C/min, in a spitless injection system with helium carrier gas, 1 mL/min, at a constant flux and volume (0.2 or 1 μL). The relative abundance of the compounds in the samples was obtained using the Hewlett Packard gas chromatograph. Results were analyzed by non parametric *Friedman's* analysis of variance BioEstat 5.0 (Ayres et al., 2007) and correspondence analysis (Landim, 2011) using only the identified compounds (table 1).

RESULTS

Secretion composition: Hydrocarbons, esters, alcohols, acetones and unidentified compounds were found in all classes of individuals. The acetones are the major component of the mandibular gland secretion, produced by all classes of individuals except virgin queens. Hydrocarbons are present only in forager workers and males (table 1). Among workers the secretion of forager has the most varied composition, with eighteen different compounds, the main compounds being 2-tridecanone (31%) and two unidentified compounds (36% each).

Table 1. Volatile compounds identified by GC/MS in mandibular gland extracts produced by *Scaptotrigona postica* workers (Newly-Emerged = NE, Nurse=N, Foragers=F), virgin queens (VQ) and males (M), arranged according to substance chemical class. Retention time = RT

	NE	N	F	VQ	M	RT		NE	N	F	VQ	M	RT
Alcohols							**Acetones**						
2-nonanol				+++		1105	2-undecanone	++	++	++		++	1295
2-undecanol	++			+++		1304	2-tridecanone	+++	+++	+++		+++	1501
2-tridecanol	++	++		+++		1507	2-pentadecanone	++	+++	++		++	1693
2-pentadecanol	++	+				1696	2-heptadecanone	++	++	++			1900
2-heptadecanol	++	+				1906	Esters						
Aliphatic Hydrocarbons							Hexyl-hexanoate		+	++			1387
Tridodecane				++	+	2300	**Unidentified**						
Tetradodecane			++		++	2400		++				+	1275
Pentadodecane			++		++	2500		++					1471
Hexadodecane			++		++	2600		++					1483
Heptadodecane			++		++	2700				++			1789
Octadodecane			++		++	2800					++		1869
Nonadodecane			++		++	2900			++	+++			1964
Triacontane			+		+	3000			+	++			1990
Monoterpenes									++	+++			2164
Neral	+					1242				+			2520
Geranial	++					1271			++	++		++	2675
												++	2876

Quantification of single compounds relative to the sum of all detected peak areas: +++ = ≥10%, ++ = ≥ 1%, + = <1%.

In this individual class the least varied secretions were those produced by the newly-emerged workers, thirteen different compounds being the main 62% 2-tridecanone, 10% 2-tridecanol and 7% 2-pentadecanone, and nurse workers with twelve different compounds being 55% of 2-tridecanone and 12% of 2-pentadecanone (table 1).

After workers, the secretion produced by the males was the next of more varied composition, with fourteen different compounds, being the main component 2-tridecanone (74%) (table 1).

Only five compounds were detected in virgin queens, including 2-nonanol and one unidentified compound with a molecular weight near 254, which is exclusive of this caste. The principal compound in the other individuals, 2-tridecanone, was not detected. The most abundant compound was 2-tridecanol (51%). This alcohol is also present at low concentrations in the secretions of newly-emerged and nurse workers (table 1).

Some compounds were exclusive of a certain class of individual, although some of them were not identified. The aldehydes, neral and geranial, the alcohol 2-undecanol and two unidentified compounds were found exclusively in the mandibular gland's secretion of newly-emerged workers. The secretion produced by the foragers also contained two exclusive unidentified compounds, and the secretion produced by the males contained one unidentified compound, which was exclusive for this sex (table 1).

In spite of the pointed differences composition in secretions produced by forager workers and mature males, there is a similar pattern between their compositions, in both the more frequent compounds are long-chain hydrocarbons (table 1). The secretion produced by the virgin queen shows a distinct compositional pattern, containing many alcohols (table 1).

The *Friedman's* non parametric analysis of variance show differences among the sum of the ranks ($p \geq 0.05\%$) but did not among the individual classes (table 2, figure 1). The correspondence analysis, however separated all classes of individuals except foragers and sexually mature males (figure 2).

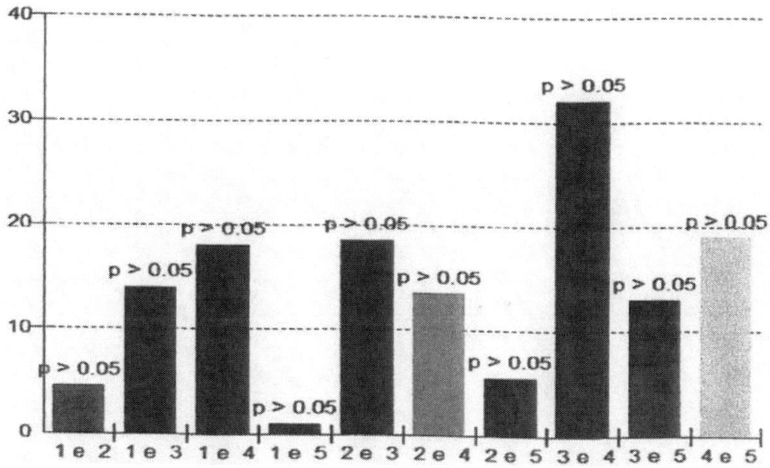

Figure 1. *Friedman test* - Differences in the ranks sum among the individuals classes. 1 = newly emerged, 2 = nurse, 3 = forager workers, 4 = virgin queen, 5 = sexually mature male, e = and.

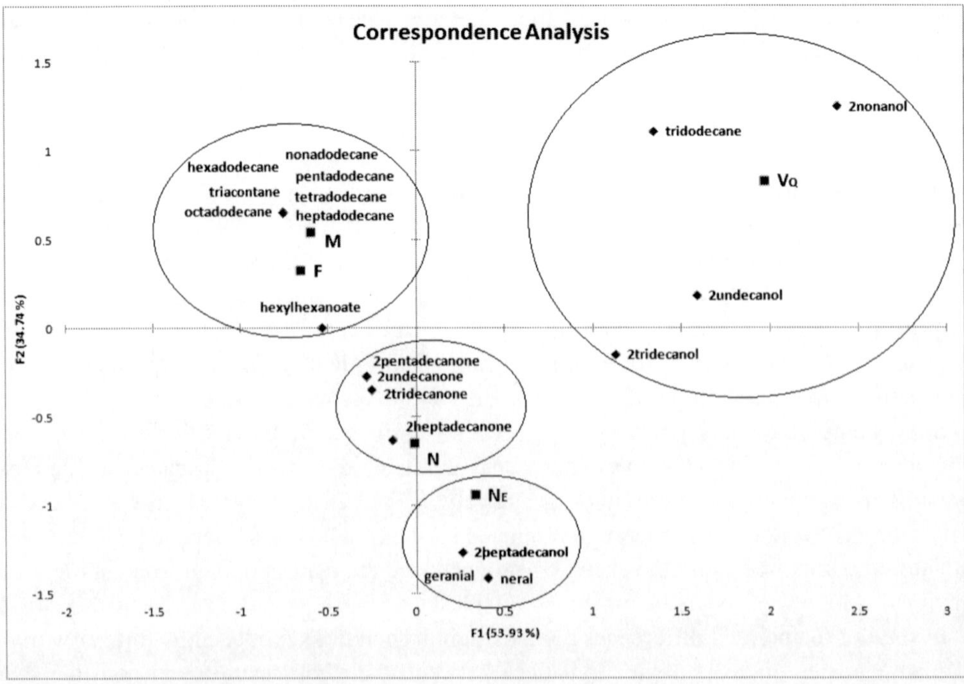

Figure 2. Correspondence analysis using only the identified compounds separates all worker classes (NE, N and F) and virgin queens (VQ). Sexually mature males (M) and forager workers (F) were not separated by their secretion composition.

Table 2. Result of *Friedman's* non parametric analysis of variance. NE=newly emerged. N=nurse, F=forager workers. VQ=virgin queen. M=sexually mature male. p≥0.05%

	NE (1)	N(2)	F(3)	VQ(4)	M(5)
Ranks sum	103.5000	99.0000	117.5000	85.5000	104.5000
Median	0.0000	0.0000	1.0000	0.0000	0.0000
Ranks mean	3.0441	2.9118	3.4559	2.5147	3.0735
Values mean	0.7647	0.6471	1.0882	0.3824	0.7647
Standard deviation	1.0168	0.9811	1.1110	0.9539	0.9865
Friedman (Fr)	6.2353				
D.F.	4				
(p) =	0.1823				

Discussion

Chemical composition of mandibular gland secretion: Most of the studies of the composition of bee's glands secretion focused on workers. However, the gland's morphologies differ among castes and sex as well as along the individual's age/life stage (Crewe 1976; Crewe and Velthuis 1980; Salles and Gracioli 2002; Gracioli-Vitti and Abdalla

2006; Cruz-Landim et al., 2005; Cruz-Landim et al., 2006), suggesting that there might be corresponding composition differences in the secretion produced by them.

In fact the composition of the mandibular gland extracts of *Melipona bicolor* are different between workers performing different tasks in the colony as well as between queens and males (Gracioli-Vitti et al., 2004). In the same way, given the diversity of the bee specie's habits, it is not wise to expect that the secretion of the mandibular gland has the same function in all them. As some compounds found in *S. postica* are also present in several other species of Meliponines (Francke et al., 2000), we suggest that the use or the perception of these compounds by different species may be plastic, depending upon the precise compounds blend, individual phase and the behavioral context.

The composition of the mandibular gland secretion produced by *S. postica* workers was previously studied by Blum (1974) and in *S. aff. depilis* by Schorkopf et al. (2009). The class of worker used in these studies is not mentioned but probably was foragers, which are easier to capture. Almost all alcohols and acetones detected in the present work are common to the secretions of the previous studies, but the secretion of *S. aff. depilis* contains a larger variety of aldehydes than the other species, and the secretion produced by *S. postica* contains only one ester. Blum (1974) did not detect hydrocarbons, but by comparing the present data and that of Schorkopf et al. (2009), it can be seen that the aliphatic hydrocarbon composition differs greatly between forager workers of *S. postica* and *S. aff. depilis.* Even taking in account that the environment might have some influence in the secretion composition, assuming that the data of Schorkopf et al. (2009) are from the plasticity of gland's secretion among species seems demonstrated.

Secretion use by workers: The function of the scent trail attributed to the mandibular gland secretion produced by Meliponines, specifically Trigonina (Lindauer and Kerr 1958, 1960; Kerr 1969; Kerr et al., 1981), has been contested by several papers (Jarau et al., 2004a, 2004b; Schorkopf et al., 2009). In fact, several studies in a wide variety of different stingless bees species provide evidence that the function of mandibular gland secretion as alarm or defense substances is much more disseminated than previously thought (Johnson 1980; Smith and Roubik 1983; Engels et al., 1997; Schorkopf et al., 2009). In *A. mellifera* Malka et al. (2008) reports similar function for the mandibular gland secretion in workers from orphanage colonies, demonstrating its action in establishing dominance between them through a competitive/aggressive behavior.

The heptanol and heptanone were the compounds identified by Schorkopf et al. (2009) as responsible for the alarm function and heptanone produced in the mandibular gland of honeybee workers, particularly in foragers, was also identified as an alarm pheromone to defend the colony (Winston et al., 1989, 1990; Pankiw et al., 1994, 1996) which are present in the mandibular gland secretion. In the same way, 2-heptanol and 2-heptanone in the mandibular gland secretion of some Meliponines species elicit defensive or aggressive behavior in forager workers (Kerr 1969; Smith and Roubik 1983). Therefore, these compounds, especially 2-heptanol, seem to function as allomones in intra- and inter-specific defense (Schorkopf et al., 2009).

In the present work, 2-heptanone and 2-heptanol were not detected in the gland extracts of *S. postica*, although the presence of 2-heptanone was reported by Blum (1974). Instead, tridecanone was the most common compound in the gland secretions produced by all classes of workers and mature males. This acetone was not detected in the secretion produced by the virgin queens. The corresponding alcohol, tridecanol, was found in workers, with intermediate concentrations detected in the secretion of newly-emerged bees and a very small amount detected in the secretion of nurses, but was not detected in the secretion of foragers or males. In the secretion produced by the virgin queens, tridecanol is the compound of highest concentration (table 1).

If acetones and alcohols are the compounds responsible for colony defense and aggression, in the absence of 2-heptanone and 2-heptanol, tridecanone and tridecanol could instead be responsible for those behaviors. In this case, it was expected that great concentrations of these compounds would be found in the individuals responsible for colony defense. In accordance with this idea, the workers show high concentrations of tridecanone; however, the highest concentration of this compound appears in newly-emerged workers, rather than in foragers, and tridecanol is present in significant amounts only in newly-emerged workers. However the compounds eliciting specific behaviors are not necessarily the compounds present in the highest concentrations. Although the defensive behavior in *S. aff. depilis* is elicited by 2-heptanone, tridecanone is the acetone present in the highest concentration in the extracts of this species (Schorkopf et al., 2009) and heptanone and heptanol may be among the unidentified compounds in the extracts of *S. postica*.

Comparison of secretion composition and it use between workers, mature males and virgin queens: Although the statistical analysis of the results show no differences among these colonial members, the table 1 shows that newly-emerged workers and virgin queens share 2 out of 13 compounds whereas forager workers shares none and secretion produced by newly emerged workers are more similar to those of the sexually-mature males. Secretions produced by the males do not contain many alcohols, and secretions produced by the newly-emerged workers and virgin queens do not contain many hydrocarbons. Besides, the correspondence analysis showed that only foragers and sexually mature males are indistinct. Therefore, seem clear that the final "bouquet" of the whole secretion must be different in those individuals, possibly resulting in specific functions linked to their colony "status".

The fact that the secretion of some glands might be spread over the body surface and function as a recognition pheromone, combined with the fact that hydrocarbons are the main cuticular pheromones in insects, including bees, could explain why those compounds are almost absent from newly-emerged workers and virgin queens. These classes of individuals must not have yet a defined identity within the colony population or even they may not need many identification clues, since they do not leave the colony as was the case of the individuals with tasks out of the colony, as foragers and mature males. In fact, for the virgin queen, recognition could be damaging.

CONCLUSION

It must be considered that the glands do not have single functions and instead are very versatile (Cruz-Landim 1994; Cruz-Landim et al., 2005, 2006). The same gland secretion might perform different functions depending on the species, class or life stage of the individual, as suggested by the changes in gland cell's morphology (Callow et al., 1959; Crewe and Velthuis 1980; Francke et al,. 1983; Francke et al., 2000; Abdalla and Cruz-Landim 2001; Katzav-Gozansky et al., 2003; Gracioli-Vitti et al., 2004). The same occurs with the secretion compounds. Blum (1974) show that 2-heptanone and 2-nanonone found in the mandibular gland secretion of *A. mellifera,* when in moderate concentrations, function as attractive pheromones, and when in high concentration as alarm signals. Acetones with eleven and fifteen carbon atoms have low volatility and therefore only enhance the action of the more volatile compounds. Therefore, it can be suggested that the mandibular gland secretion of *S. postica* is used for different functions according to the individual's caste, sex, age or physiological state and the colony circumstances being an excellent example of Evo-Devo.

ACKNOWLEDGMENTS

The authors would like to thank Fundação de Amparo à Pesquisa do Estado de São Paulo (FAPESP), Conselho Nacional de Pesquisa Cientifica e Tecnológica (CNPq) for financial support and Dr. Paulo Milton Barbosa Landim for the AID with statistics.

REFERENCES

Abdalla, F. C. and Cruz-Landim, C. (2001). Dufour gland in the Hymenoptera (Apidae, Formicidae, Vespidae): a review. *Rev. Brasil. Biol.,* 61(1), 95-106.

Abdalla, F. C. (2006). Morphological, chemical and developmental aspects of the Dufour gland in some eusocial bees (Hymenoptera, Apidae): a review. *Rev. Bras. Entomol.,* 50(4), 153-162.

Ayres, M., Ayres, M. J., Ayres, D. L. and Santos, S. A. (2007). Bioestat: Aplicações estatísticas nas áreas das ciências biomédicas. Versão 5.0. Belém: Mamirauá/CNPq, 324pp.

Blum, M. S. (1974). Pheromonal sociality in Hymenoptera. In: M. C. Birch (Ed.). *Pheromones,* pp. 222-249. North Holland Publishing Co. NY.

Blum, M. S. (1992). Honey bee pheromones. In: J. M. Graham (Ed.). *The hive and the honey bee*, pp. 373-400. Dadant and sons Hamilton, Illinois.

Butler, C. G. (1954). The method and importance of the recognition in a colony of honeybees (*Apis mellifera*) of the presence of its queen. *Proc. Roy. Entomol. Soc.,* London, 105, 11-19.

Butler, C. G. (1967). Insect Pheromones. *Biol. Rev.,* 42, 42-87.

Callow, R. K., Johnston, N. C. and Simpson, J. (1959). 10-hydroxy 2-decenoic acid in the honeybee (*Apis mellifera*). *Experientia,* 15, 421.

Costa-Leonardo, A.M. (1980). Estudos morfológicos do ciclo secretor das glândulas mandibulares de *Apis mellifera* L. (Hymenoptera, Apidae). *Rev. Bras. Biol.,* 24 (2), 143-151.

Costa-Leonardo, A.M. (1981). Ultra-estrutura do ciclo secretor das glândulas mandibulares de operárias de *Apis mellifera* L. (Hymenoptera, Apidae). *Rev. Bras. Zool.,* 41(2), 307-316.

Crewe, R. M. (1976). Aggressiveness of honeybees and their pheromone production. *South African J. Sci.,* 72, 209-212.

Crewe, R. M. and Velthuis, H. H. W. (1980). False queens: a consequence of mandibular gland signals in worker honeybees. *Naturwissenschaften,* 67, 467-469.

Cruz-Landim, C. (1994). Polimorfismo na ocorrência de glândulas nas abelhas (Hymenoptera: Apoidea). *Anais do Encontro sobre Abelhas,* 1, 118-129.

Cruz-Landim, C. (2002). Tipos de células secretoras presentes nas glândulas exócrinas das abelhas. In: C. Cruz-Landim and F. C. Abdalla (Eds) *Glândulas exócrinas das abelhas,* pp. 1-20. FUNPEC, Ribeirão Preto, Brasil.

Cruz-Landim, C., Abdalla, F. C. and Gracioli-Vitti, L. F. (2005). Morphological and functional aspects of volatile producing glands in bees (Hymenoptera: Apidae). *Insect Science and its Application,* 12, 467-480.

Cruz-Landim, C., Abdalla, F. C. and Gracioli-Vitti, L. F. (2006). Class III glands in the abdomen of Meliponini. *Apidologie,* 37, 164-174.

Cruz-López, L., Malo, E.A., Morgan, E. D., Rincon, M., Guzmán, M. and Rojas, J. C. (2005). Mandibular gland secretion of *Melipona beecheii*: chemistry and behavior. *J. Chem. Ecol.,* 31, 1621-1632.

Engels, W., Rosenkranz, P., Adler, A., Taghizadeh, T., Lubke, G. and Francke, W. (1997). Mandibular gland volatiles and their ontogenetic patterns in queen honey bees, *Apis mellifera carnica. J. Ins. Physiol.,* 43(4), 307-313.

Francke, W., Lubke, G., Schroder, W., Reckziegel, A., Imperatriz-Fonseca, V. L., Kleinert, A., Engels, E., Hartfelder, K. and Engels, W. (2000). Identification of Oxygen Containing volatiles in cephalic secretions of workers of Brazilian stingless bees. *J. Braz. Chem. Soc.,* 11, 562-571.

Francke, W., Schroder, W., Engels, E. and Engels, W. (1983). Variation in cephalic volatile substances in relation to worker age and behavior in the stingless bee, *Scaptotrigona postica. Z. Naturforsch.,* 38c, 1066-1068.

Gary, N.E. (1962). Chemical mating attractants in the queen honeybee. *Science,* 136 (3518), 773-774.

Gracioli-Vitti, L. F. and Abdalla, F. C. (2006). Comparative ultrastructure of the mandibular gland in *Scaptotrigona* postica (Hymenoptera, Apidae, Meliponini) workers and males. *Braz. J. Morphol. Sci.,* 23(3-4), 415-424.

Gracioli-Vitti, L.F., Abdalla, F. C., Silva de Moraes, R. L. and Jones, G. R. (2004). The chemical composition of the mandibular gland secretion of *Melipona bicolor* Lepeletier, 1836 (Hymenoptera, Apidae, Meliponini): A comparative study among castes and sexes. *J. Chem. Soc.,* 15(5), 777-781.

Jarau, S., Hrncir, M., Ayasse, M., Schulz, C., Francke, W., Zucchi, R. and Barth, G. (2004a). A stingless bee (*Melipona seminigra*) marks food sources with a pheromone from its claw retraction tendons. *J. Chem. Ecol.,* 30, 793-804.

Jarau, S., Hrncir, M., Zucchi, R. and Barth, F. G. (2004b). A stingless bee uses labial gland secretion for scent trail communication (*Trigona recursa* Smith, 1863). *J. Comp. Physiol.,* 190(3), 233-239.

Johnson, L. K. (1980). Alarm response of foraging *Trigona fulviventris* (Hymenoptera: Apidae) to mandibular gland components of competing bee species. *J. Kans. Entomol. Soc.,* 53, 357-362.

Katzav-Gozansky, T., Soroker, V., Francke, W. and Hefetz, A. (2003). Honeybee egg-laying workers mimic a queen signal. *Insectes Sociaux,* 50, 20-23.

Kerr W.E. (1969). Some aspects of the evolution of social bees. *Evol. Behav.,* 3, 119–175.

Kerr W.E. (1994). Communication among *Melipona* workers (Hymenoptera: Apidae). *J. Insect Behavior,* 7(1), 123-128.

Kerr, W. E. and Costa-Cruz, C. (1961). Funções diferentes tomadas pela glândula mandibular na evolução das abelhas em geral e em "*Trigona (oxytrigona) tataira*" em especial. *Rev. Bras. Biol.,* 21(1), 1-16.

Kerr W.E. and Nielsen R.A. (1966). Evidence that genetically determined *Melipona* queens can become workers. *Genetics,* 54, 859-865.

Kerr, W. E., Ferreira, A. and Mattos, N. S. (1963). Communication among stingless bees – additional data (Hymenoptera: Apidae). *J.N.Y. Entomol. Soc.,* 71, 80-90.

Kerr, W.E., Blum, M. and Fales, H.M. (1981). Communication of food source between workers of *Trigona (Trigona) spinipes. Rev. Bras. Biol.,* 41, 619- 623.

Landim, P. M. B. (2011). *Análise estatística de dados geológicos multivariados.* Oficina de Textos, São Paulo, 208pp.

Lindauer, M. and Kerr, W. E. (1958). Die gegenseitige Verstandigung bei den stachellosen Bienen. *Z. Vergl. Physiol.,* 41, 405-434.

Lindauer, M. and Kerr, W. E. (1960). Communication between the workers of stingless bees. *Bee World,* 41(2), 29-41.

Malka, O., Shnieor, S., Katzav-Gozansky, T. and Hefetz, A. (2008). Aggessive reproductive competition among hopelessly quenless honeybee workers triggered by pheromone signaling. *Naturwissenschaften,* 95, 553-559.

Michener, C. D. (2000). *The bees of the world.* The Johns Hopkins University Press. Baltimore, USA, 913pp.

Mota, M. H. V. B. and Cruz-Landim, C. (1988). Ocorrência e morfometria de glândulas tegumentares abdominais em *Apis mellifera* (Hymenoptera, Apidae). *Rev. Brazil. Biol.,* 5, 119-154.

Nedel, J. O. (1960). Morphologie und Physiologie der mandibeldruse einiger Bienen-Arten (Apidae). *Zeitschrift fur morphologies und okologie der tiere,* 49, 139-183.

Nieh, J. C., Barreto, L. S., Contrera, F. A. L. and Imperatriz-Fonseca, V. L. (2004). Olfactory eaves dropping by a competitively foraging stingless bee, *Trigona spinipes. Proc. R. Soc. Lond. B,* 271, 1633-1640.

Pain, J. and Ruttner, F. (1963). Les extraits des glandes mandibulaires das reines d'abeilles attirent les mâles, lors du vol nuptial. *C.R.Acad.Sci. Paris,* 256, 512-515.

Pankiw, T., Winston, M. L. and Slessor, K. N. (1994). Variation in worker response to honey bee (*Apis mellifera*) queen mandibular pheromone (Hymenoptera: Apidae). *J. Insect Behav.,* 7, 1-15.

Pankiw, T., Winston, M. L., Plettner, E., Slessor, K. N., Pettis, J. S. and Taylor, O. R. (1996). Mandibular gland components of European and Africanized honeybee queens (*Apis mellifera*). *J. Chem. Ecol.,* 22, 605-615.

Pankiw, T., Page, R. E. and Fondrk, K. M. (1998). Brood pheromone stimulates pollen foraging in honey bees, *Apis mellifera* L. *Behav. Ecol. Sociobiol.,* 44, 193-198.

Plettner, E., Slessor, K. N., Winston M. L. and Oliver, J. E. (1996). Caste selective pheromone biosynthesis in honeybees. *Science,* 271, 1851-1853.

Salles, H. C.and Gracioli, L. F. (2002). Glândulas mandibulares e intramandibulares. In: C. Cruz-Landim and F. C. Abdalla (Eds.). *Glândulas exócrinas das abelhas*, pp. 71-90. FUNPEC, Ribeirão Preto, Brasil.

Schorkopf, P. L. D., Hrncir, M., Mateus, S., Zucchi, R., Schimidt, V.M. and Barth, F. G. (2009). Mandibular gland secretions of meliponine worker bees: further evidence for their role in interspecific defense and aggression and against their role in food source signaling. *J. Econ. Entomol.,* 212, 1153-1162.

Smith, B. H. and Roubik, D. W. (1983). Mandibular glands of stingless bees (Hymenoptera: Apidae): chemical analysis of their contents and biological function in two species of *Melipona. J. Chem. Ecol.,* 9, 1465-1472.

Strang, G. E. (1970). A study of honey bee drone attraction in the mating response. *J. Econ. Entomol.,* 63, 641-645.

Von Frisch, K. (1967). *The dance language and orientation of bees*. Cambridge: Harvard University Press.

Wilson, E.O. (1971). *The insect societies*. Belknap Press of Harvard University Press, Cambridge, Massachusetts.

Wilson, D. S. and Sober, E. Reviving the superorganism. *J. Theor. Biology,* 136, 337-356.

Winston, M. L. (1987). *The biology of the honey bee*. Harvard University Press, Cambridge, Massachusets.

Winston, M. L., Slessor, K. N., Willis, L. G., Naumann, K., Higo, H. A., Wyborn, M. H. and Kaminski, L. A. (1989). The influence of queen mandibular pheromone on worker attraction to swarm clusters and inhibition of queen rearing in the honey bee (*Apis mellifera* L.). *Insect Sociaux,* 36, 15-27.

Winston, M. L., Higo, H. A. and Slessor, K. N. (1990). Effect of various dosages of queen mandibular gland pheromone on the inhibition of queen rearing in the honey bee (Hymenoptera: Apidae). *Ann. Entomol. Soc. Am.,* 83, 234-238.

Winston, M. L., Higo, H. A., Colley, S. J., Pankiw, T. and Slessor, K. N. (1992). The role of queen mandibular pheromone and colony congestion in honey bee (*Apis mellifera* L.) reproductive swarming. *J. Insect Behav*., 4, 649-660.

In: Vertebrate and Invertebrate Animal Approaches
Editor: José Rosa Gomes
ISBN: 978-1-63117-102-4

Chapter 14

EFFECTS OF DIMETHONATE ON THE SURVIVAL AND GROWTH OF *(BUFO HIMALAYANAS)* TADPOLES IN LABORATORY CONDITIONS

Rujas Yonle**, *Arpita Chaudhury, Deependra Sharma, Santosh Basnet, Somyadip Ghosh and Tashi T. Bhutia
Post Graduate Department of Zoology, Darjeeling Government College,
Darjeeling, West Bengal, India

ABSTRACT

The decline of many amphibian populations is associated with insecticides, but for most insecticides we know little about their toxicity to amphibians.The short term exposure effect of insecticide (21 days) Dimethonate (commercial grade) on the survival and growth of tadpoles of *Bufo himalayanas* (Gunther) was examined in the laboratory. The insecticide was chosen due to their regular use in the tea gardens and orange orchards of the Darjeeling Himalayas. These toad tadpoles are one of the major biotic components of the ecosystem of the Darjeeling Himalayas. Forty-eight hours LC 50 values were found to be >0.617 mg/L with the Dimethonate (commercial grade). It was found out that Dimethonate had some direct negative effects on the tadpole larvae, reducing the tadpole survival and biomass substantially.

Keywords: Darjeeling Himalayas; Bufo himalayanas; Dimethonate

INTRODUCTION

The tadpoles of the *Bufo himalayanas* inhabit the fresh waters of Darjeeling Himalayas (26.52'48'' N and 88.16'48''E) freely. Dimethonate has been available in the open market and also has been an insecticide of choice for a variety of applications. Anthropogenic chemicals are pervasive in nature and biologist are faced with the challenge of understanding

* Corresponding author e- mail: rujasyonle@gmail.com.

how these chemicals impact ecological communities (Relyea et al., 2005). Recent toxicological studies have shown that environmentally relevant level of insecticides can affect the population dynamics through altered abundance from differential tolerance and by reduced growth, which may affect fitness (Fairchild et al., 1994, Diana et al., 2000, Semlitsh et al., 2000, Boone and James 2003). Laboratory toxicity tests are designed to access the direct impact of insecticides under controlled environmental conditions to determine the direct effects of contaminants on individuals or populations. Other environmental cues also play an active role in controlling the population dynamics however which are excluded in this laboratory studies. Insesticides are important in controlling domestic, agriculture, and industrial pests, including vector organisms. Insesticides adversely affect nontarget organisms and are widely used in agricultural ecosystems, wherein amphibians, the nontarget species, are major components of the wetland biota. (Gurushankara, Krishnamurthy and Vasudev 2003) Thus, amphibians living in agricultural ecosystems and wetlands are exposed to these pesticides (Kaur and Chemma 1986; Lambert 1997). While insecticides have the potential to affect many aquatic taxa, the, the impacts on amphibians are of particular concern in the past decade because of the apparent global decline of many species (Blaustein and Wake 1994, Alford and Richards 1999, Houlihan et al., 2001, Kiesecker et al., 2001). The list of possible causes of amphibian declines are numerous, and pesticides have been implicated in at least in some of the declines. Pesticides occur in amphibian habitats (Harris et al., 1998, McConnel et al., 1998, LeNoir et al., 1999), amphibians living with insecticide in these habitats exhibit physiological signatures of these pesticides (i.e., reduced acetyl-choline esterase activity; (Sparling et al., 2001), and declining populations are correlated with greater amounts of upwind agriculture where pesticide and insecticide use are common(Davidson et al., 2001, 2002).

To understand the effects of insecticides on amphibians, the natural first step is to understand the direct toxicity of the insecticides. There is growing number of excellent laboratory –based, single species toxicology experiments that can inform us about the LC50 values of the different insecticides concentrations (the concentrations expected to kill 50% of a test population); e.g., (Relyea et al., 2005). Studies have revealed the indiscriminate input of pesticides and insecticides in the Tea gardens of Darjeeling Hills and Dooars Terai. (Bishnu et al., 2008). Tadpoles of *Bufo himalayanas* comprises as one of the chief biotic components of the fresh water ecosystem of Darjeeling hills. The present study is thus aimed to examine the direct effects of the insecticide Dimethonate on the survival and growth of tadpoles of *Bufo himalayanas*.

MATERIALS AND METHODS

The experiments were conducted at Darjeeling Government College, Department of Zoology, laboratory of Environmental biology. We began by filling the stocking tanks (5 X 2.5 X 2ft) with fresh water. During the next week, dry leaves (primarily of *Quercus* sp.) were added along with 10 liters of water collected from the nearby seasonal pond to serve the source of algae and zooplankton for prior usage of the experiment (Relyea et al., 2005). After allowing the tank algal and zooplankton communities to establish for a week the tadpoles collected from the side pools of torrents in and around the campus area were released. All the tadpoles were still in the early developmental stages and were left for acclimatization. Static

LC50 was performed for determining the 48 hour LC50 values of dimethonate on tadpoles of *Bufo himalayanas* after acclimatizing the tadpoles for 4 days in aquarium conditions. Tadpoles were fed with fish food but not 24 hours prior to the experiment. Before the experiment all the 30 liter capacity aquarium were cleaned and filled with 15 liters of rain water (temp was maintained at 18°C-19°C, pH 7 - 6, and hardness 8.2 - 9.4 mg/L as $CaCO_3$).Six different concentration (0.4, 0.5, 0.6, 0.7, 0.8, 0.9 mg/lit) were selected in triplicates. The aquariums were continuously aerated for an hour to get a homogenous concentration of toxic compound and 10 individuals were transferred to each aquarium. Mortality was assessed for 24 and 48 hours with the dead being quickly removed from the test aquarium.

After the 48 hours LC 50 values of dimethonate on tadpoles of *Bufo himalayanas* were ascertained, ten tadpoles each were treated to $1/3^{rd}$, $1/5^{th}$, $1/8^{th}$ concentration of the48 hour LC 50 in triplicates in glass aquaria measuring 90x35x35 cm for a period of 21 days. Body weights of tadpoles and temperature, free carbon dioxide, total hardness and pH of water were measured at an interval of 7 days (Table 1). A summer photo period was maintained with a light dark cycle of 16:8 h.

RESULTS

The 48 hr LC 50 values of 3 weeks old tadpoles of *Bufo Himalayanas* was calculated to be 0.617 mg/L. Reduction in body weight was observed in all the three exposures doses (lesser than 48 hr LC 50 value). Maximum loss of body weight was observed in $1/3^{rd}$ dose of LC 50 ie.0.206 mg/L and minimum in $1/8^{th}$ dose of LC 50 (0.077mg/L) (figure1); (table 1). Mortality also increased with increase in insecticide concentration (figure 2).

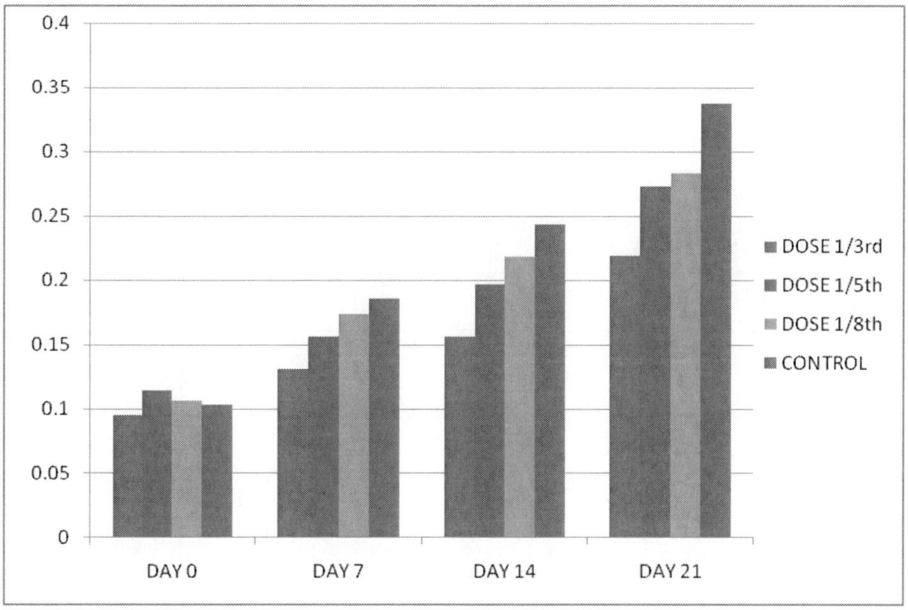

Figure 1. Influence of different concentrations of dimethonate on body weight of tadpoles of *B. himalayanas* recorded during 21 days.

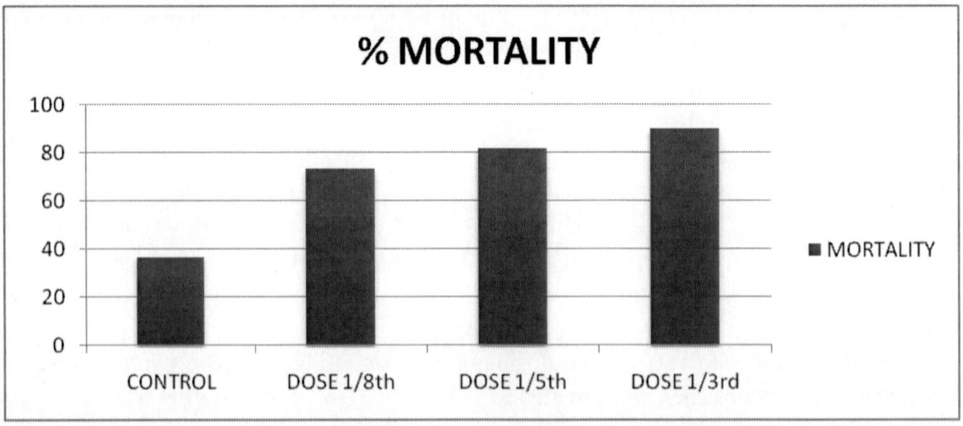

Figure 2. Percentage mortality of *B. himalayanas* tadpoles recorded over a period of 21 days.

The change in the body weight and mortality of tadpole of *Bufo himalayanas*, brought about by three different doses of insecticide dimethonate differed significantly as revealed by *t* - tests (t=5.9732 $p<0.05$, df=3).

Table 1. Effects of Dimethonate on body weight of tadpoles of *Bufo himalayanas* exposed for 21 days

DOSE	0 DAY MEAN± SD n = 10	7 DAYS MEAN± SD n = 10	14 DAYS MEAN± SD n = 10	21 DAYS MEAN± SD n = 10
CONTROL	0.105±0.015	0.18±0.006	0.23±0.001	0.31±0.025
DOSE 1/3rd	0.105±0.013	0.131±0.04	0.1577±0.009	0.232±0.020
DOSE 1/5th	0.113±0.003	0.156±0.013	0.198±0.003	0.272±0.120
DOSE 1/8th	0.112±0.007	0.177±0.009	0.217±0.004	0.284±0.03

Table 2. Tests of water quality parameters conducted with tadpoles of *Bufo himalayanas* over a period of 21 days

DOSE	Temperature (°C)				Total hardness(mg/L)				Dissolved CO_2 (mg/L)				Median pH			
	DAY 0	DAY 7	DAY 14	DAY 21	DAY 0	DAY 7	DAY 14	DAY 21	DAY 0	DAY 7	DAY 14	DAY 21	DAY 0	DAY 7	DAY 14	DAY 21
CONTROL	19	19	18	18	9.4	9.2	9.2	9.1	3.0	3.0	3.1	3.2	7	7	6	6
DOSE 1/3rd	19	19	18	18	8.7	8.5	8.3	8.2	3.0	3.6	3.7	4.2	7	6	6	6
DOSE 1/5th	19	19	18	18	8.7	8.6	8.5	8.5	3.0	3.4	3.6	3.8	7	7	6	6
DOSE 1/8 th	19	19	18	18	8.6	8.5	8.4	8.4	3.0	3.1	3.2	3.2	7	7	6	6

DISCUSSION

The results of the experiment indicate that insecticides have some direct effects on the survival and body weight of the tadpoles of *Bufo himalayanas*. Most of the earlier toxicological studies in amphibian larvae revealed that increased concentration of toxicant shows varying effect of tadpole growth, survivorship, development and swimming ability (Clotfelter et al., 2004). The insecticide dimethonate caused a 60% reduction on tadpole survival and 35% reduction on the tadpole body weight. However Roundup caused a 40% reduction on total tadpole survival and biomass (Relyea et al., 2005).The 48 hour LC50 value of tadpoles of *Bufo himalayanas* was calculated to be 0.617 mg/L, where the 48 hour LC50 value ranged between 3.9 to 15.5mg/L in some species of Australian tadpoles (Perkins et al., 2000). The low LC50 of *Bufo himalayanas* clearly suggests the vulnerability of the tadpoles of *Bufo himalayanas* to the use of insecticides. However the bioaccumulation in adult has yet to be investigated. The drastic decrease in growth and development would prevent the tadpoles from reaching metamorphosis.

ACKNOWLEDGMENT

We express deep sense of gratitude to the Head, Department of Zoology, Darjeeling Government College for letting us carry out this research work. Authors are also thankful to Benoy Kishore Rai and Dawa Bhutia, SRF, University of North Bengal for their support throughout the work.

REFERENCES

Alford, R.A. and S.J Richards. (1990). Global amphibian declines: a problem in applied ecology. *Annual review of Ecology and Systematics.30*, 133-165.

Blaustein, A.R., Wake. and D.B, Sousa, W.P. (1994). Amphibian declines: Judging Stability, persistence, and susceptibility of populations to local and global extinctions. *Conserv Biol, 8*:60-71.

Boone, M.D. and S.M. James. (2003). Interactions of an insecticide, herbicide, and natural stressors in amphibian community mesocosms. *Ecolo App, 13*, 829-841

Clotfelter, E.D., Bell, A.M. and Levering, K.R. (2004). The role of animal behavior in the study of endocrine-distrupting chemicals. *Animal Behaviour, 68,* 465-476.

Davidson, C., Shaffer, H.B. and Jennings M.R. (2002). Spatial tests of the pesticide drift, habitat destruction, UV-B, and climate-change hypothesis for California amphibian declines. *Conserv Biol, 16,* 1588-1601.

Diana, S.G., Resetarites, W.J., Schaeffer, D.J., Beckmen, K.B. and Beasley, V.R.. (2000). Effects of atrazine on amphibian growth and survival in artificial aquatic communities. *Environm Toxicol Chem, 19,* 2961-2967.

Fairchild, J.F., La Point, T.W. and Schwart, T.R. (1994). Effects of an herbicide and insecticide mixture in aquatic mesocosms. *Archives of Environ Contam and Toxicol, 27,* 527-533.

Gurushankara, H.P, Vasudev, V. and Krishnamurthy S.V. (2003). Estimation of acute toxicity of malathion insecticide on tadpoles and adults of Rana (Limnonectus) limnocharis. *Indian J Compar Animal Physio, 21*, 48-54.

Harris, M.L., Bishop, C.A., Struger, J., Ripley. B. and Bogart, J.P. (1998).The functional integrity of Northern leopard frog (Rana pipiens) and green frog (Rana calamitans) populations in Orchand wetlands.II. Genetics, physiology, and biochemistry of breeding adults and young-of-the year. *Environ Toxicol Chem, 17*, 1338-1350.

Houlihan, J.E., Findlay, C.S., Schmidt, B.R., Meyers, A.H. and Kuzmin, S.L. (2001).Quantitative evidence for global amphibian population declines. *Nature*, *404*, 752-755.

Kaur., K. and Chemma, G. (1986). Acute toxicity of some pesticides to the eggs and larvae of the common carp, Cyprinus carpio. *Zool Orient*, *2*, 52-58.

Kiesecker, J.M., Blaustein, A.R. and Belden, L.K. (2001).Complex causes of amphibian population declines. *Nature 410*, 681-684.

Lambert, M.R.K. (1997). Effects of pesticides on amphibians and reptiles in sub-saharan Africa. *Rev Environ Contam Toxicol, 150,* 31-73.

Lenoir, J.S., Mc Connell, L.L., Fellers, G.M., Cahill, T.M. and Seiber, J.N. (1999). Summertime transport of current- use pesticides form California's central valley to the Sirra Nevada Mountain range, USA. *Environ Toxicol Chem, 18,* 2715-2722.

Mc Connell, L.L., Lenoir, J.S., Datta, S. and Seiber, J.N. (1998). Wet deposition of current-use pesticides in the Sierra Nevada Mountain range, Califoirnia, USA. *Environ Toxicol Chem, 17,* 1908-1916.

Relya, R.A., Schoeppner, N.M. and Hoverman, J.T. (2005). Pesticides and amphibians: The importance of community context. *Ecol App, 15,* 1125-1134.

Semlitsch, R.D., Bridges, C.M. and Welch, A.M. (2000). Genetic variation and a fitness tradeoff in the tolerance of gray tree frog (Hyla versicolor) tadpoles to the insecticide carbaryl. *Oecologica, 125*, 179-185.

Sparling, D.W., Fellers, G.M. and Mc Connell, L.L. (2001). Pesticides and amphibian population declines in California, USA. *Environ Toxicol Chem, 20*, 1591-1595.

In: Vertebrate and Invertebrate Animal Approaches
Editor: José Rosa Gomes
ISBN: 978-1-63117-102-4

Chapter 15

DEVELOPMENTAL STAGES OF *TETRANYCHUS LUDENI* ZACHER (ACARI: TETRANYCHIDAE) AND *BREVIPALPUS DELEONI* PRITCHARD AND BAKER (ACARI: TENUIPALPIDAE) ON NEW HOST PLANT

Indranil Roy and Goutam Kumar Saha*
Entomology and Wildlife Biology Research Laboratory
Department of Zoology, University of Calcutta, Kolkata, India

ABSTRACT

The present report represents new record of medicinal host plants *Abutilon indicum, Ocimum gratissimum, Tinospora cordifolia* for *Tetranychus ludeni* Zacher and medicinal host plant *Ocimum gratissimum* for *Brevipalpus deleoni* Pritchard and Baker along with a note of their developmental stages at laboratory condition. The mean duration of egg to adult period was 14.4±1.34 days in *T. ludeni* on *Tinospora sinensis* (Lour.) Merrill. In *B. deleoni* the duration of egg to adult period was 21.3±0.67 on *Ocimum gratissimum* Linn.

Keywords: Phytophagous mite, new record, medicinal plant, developmental stages

INTRODUCTION

Tetranychus ludeni Zacher, a phytophagous mite, was responsible for infesting a wide variety of medicinal plants viz. *Cajanus cajani, Citrus* sp., *Coccinia grandis, Clitoria tematea, Cucumis sativus, Datura metel, Dolichos biflorus, Ficus carica, Ricinus communis, Pterocarpus marsupium, Phyllanthus fraternus, Gossypium arboretum, Pavonia odorata* (Gupta, 2005). Another important phytophagous mite Brevipalpus deleoni Pritchard and Baker was also reported for infesting medicinal plant like *Datura metel, Strychnos nux-vomica, Vallaris solanacea, Vitex negundo* (Gupta, 2005). During the course of present

* Corresponding Author: Email: indranilzoology@gmail.com; Email: gkszoo@gmail.com.

Eggs

The eggs of *T. ludeni* were spherical, about 80μ in length and were yellowish transparent and were laid singly on the undersurface of leaf. The mean incubation period was 1.57±0.22 days with minimum of 1 day and maximum of 3 days. In case of *B. deleoni* the eggs were elliptical, about 81μ in length and red in colour. Minimum and maximum incubation period of 1 and 2 days (mean 1.4±0.41 days) were recorded.

Larva

The larva of *T. ludeni* was white in colour with three pairs of legs. The mean larval period was 2.38±0.14 days. The minimum and maximum larval period of 2 and 3 days were noticed. The larva of *B. deleoni* was red in colour with three pairs of legs. The mean larval period was 2.6±0.30 days. Minimum of 2 days and maximum of 3 days of larval period were observed.

Protonymph and Deutonymph

The protonymph of *T. ludeni* was also white in colour, having four pairs of legs. The protonymph was more active than the larva and this period on an average lasts for 3.8±0.20 days, with the minimum of 3 days and maximum of 5 days duration. On the other hand the mean duration of protonymph of *B. deleoni* was 1.8±0.38 days with minimum of 1 day and maximum of 3 days duration.

Deutonymph of *T. ludeni* was bigger in size than protonymph with much pigmentation. The mean deutonymphal period was 6.8±0.37 days with the minimum of 6 days and maximum of 8 days duration.

In case of *B. deleoni*, the mean duration of deutonymphal period was 3.2±0.55 days with minimum of 2 days and maximum of 5 days observed.

Adult longevity

Adults of *T. ludeni* were red in colour. The adult longevity was 9.8±1.39 days with the minimum of 5 days and maximum of 12 days.

Adult longevity of *B. deleoni* was 12.4±0.89 days. Minimum and maximum longevity period of 9 days and 19 days were recorded.

Egg to Adult

The mean duration of egg to adult period was 14.4±1.34 days with the minimum of 21 days and maximum of 25 days in *T. ludeni.*

In *B. deleoni* the duration of egg to adult period was 21.3±0.67 days with minimum and maximum of 18 days and of 28 days.

Fecundity

The average fecundity of *T. ludeni* was 16.25±1.20 eggs per day per female in the species studied whereas in *B. deleoni* the average fecundity was 1.75±0.25 eggs per day per female.

DISCUSSION

It is to be noted that the life cycle of *Tetranychus ludeni* on *Tinospora sinensis* (under controlled laboratory condition) at a mean temperature and relative humidity of 31°C and 72 percent differ from that of the earlier studies. In the present study, the mean duration of egg to adult period was 14.4±1.34 days. Earlier Puttaswamy and Channa Basavanna (1981) studied the effect of temperature and relative humidity on development of *T. ludeni*. They reported that the optimum condition for development and maximum survival of eggs were between the temperatures of 32 $\pm 1^0$C with 65 ±3 percent relative humidity and 75 ±3 percent relative humidity. Immature stages developed best at 30 $\pm 1^0$C and 35 $\pm 1^0$C. Earlier, Mallik and ChannaBasavanna (1983) studied the life history and life tables of *T. ludeni* and its predator *Amblyseius longispinosus* and found that duration of egg, larva, protonymph and deutonymph of *T. ludeni* were 10.6, 32.5, 34.5 and 49 hours respectively which differ markedly from the present study. Total development time for male and female were 200 hrs. and 222 hrs., respectively. Malaviya and Rai (1995) worked out the life cycle of *T. ludeni* at a temperature range of 20.24^0C-22.33^0C at relative humidity of 40-55%. They reported that fecundity of mated females was higher (4.96 ±0.77 egg/day) that of unmated females.

Life cycle of *Brevipalpus deleoni* was studied for the first time from India. Developmental stages of other species of *Brevipalpus* were recorded earlier by different authors. Goyal and Sadana (1986) studied development stages of *B. obovatus*. Sadana and Sharma (1989) studied the influence of temperature on development of *B. rugulosus* on *Luffa acutangunla* Roxb and found that 30 ± 1^0C was most favorable temperature. Sadana and Kumari (1993) in their studies on development of *B. phoenicis* at 20, 25, 30^0C in combination with 50, 70 and 90% RH, reported that 25^0C in combination with 70% RH was most suitable for the development of the mite. Considerable variation in the life cycle data of *T. ludeni* and, *B. deleoni* were noted contrast to different geographical situations reported. The life cycle data of these mite species will help to compare the biology of these mites to predict possible abundance pattern in nature.

ACKNOWLEDGMENTS

We thank, Dr. S. K. Gupta, Former Joint Director, Zoological Survey of India, Kolkata, for identification of mite species. The authors are grateful to the Head, Department of Zoology, University of Calcutta, Kolkata and Assistant Secretary, Ramakrishna Mission Ashram, Narendrapur, West Bengal, India, for providing facilities to carryout research work. The financial assistance from Ministry of Environment and Forests, Government of India, through the Research Project (Sanction No. 14/26/2004 – ERS/RE) is thankfully acknowledged.

REFERENCES

Ghoshal, S., Gupta, S.K. and Choudhury, A. (2004) Life cycle of *Amblyseius multidentatus* (Swirski and Shechter) at room temperature feeding upon *Eotetranychus fremonti* Tuttle and Baker on *Avicennia albae* Blume. *Rec. zool. Surv. India.,* 102 (3and4): 47-52.

Goyal, N. and Sadana, G. L. (1986) Developement, fecundity and longevity of *Brevipalpus obovatus* (Acari: Tenuipalpidae) on different host plants. *Indian J. Acar.,* 11: 41-50.

Gupta, S. K. (2005) *Insects and mites infesting medicinal plants in India,* Ramakrishna Mission Ashrama, Narendrapur, Kolkata, 214 pp.

Gupta, S. K., Ghoshal, S. and Choudhury, A. (2003) Life cycle of Amblyseius alstoniae Gupta, at room temperature feeding upon Eotetranychus hicoriae McGregor on guava (Psidium guajava). *Rec. zool. Surv. India.,* 101(1 and2): 93-99.

Malaviya, M. D. and Rai, A. B. (1995) Studies on biology of red spider mite *Tetranychus ludeni* on Indian bean. *Abst. Vth Nat. Symp. Acarol. Sept., Bangalore*: 43 pp.

Mallik, B. and ChannaBasavanna, G. P. (1983) Life history of *Tetranychus ludeni* and its predator, *Amblyseius longispinosus* (Acari: Tetranychidae; Phytoseiidae). *Indian J. Acar.,* 8: 1-12.

Puttaswamy and ChannaBasavanna, G. P. (1981) Influence of weather factors and predators on the populations of the spider mite *Tetranychus ludeni* (Acari: Tetranychidae). *Indian J. Acar.*, 5: 69-79.

Sadana, G. L. and Kumari, M. (1993) Seasonal history of *Brevipalpus phoenicis* (Geiesks) (Acari: Tenuipalpidae) on *Psidium guajava* cv seedless guava. In: *Progress in Acarology* (Ed. ChannaBasavanna, G. P. and Viraktamath, C. A.)*Vol.* 2, Oxford and IBH (Publ.): 93-100 pp.

Sadana, G. L. and Sharma, R. (1989) Influence of temperature on the development of *Brevipalpus rugulosus* (Acari: Tenuipalpidae). In: *Progress in Acarology* (Ed. ChannaBasavanna, G. P. and Viraktamath, C. A.)*Vol.* 2, Oxford and IBH (Publ.): 95-99 pp.

In: Vertebrate and Invertebrate Animal Approaches
Editor: José Rosa Gomes
ISBN: 978-1-63117-102-4

Chapter 16

ANTIOXIDANT AND DETOXIFICATION ACTIVITIES OF GILL AND DIGESTIVE GLAND OF *LAMELLIDENS MARGINALIS* (MOLLUSCA: BIVALVIA) EXPOSED TO AZADIRACHTIN

Suman Mukherjee[1], Mitali Ray[2] and Sajal Ray[2*]
[1]Post Graduate Department of Zoology, Immunology Laboratory,
A.B.N. Seal College, Coochbehar, West Bengal, India
[2]Department of Zoology, Aquatic Toxicology Laboratory,
University of Calcutta, Kolkata, West Bengal, India

ABSTRACT

L. marginalis is an important biological component of freshwater ecosystem and an efficient bioaccumulator of toxin residues. Being a filter feeder, this species performs biological filtration of ambient water and keeps the water body biologically safe for other species.

Natural habitat of the species is under risk of contamination by multineeem, a newly introduced azadirachtin (limonoid) based pesticide. During monsoon, natural habitat of *L. marginalis* received residues of diverse xenobiotics along with azadirachtin. Reactive oxygen intermediates are major cytotoxic factor which is generated in invertebrates in diverse physiological conditions.

The stimulation of superoxide anion production and resulting physiological damage is modulated by the presence of antioxidant enzyme namely superoxide dismutase (SOD) and catalase (CAT).

Further the toxin residues are conjugated in gluthathione pool through glutathione-S-transferase (GST). This is a protective phenomenon in bivalves. In *L. marginalis*, the activity levels of antioxidant enzymes namely superoxide dismutase and catalase activites were significantly low than control in gill and digestive gland under toxin exposure respectively. In case of glutathione-S-transferase, azadirachtin moderately suppress the activity of the enzyme in selected tissues of *L. marginalis*.

[*] E. mail: raysnailmail@rediffmail.com.

Keywords: Bivalve, Azadirachtin, Antioxidant and detoxification enzymes

INTRODUCTION

Molluscs are of interest not only to the farmers and horticulturists but to the ecotoxicologist as biomonitor species for environmental pollution (Ebenso, 2004; Chakraborty et al., 2010). They encounter toxic materials either by contact through medium or during feeding. In recent years there has been a considerable interest in the use of biochemical indicators within bivalve mollusc in contrast to previous monitoring studies, which mainly focus on the level of contaminants. *L. marginalis* has been identified as an efficient filter feeder bivalve (Das and Jana, 1999) and is widely distributed along the floors of the freshwater bodies. Azadirachtin is a natural bioactive agent derived from the seeds of the neem tree *Azadirachtin indica* (Schmutterer, 1990). Azadirachtin is a triterpenoid compound and expresses insecticidal activity against a broad range of insect pests (Singh and Singh, 1995). The insecticidal activity involves pest repellence, feeding and ovipositional deterrence and insect growth regulation (Schmutterer, 1990). 'Multineem' is an azadirachtin based biopesticide which is being recently used by the Indian farmers for the purpose of protection of crops from insect pest attack. Indiscriminate application of insecticide to agricultural field may result in contamination of adjacent freshwater bodies leading to exposure of nontarget organisms like *L. marginalis* to toxic azadirachtin (Isman, 1997).

An important body of investigation has focused on biomarkers of oxidative stress within organisms. Antioxidants enzyme systems (superoxide dismutase (SOD), catalase (CAT)) and detoxification enzyme like glutathione-S- transferase (GST) are efficient protective mechanisms against chemical reactive species produced by endogenous metabolism or by the biotransformation of xenobiotics. The reactive oxygen species (ROS) are generated in haemocytes and tissues in molluscs as superoxide anion (O_2^-) having microbicidal activity and are rapidly dismutated by the action of metalloenzyme superoxide dismutase to hydrogen peroxide and oxygen. Hydrogen peroxide is continuously generated by several enzymes including SOD which must be degraded to prevent oxidative damage (Gamble et al., 1995; Isani et al., 2000). The antioxidant catalase, is a hematin-containing enzyme based in the peroxisomes of cells and is an important component of intracellular and antioxidant defences of aquatic organisms (Jamil, 2002). It reduces the hydrogen peroxide into water and oxygen to prevent oxidative stress and in maintaining cell homeostasis (Arun and Subramanian, 1998; Arun, 2000). The decomposition of hydrogen peroxide is directly proportional to both the concentration of enzyme and the concentration of substrate (Di Giulio et al., 1989). Peroxide, including hydrogen peroxide is one of the main reactive oxygen species (ROS) leading to oxidative stress (Gamble et al., 1995). The cytotoxic effect of H2O2 is thought to be caused by hydroxyl radicals generated from iron catalyzed reactions causing subsequent damage to DNA, proteins and membrane lipids. In molluscs, catalase catalyzes the conversion of H2O2 to H2O and O2 (Scott et al., 1991; Arun, 2000). Antioxidant and detoxification responses have evolved as adaptive response in animal series. GST is an effective enzyme involved in conjugation phase of detoxification response. Glutathione-S transferase catalyzes the conjugation of reduced glutathione to nucleophiolic xenobiotics (Ishizuka et al., 1998; Nielsen et al., 1999). GST primarily is involved in the chemical disposition of toxic substances and has the ability to catalyse the conjugation of GSH to various toxic electrophiles and inactivate toxic compounds by non catalytic binding (Lamoureux and Rusness, 1987) and elimination from the system. The present study evaluates the influence of a biopesticide on the antioxidant and detoxification enzyme activities in the gill and digestive gland of the freshwater bivalve, *L. marginalis.*

MATERIAL AND METHOD

Collection and Treatment of Animals

The adult healthy *L. marginalis* with shell size of 7-8 cm were manually collected from the selected wetlands of the district of South 24 Parganas of West Bengal. Animals were transported to the laboratory in rectangular plastic containers with a dimension of 12'x18'x 6' at a density of 4-6 individuals per box in moist condition. Prior to experimentation, animals were acclimatized for 15 days in the laboratory.

During acclimatization, *L. marginalis* were maintained in aquaria with fresh supply of pond water with temperature of 29°C±3°C and the animals received uniform ration of illumination. During the course of acclimatization and experiment, the animals were fed with chopped *Hydrilla* sp. and common aquatic weeds (Raut, 1991).

Routine replenishment of water was carried out in every 12 hours to avoid residual toxicity. Aqueous solutions of Multuneem (Multiplex, India Private Limited, Azadirachtin E.C. 0.03%) formulations were prepared in Borosilicate glass containers with azadirachtin concentrations of 0.006, 0.03, 0.06, and 0.09 ppm. The pH of the solution was maintained at 7.2. Each experimental set consisted of 10 animals of same shell length. Animals were exposed to a volume of 5 litre of pesticide solution for varied span of exposure i.e. 1, 2, 3, 4, 7, 15 and 30 days. For control, a set of animals were kept in identical volume of pesticide free analytical grade water. The experiments were carried out in static water environment and fresh solutions of pesticide were replenished in every 12 hour.

Preparation of Tissue Lysate

Gills and digestive gland of *L. marginalis* was dissected in cold sterile saline and tissue of uniform weight (100 mg) was homogenized in 0.25 M sucrose solution (3%w/v) in mechanical glass homogenizer at 4°C and was centrifuged at 100x g for 15 minutes at cold (Vijayavel and Balasubramanian, 2006). Estimation of protein and enzyme were carried out from supernatant.

Super oxide Dismutase (EC1.15.1.1)

Activity of super oxide dismutase was determined in tissue lysate following the method of Krishnan et al., (2002). Assay involved reaction of gill and digestive gland suspension to Griess reagent followed by absorbance at 560 nm. One unit of enzyme activity is defined as the amount of SOD capable of inhibiting 50% of nitrite formation under assay condition.

Activity of superoxide dismutase was calculated and expressed as unit of SOD/mg of protein/min.

Catalase (EC1.11.1.6.)

Catalase activity was estimated according to Prakash and Rao (1995). For test sample, the reaction mixture consists of 5 ml of hydrogen peroxide (6mM) as substrate and 0.5ml of enzyme (0.1% Triton-X lysed gill and digestive gland). For standard, the reaction mixture consists of 5.5ml of phosphate buffer (0.01M). For both test and standard sample, the reaction was stopped by adding 1ml of sulphuric acid (6N) followed by 7ml of potassium permanganate (0.01N) after 3 mins of reactions. The final reaction mixture was monitored at 480 nm for 30-60 secs after the stoppage of the reaction. The activity was expressed as μl of catalase/mg protein/min.

Glutathione -S Transferase (EC 2.5.1.13)

Glutathione-S transferase activity was estimated in the gill and digestive gland following the method of Aceto et al., (1991). One unit of GST refers to the enzyme amount needed to conjugate 1μM of 1 chloro 2, 4dinitrobenzene per min, per mg of total protein present in the homogenate. The enzyme activity was expressed as μM GST/ mg protein/ min.

RESULTS

Antioxidant Activity

The activity of SOD was highest in digestive gland followed by gill (Figure 1). The degree of inhibition of SOD in the tissues expressed a dose dependent response and considerable inhibition occurred even at low concentrations (0.006ppm) of azadirachtin exposure. Maximum inhibition in SOD activity was observed as 0.5 unit/mg of protein/min in the digestive gland as compared to control upon 0.09 ppm of azadirachtin exposure for 7 days (Figure 2).

Oxygen toxicity, referred to in earlier days as the oxygen toxicity is currently in focus in stress related environmental effect (Arun and Subramanian, 1998). The toxic effects of oxyradicals include lipid peroxidation, nucleic acid damage and enzyme inactivation. Self scavenging of these oxyradicals is an adaptive response which is studied by estimating catalase which is involved predominantly in the removal of hydrogen peroxide. Catalase has been used in aquatic organisms as a biomarker for monitoring environmental pollution and oxyradical induced damage.

In tissues of control animal, the activity of catalase was highest in the digestive gland followed by gill (Figure 4 and 3). The degree of inhibition of catalase in the tissues expressed a dose dependent response and considerable inhibition occurred at low concentrations of azadirachtin exposure (Figure 3 and 4).

Maximum inhibition in catalase activity was observed as 0.35 μm /mg protein/min in the digestive gland as compared to control for 0.09 ppm/7days exposure (Figure 4).In the gill, the maximum inhibition of the activity of catalase was recorded as 0.4μm /mg protein/min against 0.09 ppm of azadirachtin exposure for 7 days (Figure 3).

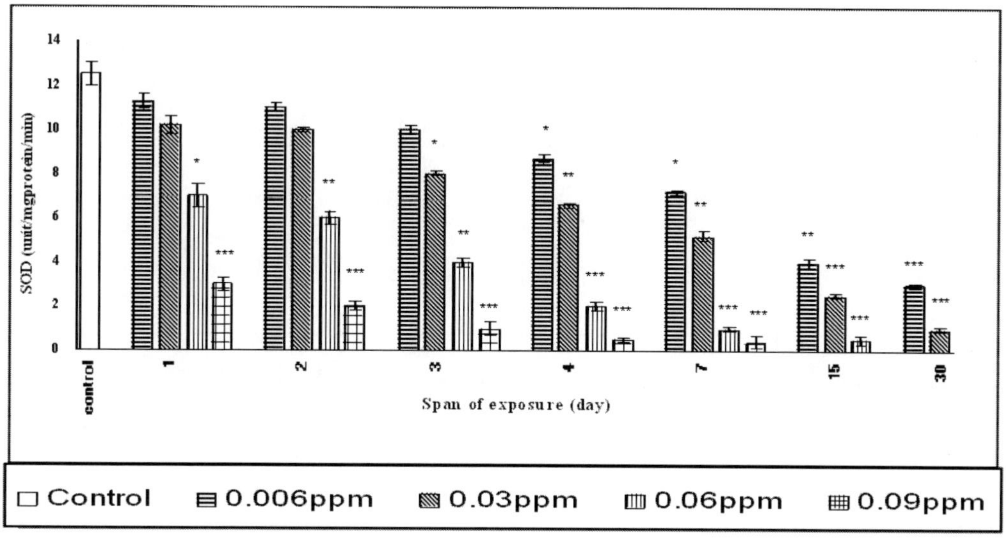

Figure 1. Activity of superoxide dismutase of gill of *L. marginalis* (7-8 cm shell length) exposed to azadirachtin *in vivo*. Data is represented as Mean±S.D. Statistical significance is shown at P<0.05*, P<0.01**, P<0.001***. (n=5).

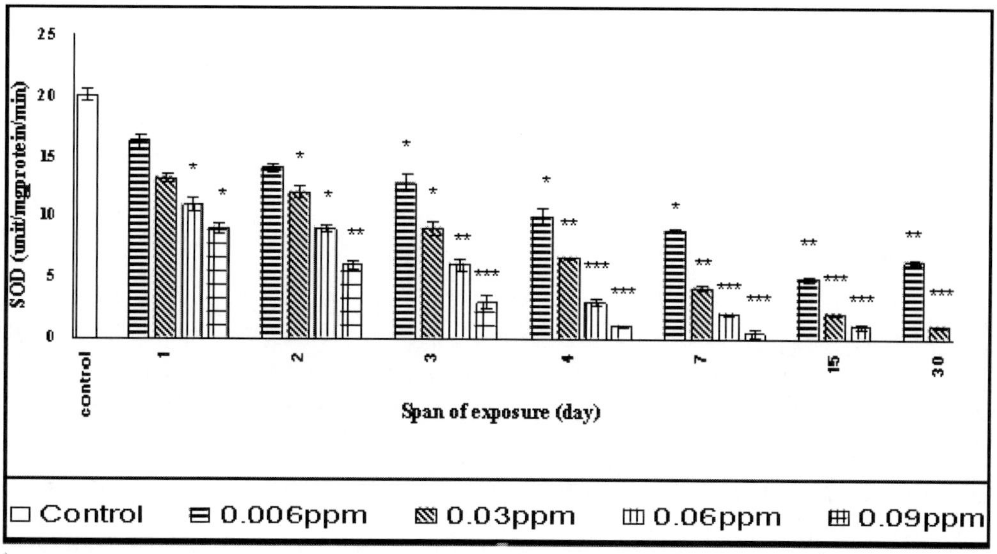

Figure 2. Activity of superoxide dismutase of digestive gland of *L. marginalis* (7-8 cm shell length) exposed to azadirachtin *in vivo*. Data is represented as Mean±S.D. Statistical significance is shown at P<0.05*, P<0.01**, P<0.001***. (n=5).

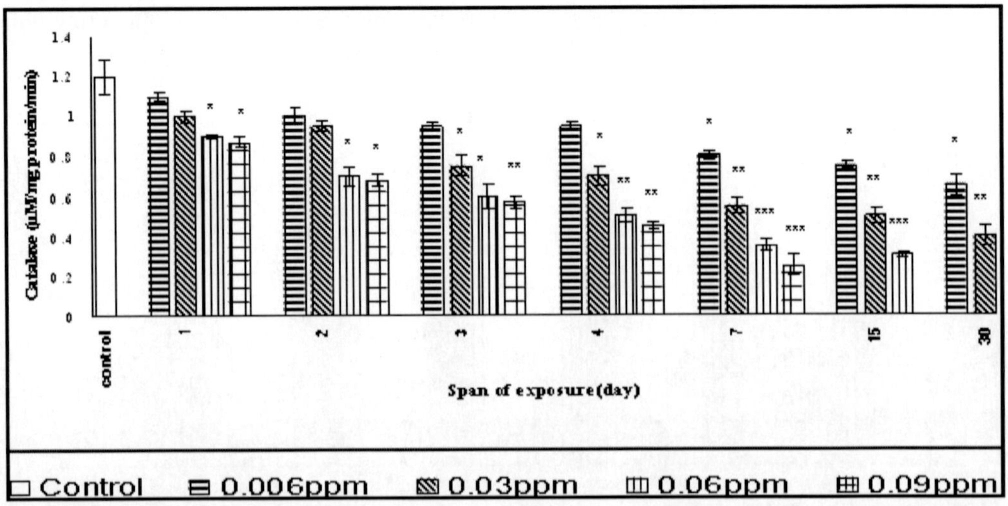

Figure 3. Activity of catalase of gill of *L. marginalis* (7-8 cm shell length) exposed to azadirachtin *in vivo*. Data is represented as Mean±S.D. Statistical significance is shown at P<0.05*, P<0.01**, P<0.001***. (n=5).

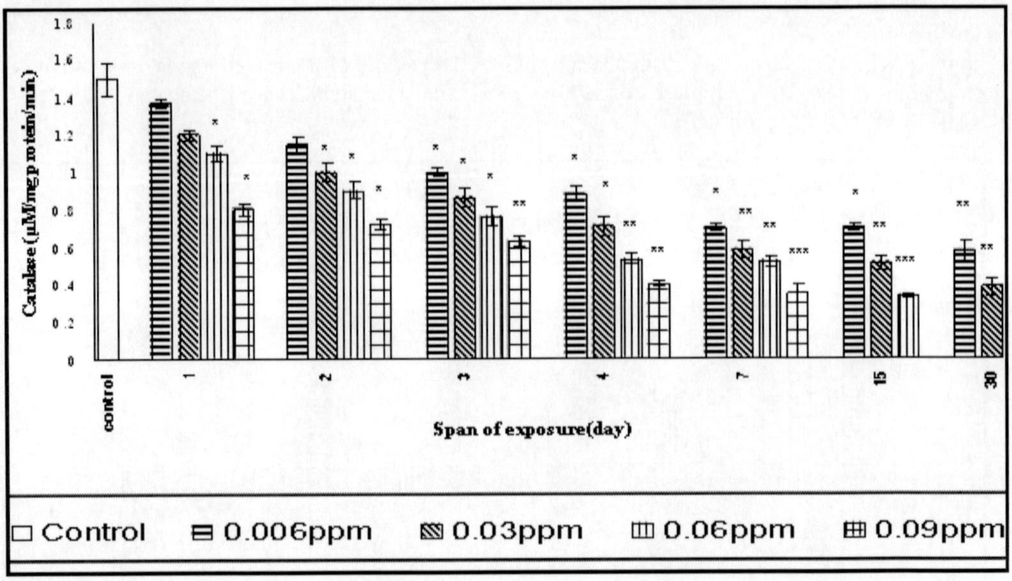

Figure 4. Activity of catalase of digestive gland of *L. marginalis* (7-8 cm shell length) exposed to azadirachtin *in vivo.* Data is represented as Mean±S.D. Statistical significance is shown at P<0.05*, P<0.01**, P<0.001***. (n=5).

Detoxification Activity

Glutathione-S-transferase is involved in the detoxification reaction process of various xenobiotic chemicals. The activity level of GST was relatively lower than that of control throughout the study in all the tissues for azadirachtin exposure. In tissues of control animal, the activity of GST was highest in digestive gland followed by the activity in gill (Figure 5 and 6). The degree of inhibition of GST in the tissues expressed a dose dependent response and considerable inhibition occurred at low concentrations of pesticide exposure (Figure 5 and 6). Azadirachtin of 0.09 ppm/7 days exposure resulted in maximum inhibition in GST activity which was determined as 1μm /mg protein/min in the digestive gland as compared to control (Figure 6).In the gill, the maximum inhibition of the activity of GST was recorded as 1.5 μm /mg protein/min against 0.09 ppm of azadirachtin exposure for 7 days (Figure 5).

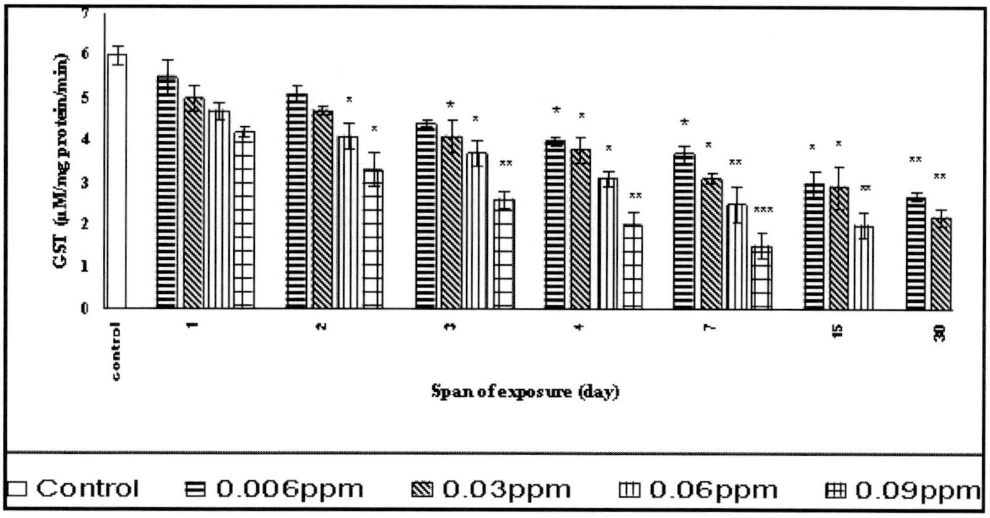

Figure 5. Activity of GST of gill of *L. marginalis* (7-8 cm shell length) exposed to azadirachtin *in vivo*. Data is represented as Mean±S.D. Statistical significance is shown at P<0.05*, P<0.01**, P<0.001***. (n=5).

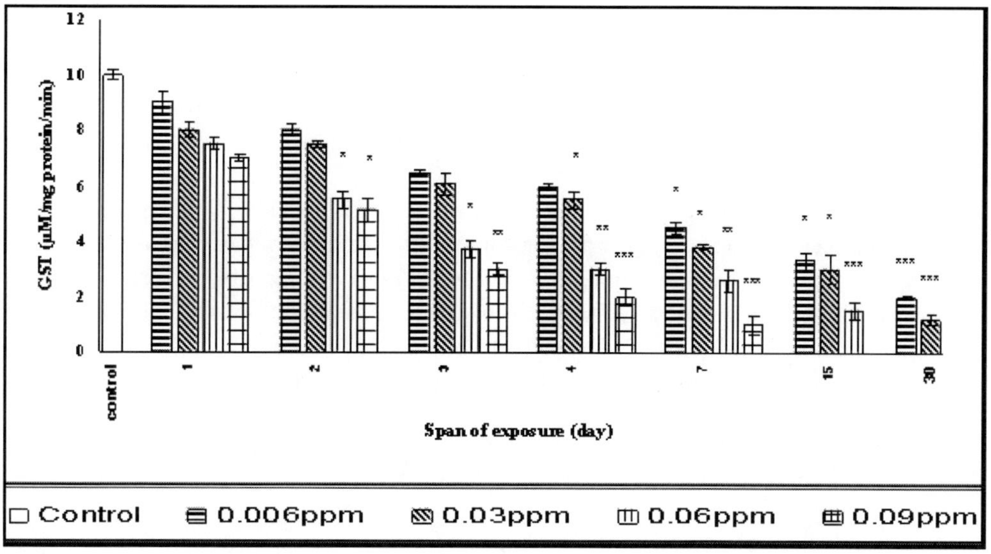

Figure 6. Activity of GST of digestive gland of *L.marginalis* (7-8 cm shell length) exposed to azadirachtin *in vivo*. Data is represented as Mean±S.D. Statistical significance is shown at P<0.05*, P<0.01**, P<0.001***. (n=5).

Discussion

Molecular oxygen is an essential element of life and reactive oxygen species (ROS) are generated in all aerobes (Krishnan et al., 2002). Strategy of oxygen radical mediated killing is based on the premises of toxicity evolved due to high concentrations of molecular oxygen. Superoxide anions undergo reaction with hydrogen to produce hydrogen peroxide, hydroxyl radicals and finally water (Rodriguez and Moullac, 2000). Superoxide anions are extremely toxic, powerful and hyperactive killing agents which are capable of creating damage of the cells and tissues of self. Most ROS are generated as superoxide anion (O_2^-) having microbicidal activity and are rapidly dismutated by the action of superoxide dismutase to hydrogen peroxide and oxygen. In this study, the activity of antioxidant enzyme namely superoxide dismutase is higher in digestive gland followed by in gills (Figure 2 and 1). Exposure to azadirachtin decreased the generation of superoxide dismutase for all four experimental concentrations in gill and digestive gland which expressed a dose dependent response. Since xenobiotic induced production of ROS vary significantly against environmental factors, necessary adjustments in antioxidant defences are required to maintain the steady state concentration of ROS for prevention and minimization of oxidative stress and cellular damage (Lesser, 2006) and is considered as a biochemical adaptive response. A decrement of superoxide dismutase production was observed against increasing azadirachtin concentrations. Catalase activity decreased significantly both in gill and digestive gland (Figure 3 and 4). In conclusion, present data demonstrate the sensitivity of antioxidant enzymes of *L. marginalis* under the exposure of azadirachtin. Azadirachtin induced decrement of antioxidant enzymes is indicative to a possible onset of physiological stress in the specimen distributed in water bodies contaminated with neem pesticide. Glutathione-S transferase which catalyzes the conjugation of reduced gluthathione with nucleophiolic xenobiotics or cellular components damaged by oxyradical attack in their detoxification reaction (Ishizuka et al., 1998). Azadirachtin moderately suppressed the activity of the GST in haemocytes and selected tissues of *L. marginalis* (Figure 5 and 6) which is indicative of possible impairment of detoxification machinery. Toxin induced inhibition of GST activity would increase the magnitude of residual toxicity of azadirachtin accumulation in the selected tissues and organs of *L. marginalis*. Being a filter feeder, this species performs biological filtration of ambient water and maintain the water body biologically safe for other species. Molluscs, in recent years are gaining enormous interest among toxicologists as suitable biomarker of toxicity. This slow moving biofilter organism is capable of providing useful information of the health of waterbodies as suitable indicator of water pollution. Present investigation would help to establish a suitable biomonitor species of aquatic pollution with reference to toxicity of neem based pesticide and allied xenobiotics.

References

Aceto, A., Di, Ilio., C., Bucciarelli, T., Pantani, C., Dell'Agata, M., Pannunzio, G. and Frederic, G. 1991.Characterization of glutathione transferase from *Gammarus italicus*. *Comp. Biochem. Physiol.*, 99B: 523–527.

Arun, S. and Subramanian, P. 1998. Antioxidant enzymes in fresh water prawn *Macrobrachium malcolmsonnii lamarrci* during embryonic and larval development. *Comp. Biochem. Physiol.*, 121: 273-277.

Arun, S. 2000. Biotransformation and antioxidant enzymes in fresh water prawns *Macrobrachium malcolmsonnii* and *M. lamarrei. lamarrei* Ph.D. thesis. Bharathidasan University. India.

Chakraborty, S., Ray, M. and Ray, S. 2010. Toxicity of sodium arsenite in the gill of an economically important mollusc of India. *Fish Shellfish Immunol.*, 29: 136-148.

Das, S. and Jana, B.B.1999.Dose-dependent uptake and *Eichhornia*-induced elimination of cadmium in various organs of the freshwater mussel, *Lamellidens marginalis* (Linn.). *Ecol. Eng.*, 12: 207–229.

Di Giulio, R.T., Washburn, P.C. and Wenning, R.J. 1989. Biochemicalresponses in aquatic animals: A review of determinants of oxidative stress. *Environ. Toxicol. Chem.,* 8: 1103-1123.

Ebenso, I.E. 2004. Molluscicidal effect of neem *Azadirachta indica* extracts on edible tropical land snails. *Pest Managem. Sci.*, 60: 178-182.

Gamble, S C., Goldfarb, P. S., Porte, C. and Livingstone, D. R. 1995. Glutathine peroxidase and other antioxidant enzyme function in marine invertebrate (*Mytilus edulis, Pecten maximus, Carcinus maenas* and *Asterias rubens*). *Mar. Environ. Res.*, 39(1-4): 191 – 195.

Isani, G., Andreani, G., Kindt, M. and Carpene, E. 2000. Metallothioenes (MTS) in marine molluscs. *Cell. Mol. Biol.*, 46: 311-330.

Ishizuka, M., Sakiyama, T., Iwata, H., Fukushima, M., Kazusaka, A. and Fujita, S.1998. Accumulation of halogenated aromatic hydrocarbons and activities of cytochrome P450 and glutathione S-transferase in crabs (Eriocheir japonicus) from Japanese rivers. *Environ. Toxicol. Chem.*, 17:1490–1498.

Isman, M.B. 1997. Neem insecticides. *Pest Outlook.,* 12: 32-38.

Jamil, K. 2002. Bioindicators and biomarkers of environmental pollution and risk assessment. Science Publishers, Enfield: 4, 15, 56.

Krishnan, N., Chattapadhyay, J. K. and Chaudhuri, A. 2002. Superoxide dismutase activity in hemocytes and hemolymph of *bombyx mori* following bacterial infection. *Curr. Sci.*, 83(3): 321 – 325.

Lamoureux, G.L. and Rusness, D.G. 1987. Synergism of diazinon toxicity and inhibition of diazinan metabolism in the housefly by Tridhiphane inhibition of gluthathione-S-transferase activity. *Pest Biochem. Phys.*, 27: 318-329.

Lesser, M. P. 2006. Oxidative stress in marine environment: biochemistry and physiological ecology. *Annu. Rev. Physiol.*, 68: 253 – 278.

Nielsen, S. A., Toft, S. and Clausen, J. 1999. Cypermethrin effects on detoxification enzymes in active and hibernating wolf spiders (*Pardosa amentata*). *Ecol. Appl.*, 9: 463–468.

Prakash, N. T. and Rao, K. S. J. 1995. Modulations in antioxidant enzymes in different tissues of marine bivalve *Perna viridis* during heavy metal exposure. *Mol. Cell. Biochem.*, 146: 107 – 113.

Raut, S.K. 1991. Laboratory Rearing of Medically and Economically Important Molluscs. In *"Snails, Flukes and Man"*. (M.S. Jairajpuri, Ed.), pp.79- 83. Zoological Survey of India Pub.

Rodriguez, J. and Moullac, G.L. 2000. State of the art of immunological tools and health control of penaeid shrimp. *Aquacult.*, 191: 109-19.

Schmutterer, H. 1990. Properties and potential of natural pesticides from the neem tree *Azadirachta indica*. *Annu. Rev. Entomol.*, 35: 271-297.

Scott, M.D., Zuo, L., Lubin, B.H. and Chiu, T.Y. 1991. NADPH, not glutathione, status modulates oxidant sensivity in normal and glucose-6-phosphate dehydrogenase- deficient erythrocytes. *Blood.*, 77: 2059-2064.

Singh, K. and Singh, D.K. 1995. Molluscicidal activity of *Azadirachta indica* (neem) on the biochemical parameters in the ovotestis of *Lymnea stagnalis*. *Malaysian Applied Biol.*, 24: 7-11.

Vijayavel, K. and Balasubramanian, M. P. 2006. Fluctuations of biochemical constituent and marker enzyms as a consequence of naphthalene toxicity in the edible estuarine crab, *Scylla serrata*. *Ecotoxicol. Environ. Saf.*, 63(1): 141 – 147.

In: Vertebrate and Invertebrate Animal Approaches
Editor: José Rosa Gomes
ISBN: 978-1-63117-102-4

Chapter 17

CYPERMETHRIN INDUCED DYNAMICS OF HEMOCYTE DENSITY OF INDIAN MOLLUSC *LAMELLIDENS MARGINALIS*

Krishnendu Das, Mitali Ray and Sajal Ray*
Department of Zoology, Aquatic Toxicology Laboratory,
University of Calcutta, West Bengal, India

ABSTRACT

Hemocytes the blood cells of *Lamellidens marginalis* (Mollusca:Bivalvia) capable of performing diverse immunological functions, including phagocytosis, encapsulation, nonself surface adhesion and cytotoxicity. Cypermethrin a type II pyrethroid pesticide is usually widely used in agricultural field which contaminates the freshwater habitat of this species during monsoon. Toxicity of this pesticide is screened in *L. marginalis* in respect to hemocyte density and subpopulation. The species were exposed to 0.03, 0.05, 0.07 and 0.09 ppm of cypermethrin for 24, 48, 72, 96 hours, 7, 15 days and total counts of hemocytes were determined. Initial increase in total hemocyte count followed by suppression at the later stages was observed. Different hemocyte subpopulations were determined and enumerated microscopically. Five hemocyte classes were identified: granulocytes, agranulocytes, asterocytes, hyalinocytes and blast like cells. Decline in total count of agranulocytes, hyalinocytes and asterocytes and elevation of relative density of granulocytes and blast like cells were observed under exposure of cypermethrin in a dose dependent manner, Cypermethrin induced alteration of density of total hemocytes and different subpopulations indicates a disruption of cellular functions, homeostasis and immunological states of the species. Prolonged sub lethal exposure of cypermethrin may lead to population decline of this economically important species in its natural habitat.

Keywords: Biomarker; Hemocyte; *Lamellidens marginalis*; Cypermethrin; Neurotoxin

* Corresponding author: Fax: 913324614849. Telephone:0913324615445, E-mail: raysnailmail@ rediffmail.com.

INTRODUCTION

L. marginalis a fresh water wetland bivalve mollusk is distributed in India, Bangladesh, Srilanka and Myanmar (Rao, 1989) and favours shallow or slow flowing water bodies, like pond, tank, lakes and reservoirs. This is economically, nutritionally and economically important species of India. Different body parts of *L. marginalis* are a source of animal protein for rural Indian population (Prabhakar et al., 2009; Baby, 2010) and poultry. This species is considered as indigenous source of dietary protein in shrimp farms, fish hatcheries and for poultry production (Misra et al., 2008). Selected body parts of *L. marginalis* are used as medicine to cure a number of diseases. Bivalves are stationary filter-feeding organisms able to bioaccumulate and concentrate most pollutants even if they are present in fairly low concentrations (Niyogi et al., 2001).

Cypermethrin a pyrethroid pesticide (Jin, 1988) now being widely used by the Indian farmers and also other part of the world to control the agricultural pests and plant diseases. During monsoon, this pesticide contaminates the natural habitat of *L. marginalis*. Cypermethrin is a reported neurotoxin for aquatic and terrestrial life forms (Curtis et al., 1995). Cypermethrin may leads to malfunctioning of neurotransmitters, followed by hyper and hypo opercular activity, imbalanced swimming activity (Marigoudar et al., 2009). Cypermenthrin can produce toxic effects in mammals (Nair et al., 2011; Wang et al., 2010; Grewall et al., 2010) and human (Saxena, 2010). Soils and sediments act as main environmental reservoirs of cypermethrin (Bacci et al., 1987). During monsoon the freshwater aquatic fields, the natural habitat of *L. marginalis* are contaminated by agricultural runoff containing cypermethrin leading to their population decline. Thus the survival of the species is under threat due to cypermethrin toxicity (Beth et al., 2004; Subramaninan et al., 1999).

Hemocytes are the circulating blood cells of hemolymph of *L. marginalis* and a major cellular component of innate immunity (Auffret et al., 2002; Mukherjee et al., 2006).Classification and morphology of gastropods (Martin, 2007; Mahilini, 2007) and bivalve (Chang, 2005) has been done in detail. Effects of toxicity of arsenic and azadirachtin on hemocyte density and subpopulation are reported (Mukherjee, 2006; Chakraborty, 2008). However limited information is available of the toxicity of cypermethrin on total hemocyte count and in Indian subpopulation for bivalve, *L. marginalis*.

In the present study total hemocyte density and its morphological variations has been investigated thoroughly under exposure of sub lethal concentrations of cypermethrin. All data were analyzed to establish its suitability as a biomarker of cypermethrin in field ecosystem.

MATERIALS AND METHODS

Fresh specimens of *L. marginalis* were collected from some selected unpolluted fresh water ponds of the district of 24 parganas (South) of West Bengal. They were brought to laboratory immediately after collection in water taken from the specific habitat. They were kept in well aerated fiber-glass tanks for 5days in room temperature to acclimate and care was taken to avoid crowding. Continuous flow of tap water was maintained so that oxygen did not act as stress factor.

Along with some modification here in this experiment the rearing and maintenance of *L. marginalis* were carried out according to (Raut, 1991). *L. marginalis* of group of 5-6 cm in length was selected for this experiment. An aqueous solution of cypermethrin (United Phosphorus Limited, India; 10% EC; CAS number 52315-07-8) was prepared in borosilicate glass containers at concentrations of 0.03, 0.05, 0.07 and 0.09 ppm. Cypermethrin free analytical grade water was used to prepare desired solutions for the experiment. The pH of the solution was maintained at 7.2. A set of ten animals of the same size was taken during each experiment.

Along with the control the animals were exposed to a volume of 5 L of cypermethrin solution. During the experiment the exposed medium was changed at every 12h and the static water environment was maintained at 24-26^{0}C temperature. Hemolymph was withdrawn from the posterior adductor muscle (Adema et al., 1991) slowly to avoid bubbling along with blood in prechilled glass vials.

The collected hemolymph samples were resuspended in sterile snail saline (5mM HEPES, 3.7mM NaOH, 36 mM NaCl, 2mM KCL, 2mM $MgCl_2.2H_2O$ and 4mM $CaCl_2.2H_2O$ at pH 7.8) (Chakraborty et al., 2008) and stored at 4^{0}C. A part of collected hemolymph was placed on previously cleaned, dried and sterilized glass slides. Slides were incubated to facilitate adherence and spreading. Glass made petri dish with fitting cover was used as the chamber. Moistened filter papers were used to keep hemocytes in action mode.

The adherent hemocytes were fixed with 1% glutaraldehyde (Sigma, USA, synthesis grade) for 5 minutes before rinsed with sterile snail slaine. After the fixation, the monolayer of the hemocytes was stained with Giemsa's stain (Himedia, India). To carry out differential hemocyte count field counts were performed in under a light microscope (Olympus BH-2). The total hemocyte counts were measured with the help of Neubauer hemocytometer (Brousseau et al., 1999). 300 cells were examined from each slide for the differential hemocyte count. The data were expressed in percentage abundance of the respective hemocyte variants. Data was presented in mean ± standard deviation (S.D.). The statistical analysis was carried out using the Student's t-test. Differences were considered significant at $P<0.05$.

Results

a) Enumeration of Hemocyte Density

The total hemocyte count (THC) and differential hemocyte count (DHC) of hemocyte subpopulation are most widely used parameters to assess bivalve health status between these parameters have been shown to change under stressful condition (Koprucu, K., 2010).

A significant steady rise up to 96h of exposure followed by fall in total count of hemocytes were observed in specimens exposed to different sub lethal doses of cypermethrin compared to the control (Figure-1).

The highest total count in the exposed condition was recorded for the specimens exposed to 0.03ppm of cypermethrin for 48h. The lowest total count was observed for the specimens exposed to 0.09ppm of cypermethrin for a time span of 15 days.

b) Morphological Variation of Hemocyte Subpopulation under Cypermethrin Exposure

In this study five types of hemocytes were clearly distinguished under light microscope (Figure-2).

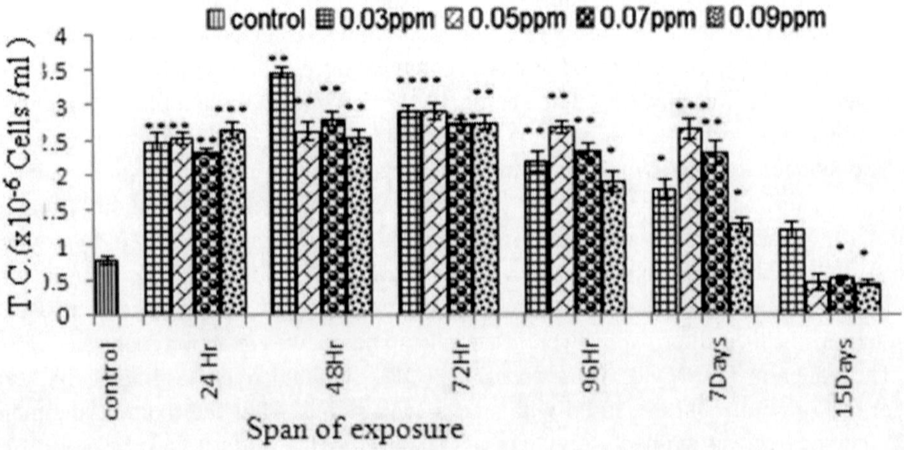

Figure 1. Total count (T.C) of hemocytes at 24, 48, 72, 96h, 7, 15 days of exposure to0.03, 0.05, 0.07, 0.09 ppm cypermethrin. Data are mean ± standard deviation (n = 10) replicates. The asterisk (*) indicate the values that are significantly different ($^{*}P<0.05$, $^{**}P<0.001$, $^{***}P<0.001$).

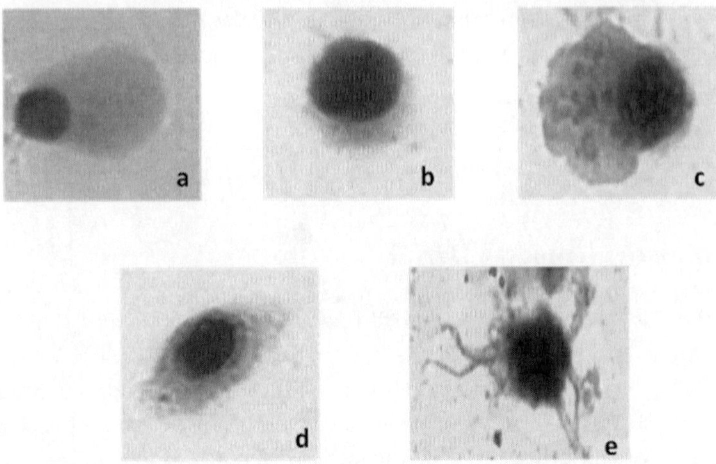

Figure 2. Light micrograph of the hemocytes of *Lamellidens marginalis* stained with Giemsa's stain. (a)Agranulocytes;(b) Blast-like cells ;(c) Granulocyte, (d) Hyalinocytes;(e) Asterocytes (x1000).

Agranular hemocyte is the most abundant among them .Cytoplasmic granules are fewer in agranulocytes (Figure-2a). The Blast cells (Figure-2b) were characterized by their high nuclear to cytoplasm ratio. No sign of granular structures were found in their basophilic cytoplasm. Granulocytes (Figure-2c) were characterized by the presence of eosinophilic cytoplasmic granules. Presence of cytoplasmic projections were occasional .Eccentrically or centrally placed nucleus were mostly round shaped. Hyalinocytes (Figure-2d) were rare and characterized by the spindle shaped conformity, dense central nuclei and sharp pseudopodial projections. Nucleus present with stippled chromatin .Few granular structures were observed in the slightly basophilic cytoplasm. Condensed nuclei bearing asterocytes (Figure-2e) found agranular.

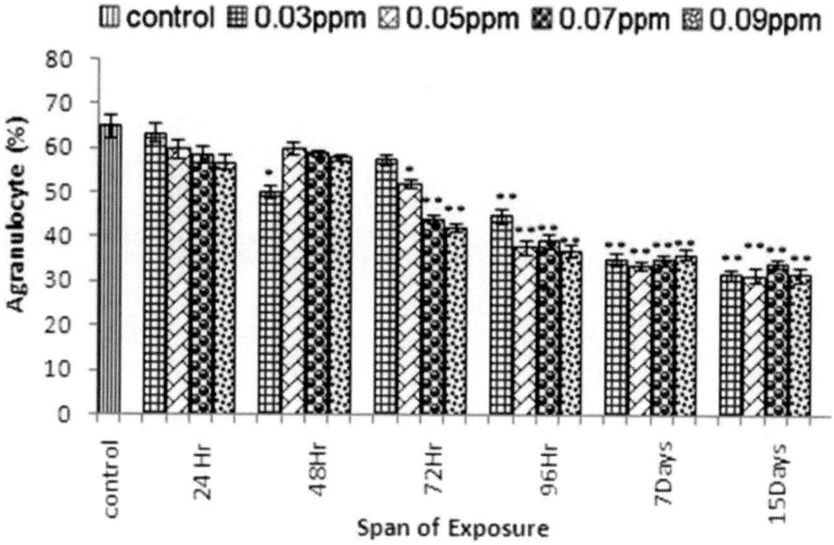

Figure 3. Frequency of agranulocytes of *L.marginalis* at 24, 48, 72, 96h , 7 and 15 days exposure to 0.03, 0.05, 0.07, 0.09 ppm of cypermethrin. Data represented is the mean ± standard deviation (n = 10).$^{*}P<0.05$, $^{**}P<0.01$, $^{***}P<0.001$.

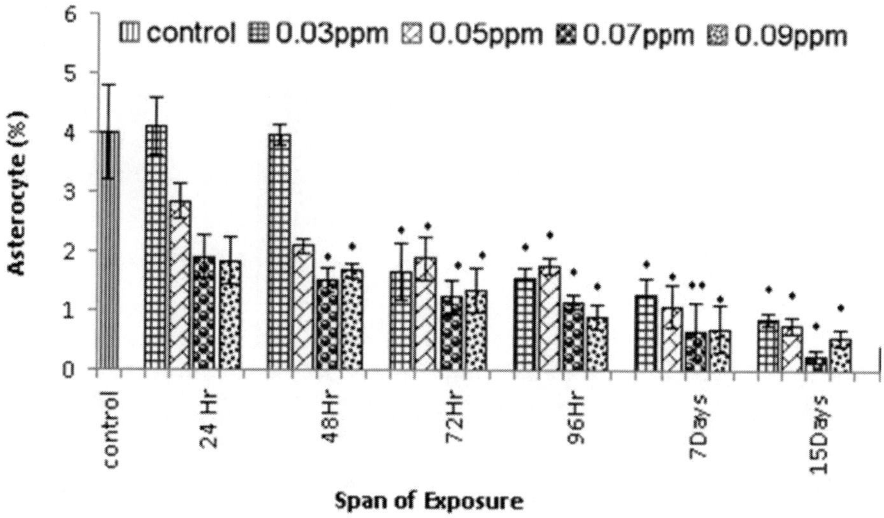

Figure 4. Frequency of asterocytes of *L.marginalis* at 24, 48, 72, 96h , 7 and 15 days exposure to 0.03, 0.05, 0.07, 0.09 ppm of cypermethrin. Data represented is the mean ± standard deviation (n = 10).$^{*}P<0.05$, $^{**}P<0.01$, $^{***}P<0.001$.

On being exposed to sub lethal doses of cypermethrin, the population of agranulocytes (Figure-3), asterocytes (Figure-4) exhibited steady decline, whereas the blast like cells (Figure-5) and granulocytes (Figure-6) showed a remarkable rise in count.

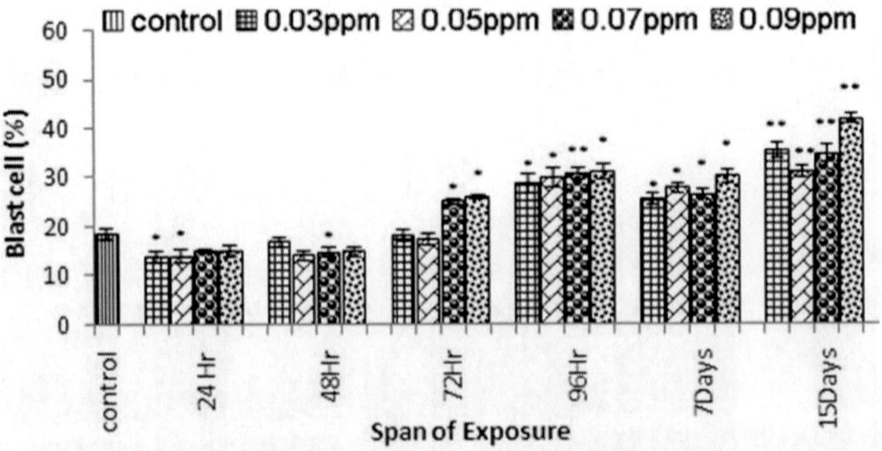

Figure 5. Frequency of blast-like cells of *L.marginalis* at 24, 48, 72, 96h , 7 and 15 days exposure to 0.03, 0.05, 0.07, 0.09 ppm of cypermethrin. Data represented is the mean ± standard deviation (n = 10).$^{*}P<0.05$, $^{**}P<0.01$, $^{***}P<0.001$.

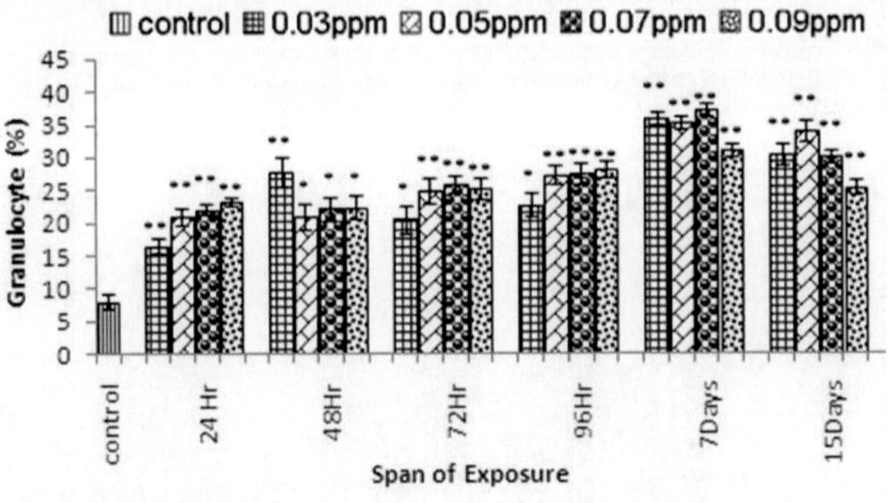

Figure 6. Frequency of granulocytes of *L.marginalis* at 24, 48, 72, 96h , 7 and 15 days exposure to 0.03, 0.05, 0.07, 0.09 ppm of cypermethrin. Data represented is the mean ± standard deviation (n = 10). $^{*}P<0.05$, $^{**}P<0.01$, $^{***}P<0.001$.

A significant decline in the hyalinocytes was observed after an initial rise at 72 hour of exposure (Figure-7). In addition to round shaped nucleus some elongated, kidney shaped and eccentrically placed granulocytes with irregular cell surfaces were recorded during this study. Vacuoles were found frequent along with decline in pseudopodia like projections in case of agranulocyte. Centrally located translucent vacuoles after 96h of exposure were positioned in the peripheral zone with transparent appearance and few of them found closely fused with the membrane. Cell surfaces were heavily damaged at 15th day of exposure. Basophilic (violet colour) cytoplasm in granulocytes were recorded at the initial stages of exposure became eosinophilic (pink colour) (Parisi, M.G., 2008).

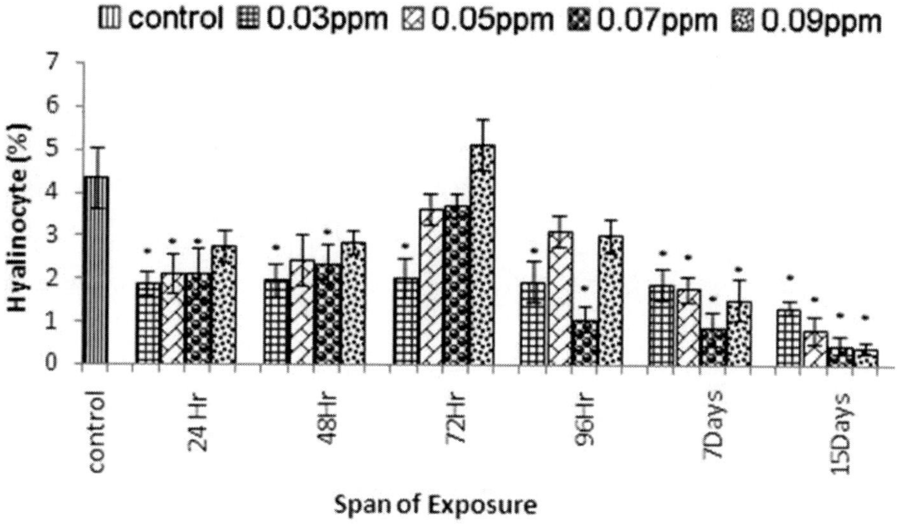

Figure 7. Frequency of hyalinocytes of *L.marginalis* at 24, 48, 72, 96h , 7 and 15 days exposure to 0.03, 0.05, 0.07, 0.09 ppm of cypermethrin. Data represented is the mean ± standard deviation (n = 10). $^{*}P<0.05$, $^{**}P<0.01$, $^{***}P<0.001$.

DISCUSSION

Rising population of granulocyte in this experiment is related to defense mechanism of the animal. Swollen irregular cell surfaces reflect the functional changes like increased pinocytic activity or intracellular transport. Immunoresponse through phenol oxidase, acid phosphatase, superoxides and peroxides are mediated by granulocytes (Aladaileh et al., 2007). Granulocytes performed as the main phagocytic hemocytes (Donaghy, 2009). Eosinophilic granule containing granulocytes supposed to play major role for the activity of peroxidase, phenoloxidase and lysomal enzymes (Parisi, M.G. et al., 2008). Spreading pseudopodia like structures and heterogeneous cytoplasm of granulocytes suggests phagocytic, chemokinetic or cytotoxic activity of granulocytes (Friebel et al., 1995). After releasing the heterogeneous cytoplasmic content full of immunocytotoxic chemicals translucent vacuoles become transparent. In some cases cytoplasmic inclusions may represent the non degradable residues of the phagocytic activity (Ottaviania et al., 1998). Granulocytes and Hyalinocyte are well known for their performance on non-self phagocytosis, encapsulation, cytotoxicity (Smith et al., 1983; Soderhall et al., 1985). Hyalinocyte supposed to play central role in hemocyte aggregation in the area of lesions to form clot and their role in wound repair was reported (Matozzo, 2007; Allam et al., 2002). With their extracellular matrix like products hyalinocytes fasten the process of wound healing (Nakayama et al., 1997). Blast-like cells are considered as hemocyte progenitors (Hine, 1999; Cima, 2000) to other cell types. Basophilic cytoplasm suggests the presence of free ribosomes and immaturity in blast cell (Chang et al., 2005). Alteration in the total count may be related to the haemopoietic activity of the bivalves (Mello et al., 2010).Increasing energy demand result from homeostasis process (e.g., detoxification) forced the specimens towards physiological stress (Uckan, 2010). Prolonged exposure of widely used cypermethrin may leads to depression in defense activity (Fisher et al., 1999). Decline in total hemocyte count at the later stages of exposure indicates the reduced mitotic activity by cypermethrin in a dose dependent manner. In this study hemocyte density of L.marginalis was referred as a possible potential biomarker of the toxicity of cypermethrin for the first time in India. Significant changes in total and differential count of hemocytes in response to different sub lethal doses of cypermethrin justify the credibility of the parameter to select as a biomarker. The present findings are of importance for the society in relation to aquaculture and public health safety (Rao et al., 2011), genotoxicity in human (Chakravarthi, 2007) and assessment of the condition of freshwater ecosystems (Oliver, 1999). Severe harmful effects on freshwater food chain were reported (Cox, 1996; Sreenivasan, 2011). This study will be helpful for the formulation of a sustainable strategy for the conservation of this fresh water bivalve in its natural habitat.

ACKNOWLEDGMENTS

The authors thankfully acknowledge the grant support received from UGC DRS SAP and the departmental instrument facility by DST FIST programme.

REFERENCES

Adema, C.M., van der Knaap, W.P.W. and Sminia, T. (1991). Molluscan hemocyte mediated cytotoxicity: The role of releasing oxygen intermediates. *Rev. Aquat. Sci.*, *4*, 201-223.

Aladaileh, S., Nair, S.V., Birch, D. and Raftos , D.A. (2007). Sydney rock oyster (*Saccostrea glomerata*) hemocytes: Morphology and function. *Journal of Invertebrate Pathology*, *96,* 48-63.

Allam, B., Alcox, K.A.A. and Ford, S.E. (2002). Flow cytometric comparison of haemocytes from three species of bivalve mollusks.*Fish and Shellfish Immunology*, *13*, 141–158.

Auffret, M., Mujdzic, N., Corporeau, C. and Moraga, D. (2002). Xenobioticinduced immunomodulation in the European flat oyster, *Ostrea edulis.Mar. Environ. Res., 54*, 585-589.

Baby, R.L., Hasan, I., Kabir K. A. and Naser, M. N. (2010). Nutrient Analysis of Some Commercially Important Molluscs of Bangladesh. *Journal of Science Research*, 2, 390–396.

Bacci, E.D. Calamari, C. Gaggi and M. Vighi. (1987). An approach for the prediction of environmental distribution and fate of cypermethrin. *Chemosphere*, 16. (7), 1373-1380.

Beth, C., Michael, S., Mac, L., Robin, O. and Jay, L. (2004). Evaluation of potential health risks to Eastern Elliptio (*Elliptio complanata*) (Mollusca: Bivalvia: Unionida: Unionidae) and implications for sympatric endangered freshwater mussel species. *Journal of Aquatic Ecosystems Stress and Recovery*, *9,* 35-42.

Brousseau, P., Payette, Y., Tryphonas, H., Blakley, B., Boernaus, H., Flipo, D. and Fournier, M. *In Manual of Immunological Methods, 29.* Boca Raton, FL: CRS Press;1999.

Chakraborty, S., Ray, M. and Ray, S.2008.Sodium Arsenite Induced Alteration of Hemocyte Density of *Lamellidens marginalis* – An Edible Mollusk from India. *Clean, 36 (2),* 195 – 200.

Chakravarthi, K., Naravaneni, B.R. and Philip, G.H. (2007). Study of Cypermethrin Cytogenesis effects on Human Lymphocytes Using In-Vitro Techniques *J. Appl. Sci. Environ. Manage, 11 (2),* 77 – 81.

Chang, S.J., Tseng, S.M. and Chou, H.Y. (2005).Morphological Characterization via Light and Electron Microscopy of the Hemocytes of two Cultured Bivalves: A Comparison Study between the Hard Clam (*Meretrix lusoria*) and Pacific Oyster (*Crassostrea gigas*). *Zoological Studies, 44(1),* 144-153.

Cima, F., Matozzo, V., Marin, M.G. and Ballarin, L. (2000).Hemocytes of the Clam *Tapes philippinarum*: Morphofunctional characterization, *Fish Shellfish Immunol, 10*, 677.

Cox, C. (1996). Insecticide fact sheet Cypermethrin. *Journal of pesticide reform, 16 (2),* 15-20.

Curtis, J. E. and Horne, P. A. (1995). Effect of Chlorpyrifos and Cypermethrin Applications on Non-Target Invertebrates in a Conservation-Tillage Crop. *J. Aust. ent. Soc, 34*, 229-231.

Donaghy, L., Hong, H.K., Lambert, C., Park, H.S., Shim, W.J. and Choi, K.S. (2009). Flow cytometry studies on the populations and immune parameters of the hemocytes of the Suminoe oyster, *Crassostrea ariakensis. Fish and Shellfish Immunology 27*, 296-301.

Fisher, W.S. et al., (1999). Decreased Resistance of Eastern Oyster (*Cassostrea virginica*) to a Protozoan pathogen (*Perkinsus marinus*) after Sublethal Exposure to Tributyltin Oxide. *Mar.Environ. Res, 47*, 185.

Friebel, B. and Renwrantz, L. (1995). Application of density gradient centrifugation for separation of eosinophilic and basophilic hemocytes from Mytitus *edulis* and characterization of both cell groups. *Comp. Biochem. Physiol., 112A (1)*, 81-90.

Grewall, K.K., Brar, R.S et al., (2010). Toxic impacts of cypermethrin on behavior and histology of certain tissues of albino rats. *toxicological international, 17(2)*, 94-98.

Hine, P.M. (1999). The Interrelationships of Bivalve Hemocytes. *Fish Shellfish Immunol, 9*, 367.

Jin, H. and G.R.B.Webster. (1998). Persistence, penetration and surface availability of cypermethrin and its major degradation products in elm bark. *J. Agric. Food Chem., 46*, 2851-2857.

Koprucu, K., Yonar, M.S. and Seker, E. (2010). Effects of cypermethrin on antioxidant status, oxidative stress biomarkers, behavior, and mortality in the freshwater mussel *Unio elongatulus eucirrus. Fisheries science, 76*, 1007–1013.

Mahilini, H.M. and Rajendran, A. (2008). Categorization of hemocytes of three gastropod species *Trachea vittata* (Muller), *Pila globosa* (Swainson) and *Indoplanorbis exustus* (Dehays). *Journal of Invertebrate Pathology, 97*, 20–26.

Marigoudar, S.R., Ahmed, R.N. and David, M. (2009). Impact of Cypermethrin on Behavioural Responses in the Freshwater Teleost, *Labeo rohita* (Hamilton) *World Journal of Zoology, 4 (1)*, 19-23.

Martin, G.G., Oakes, C.T., Tousignant, H.R., Crabtree, H. and Yamakawa, R (2007).Struscture and function of hemocytes in two marine gastropods, *Megathura crenulata* and *Aplysia californica*. *J.Mollusc.Stud, 73(4)*, 355-365.

Matozzo, V. et al., (2007).Haemocytes of the cockle *Cerastoderma glaucum*: Morphological characterisation and involvement in immune responses. *Fish and Shellfish Immunology, 23*, 732-746.

Mello, D.F., Proenca, L.A.O and Barracco, M.A. (2010).Comparative study of various immune parameters in three bivalve species during a natural bloom of *Dinophysis acuminate* in Santa Catarina Island, Brazil.*Toxins, 2,* 1166-1178.

Misra, G. and Mukhapadhyay, P.K. (2008). Mussel farming: alternate water monitoring practice. *Aquaculture Asia Magazine, July–Sept*, 32–34.

Mukherjee, S., Ray, M. and Ray, S. (2006). Azadirachtin induced modulation of total count of hemocytes of an edible bivalve *Lamellidens marginalis. Proc. Zool. Soc. Calcutta, 59(2)*,
37-39.

Nakayama, K., Nomoto, A.M., Nishijima, M. and Maruyama, T. (1997).Morphological and Functional Characterization of Hemocytes in the Giant Clam *Tridacna crocea.Journal of invertebrate pathology, 69*, 105-111.

Nair, R., Abraham, M.J. et al., (2011). A pathomorphological study of the sublethal toxicity of cypermethrin in Sprague Dawley rats. *International journal of nutrition, pharmacology, neurological diseases, 2(1)*, 179-183.

Niyogi, S., Biswas, S., Sarkers, M. and Datta, A.G. (2001). Antioxidant enzymes in brackiswater oyster, *Sacostrea cucullata* as potential biomarkers of polyaromatic hydrocarbon pollution in Hooghly Estuary (India): Seasionality and its consequences. *Sci. Total Environ, 281*, 237-246.

Oliver, L. M., Fisher, W.S. (1999). Appraisal of Prospective Bivalve immunomarkers. *Biomarkers, 4(10)*, 510.

Ottaviania, E., Franchinia, A., Barbierib, D. and Kletsasc, D. (1998).Comparative and morphofunctional studies on *mytilus galloprovincialis* hemocytes: Presence of two aging-related hemocyte stages. *Italian Journal of Zoology, 65 (4)*, 349-354.

Parisi, M.G., Li, H., Jouvet, L.B.P., Dyrynda, E.A., Cammarata, N.M. and Roch, P. (2008). Differential involvement of mussel hemocyte sub-populations in the clearance of bacteria. *Fish and Shellfish Immunology, 25*, 834–840.

Prabhakar, A.K. and Roy, S.P. (2009). Ethno-medicinal Uses of Some Shell Fishes by People of Kosi River Basin of North-Bihar, India. *Ethno-Med, 3(1)*, 1-4.

Rao, P.S., Babu, N.B. and Raju, R.R. (2011). HPLC Method for Investigations of Cypermethrin Pesticide in Edible Fish. *International Journal of Research in Pharmaceutical and Biomedical Sciences. 2(1)*, Jan-Mar.

Rao, S. et al. *Handbook on freshwater molluscs of India*.Calcutta: Zoological Survey of India; 1989.

Raut, S.K. (1991).Laboratory rearing of Medically and Environmentally Important Molluscks.in *snails, Flukes and Man* (Ed: M.S. Jairajpuri), *Zoolocial Survey of India Publication*, 79-83.

Saxena, P. and Saxena, A.K. (2010). Cypermethrin Induced Biochemical Alterations in the Blood of Albino Rats. *Jordan Journal of Biological Sciences, 3(3)*, 111-114.

Smith, V.J. and Soderhall, K. (1983). Introduction of Degranulation and Lysis of Hemocytes in Freshwater Crayfish, *Astacus astacus,* by Components of the Prophenoloxidase Activating system *in vivo*. *Cell Tissue Res., 233*, 295-303.

Soderhall, K., Wingren, A., Johansson, M.W. et al., (1985).The Cytotoxic Reaction of Hemocytes from Freshwater Crayfish *Astacus astacus. Cell Immunl, 94*, 326.

Sreenivasan, R.S., Moorthy, P.K. and Deecaraman, M. (2011). Bioaccumulation Study on the Cypermethrin Toxicity in Fresh Water Field Crab, *Spiralothelphusa hydrodroma* (Herbst). *J. Chem. Pharm. Res, 3(3)*, 397-402.

Subramanian, A.N., Lal Mohan, R.S., Karunagaran, V.M. and Babu Rajendran, R. (1999). Concentrations of HCHs and DDTs in the tissues of River dolphins *Platanista gangetica. Chemical Ecology, 16,* 143–150.

Uckan, F. and Sak, O. (2010). Cytotoxic Effect of Cypermethrin on *Pimpla turionellae* (Hymenoptera: Ichneumonidae) Larval Hemocytes. *Ekoloji, 19, 75*, 20-26.

Wang, H., Wang, Q et al., (2010). Cypermethrin exposure during puberty disrupts testosterone synthesis via downregulating StAR in mouse testes. *Arch. Toxicol., 84*, 53–61.

In: Vertebrate and Invertebrate Animal Approaches
Editor: José Rosa Gomes

ISBN: 978-1-63117-102-4

Chapter 18

HISTOPATHOLOGY OF GILL, DIGESTIVE GLAND AND ANTENNA OF *BELLAMYA BENGALENSIS* EXPOSED TO FENVALERATE

Chiranjib Mandal, Mitali Ray and Sajal Ray*
Department of Zoology, Aquatic Toxicology Laboratory,
University of Calcutta, Kolkata, West Bengal, India

ABSTRACT

Bellamya bengalensis is an important living organism of freshwater ecosystem.This animal performs biofiltration during feeding of microorganism and planktons. Fenvalerate is a type -2 pyrethroid which is exclusively used in agriculture and horticultural sectors of India for as pesticide. Gills, digestive glands and antenna are the vital organs whose morphology is adversely affected by exposure of fenvalerate. The natural habitat of *B. bengalensis* bears the risk of contamination by diverse types of pesticides including fenvalerate. During monsoon and flood, the fenvalerate laden agricultural runoff contaminate the freshwater bodies of *B. bengalensis*. *B. bengalensis* is an important bioresource and bears medicinal importance. Acute and chronic exposure of fenvalerate resulted in histological damage of gill, digestive gland and antenna of *B. bengalensis*. A detailed histopathological screening of gill, digestive gland and antenna was carried out in controlled laboratory condition under the exposure of 0.5ppm and 3ppm fenvalerate for 96 hours of duration. Structural damage of these target organs might lead to functional anomaly of the organ which might cause a gradual decrease in population density of *B. bengalensis* in its natural habitat. This study indicates the magnitude of structural damage of gill, digestive gland and antenna of *B. bengalensis* under the exposure of sublethal concentrations of fenvalerate.

Keywords: Fenvalerate, Biofiltration, *Bellamya*, Histopathology

* Corresponding author: Fax: 913324614849. Telephone: 0913324615445. E-mail: raysnailmail@ rediffmail.com.

INTRODUCTION

Bellamya bengalensis (Mollusca: Gastropoda) is an economically important organism distributed in the water bodies of eastern parts of India. *B. bengalensis* is a filter feeding organism capable of filtering enormous volume of water during feeding. This species is considered as an important member of freshwater ecosystem. Specimen is largely found in the mudwater interface of the water bodies. *B. bengalensis* constitutes a traditional item of human diet and a source of protein. This animal bears ecological and economic importance with respect to biofiltration, food value and medicinal importance. Fenvalerate, a pyrethroid group of pesticide is in use to control insect pest in different parts of India. Fenvalerate is a newly introduced formulation of pesticide which is effective in controlling pest population both in agriculture and horticultural sectors. Information of toxicity of fenvalerate in aquatic mollusc is limited. During monsoon and flood, the agricultural runoff loaded with pyrethroids contaminate the natural habitat of *B. bengalensis*. Present study is aimed to investigate the magnitude of histological damage induced by fenvalerate in tissues like digestive gland, gills and antenna of *B. bengalensis*. Digestive gland is an important organ is responsible for metabolism and detoxification of xenobiotics. Membranous structure of gill involved in respiration and collection of particulate food through filter feeding and immune-surveillance. Antenna is a tactile organ involved in various biological activities like collection of food, movement and reproduction. The experimental animals were exposed to fenvalerate at a concentration of 0.5ppm and 3ppm for 96 hours and tissue architecture were screened microscopically. Data indicates a substantial damage in the histomorphology of all three target organs.

MATERIALS AND METHOD

Collection of *B. Bengalensis* from Natural Habitat and Transportation to the Laboratory

The *B. bengalensis* with average shell size of (35-44mm) were collected from selected fresh water ponds of south 24 parganas district of West Bengal. The fresh specimens were transported to the laboratory in rectangular plastic containers with a dimension of 12inch × 18inch × 6 inch at a density 10-15 individuals per box in moist environment.

Laboratory Acclimatization and Maintenance of *B. Bengalensis*

Live healthy animals were acclimatized in pesticide free freshwater in borosilicate glass jars for 6-8 days. The maintenance of animals and entire experimentations were carried out in airy and well illuminated laboratory condition. Replenishment of water was done in every 24 hour to avoid residual toxicity. During the entire experiment, specimens were maintained according to Raut (1991).

Treatment with Fenvalerate

Animals were exposed in batches to fenvalarate in freshwater. For exposure, borosilicate glass container with a volume of 1 litre water was considered for the experiment. During experiment the animals were maintained in controlled laboratory condition with uniform light rationing. Temperature of water was maintained as 29°C ± 3°C. During the period of treatment, routine checking of mortality and morbidity were carried out. Routine replenishment of water and pesticide formulation was done throughout the study. For histopathological study of gill, digestive gland and antenna, *B. bengalensis* was treated with fenvalarate 0.5 ppm/96 hours and 3ppm/96 hours.

Autopsy of Animal

Specimens (both control and treated) were aseptically dissected for collection of gill, digestive gland and antenna which were instantly fixed in Bouin's fluid at 4°C for 12 hours. Fixed tissues were processed for histological observation after Victor (1993). The tissues were serially dehydrated by ascending grades of ethanol. Dehydrated tissues were subjected to paraffin impregnation and embedding was carried out in suitable mould at room temperature following standard protocol of micro-technique. Paraffin embedded tissues were sectioned at a thickness of 6μ using rotary microtome. The sections were thermally stretched over glass slide premeasured with Meyer's albumin. Paraffin sections were subjected to hematoxylin and eosin staining and passed through graded ethanol for dehydration.

RESULTS

The gill, digestive gland and antenna of *B. bengalensis* perform different biological functions related to respiration, digestion, sensation and immunity. A detailed histological study of these organs of the specimen under the subacute exposures of agrotoxin fenvalerate would provide an useful information about the nature and magnitude of toxicity of fenvalerate on these vital organs. The study would provide an information base about the vulnerability of the benthic freshwater species inhabiting the water bodies contaminated by fenvalerate.

Histopathology of Gill

In *B. bengalensis*, gills act as major respiratory organ. It is associated with feeding and immunological activities. Gills are present in between mantle cavity and visceral mass. Morphologically gill filaments are plate like structure. They are arranged anterio-posteriorly on either side of visceral mass. Cilia are present on the gill filaments which creates currents that transport water and food particles. Presence of hemocyte in gill lamellae is in report (Pattnaik et al., 2007; Chakraborty et al., 2010). In mussel, gills sieve the suspended food particles from water current and discriminate the ingestible and noningestible food particles through filtration (Pal et al., 1990). The transverse section (Figure 1, A, B) of gill filaments of *B. bengalensis* exhibit the elongated lamellae that are arranged equidistantly. The gill filaments showing distinct water channel are lined by epithelium. The transverse section of the control gill tissue of *B. bengalensis,* exhibit distinctive parallel gill filaments with well connected lamellary processes. Fenvalerate treated gill (Figure 2, a, b and c) exhibited membrane damage, tissue clumping within the matrix of gill filaments. It was more distinct in the tissue exposed to subacute exposure of fenvalerate. Clogging of water channel was noted in the treated gill filaments of *B. bengalensis*. The tip of lamellar epithelium exhibited the sign of tissue damage of gill exposed to 3ppm of fenvalerate for 96 hours as evident from cellular disintegration and disarray.

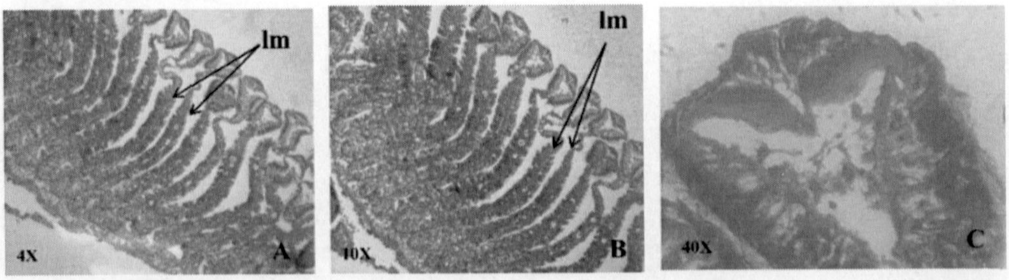

Figure 1. Transverse section of gill filaments of *B. bengalensis*: (A, B) exhibiting elongated gill lamella (lm) arranged equidistantly, (C) exhibiting intact tip of gill epithelium.

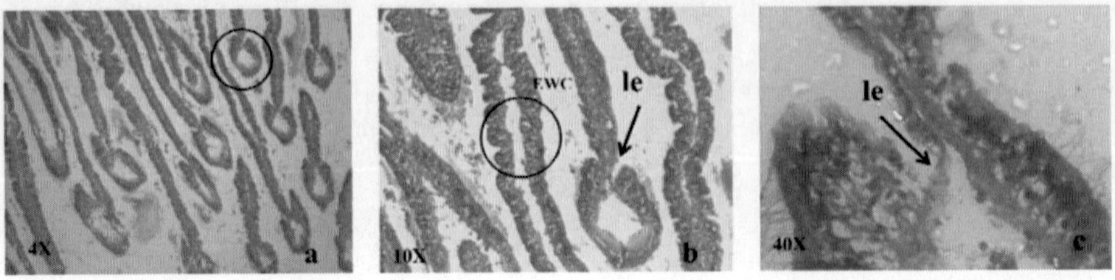

Figure 2. Transverse section of gill filaments of *B. bengalensis* (a, b, c) exposed to 3 ppm of fenvalerate for 96 hr exhibiting gill tissue damage, enlargement of water channel (EWC) damaged lamellar epithelium (le).

Histopathology of Digestive Gland

The digestive gland is an important organ of this animal involved in digestion, detoxification and metabolism. It is a large, asymmetrical organ located at the anterior portion of the body and behind the tip of whorl of *B. bengalensis*. It is also known as midgut gland or hepatopancreas. This organ performs diverse types of functions such as absorption of nutrient materials, secretion of digestive enzymes, storage of nutrient materials and release of indigestible food. In the digestive gland of gastropod, two basic types of cells are present such as digestive cells and secretory cells that undergo cyclic activity (Martoja, 1964; Merdsoy and Farley, 1973; Boghen and Farley, 1974; Wigham, 1976). Gland cells, absorbing cells, hepatic cells, secretory cells, liver cells, palisade cells, resorptive cells are the types of digestive cells whereas crypt cells, ferment cells, excretory cells, calcium secreting cells, calcareous cells are the types of secretory cells (Lutfy and Demian, 1967). The digestive gland is the food reservoir of gastropod. Cells lining the digestive gland participate in movement of the food, secreting substances in the lumina and phagocytosis (Rao et al., 1980).

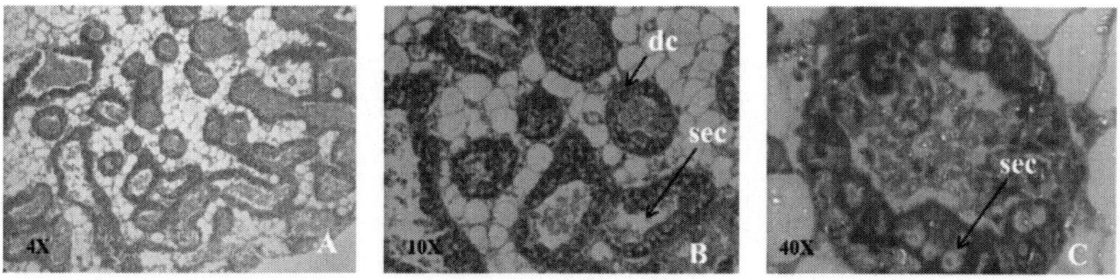

Figure 3. Transverse section of digestive gland of *B. bengalensis* (A, B, C) exibiting normal digestive cells (dc), secretory cells (sec).

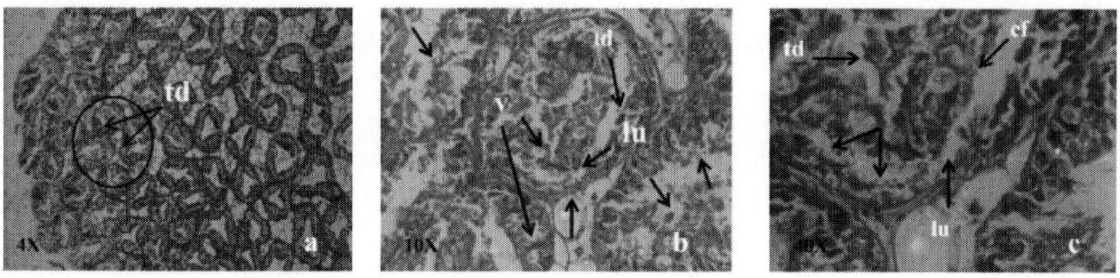

Figure 4. Transverse section of digestive gland of *B. bengalensis* (a, b, c) exposed to 3ppm of fenvalerate for 96 hr exhibiting tissue damage, disintegration of epithelium (td), channel formation (cf), large lumen formation (lu), vacule formation (v).

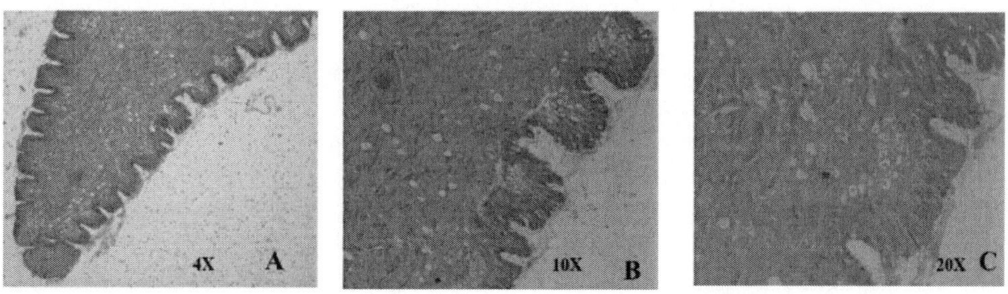

Figure 5. Transverse section of antenna of *B. bengalensis*: (A, B & C) exhibiting normal dense connective tissue.

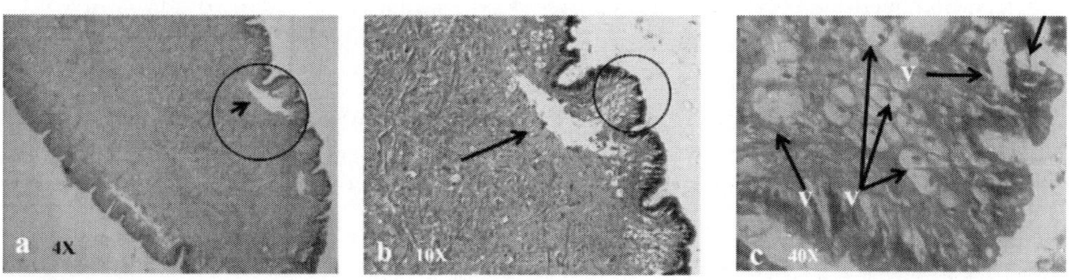

Figure 6. Transverse section of antenna of *B. bengalensis* (a, b, c) exposed to 3ppm of fenvalerate for 96 hr exhibiting epithelium of tip portion of antenna, vacule formation (v), tissue disruption.

During feeding by filtration, gastropods get into a contact with water with suspended food matter or dead or decaying leaf litter. In this present study, toxicity of fenvalerate, the contaminant of habitat of gastropod was studied and analysed in relation to histopathology.

We have reported the histology of transverse section of digestive gland of untreated control (Figure 3, A, B, C) *B. bengalensis*. The calciferous cell with nuclei were observed in the digestive gland. Control tissue of digestive gland of *B. bengalensis* exhibited tubules with narrow lumen and consisted with thick epithelium.

In unexposed digestive gland was composed of digestive, secretory cells and goblet cells. Fenvalerate exposed tissue of digestive gland of *B. bengalensis* exhibited tubules with large lumen with disrupted epithelium a sign of histological damage and disintegration.

Histopathology of Antenna

Antenna is a sensory organ and is composed of fibrous dense connective tissue. It is also known as tentacle. Tentacle of mollusc is composed of centrally placed collagen and peripheral matrix (Marshall et al., 1989). It was noted that radial nerve fibre was present around the central collagen fibre. It performs different biological activities such as movements and reproduction. Vacuolation with disruption of tissue framework was eminent in the antennae of *B. bengalensis* treated with 3ppm of fenvalerate for 96 hours.

DISCUSSION

B. bengalensis is an important aquatic mollusc which is collected from freshwater reservoir. It is collected indiscriminately for purpose of preparation of poultry feed, proteinaceous food for Indian tribal and urban community. Natural habitat of *B. bengalensis* is contaminated by different types of xenobiotics including fenvalerate. The gills are the most sensitive organs to lethal rather than sublethal levels of fenvalerate. They are composed vascularised branchial leaflets receiving hemolymph from vessels and sinuses. The histological study of the gill of fenvalerate exposed *B. bengalensis* exhibited a sign of damage and disruption of tissue architecture.

The gill of the control specimen exhibited a well and intact tissue architecture providing gill filaments with properly arranged water channels and organised lamellar epithelium. The gills of *B. bengalensis* exposed by fenvalerate exhibited tissue damage. Enlargement of water channels, damaged lamellar epithelium were noted in treated gills. This organ is involved in water filtration process. Substantial damage of the cytoarchitecture of gill was indicative to possible impairment in gaseous exchange and water filtration.

The digestive gland of mollusc is one of the vital organ which is related to detoxification, biotransformation and excretion of xenobiotics. So, it is an important organ of mollusc, subjected to toxicological insult. Histopathological changes of digestive gland of mollusc is in report. Molluscicides inhibited the process of food transport mechanism of mollusc (Wedgwood and Bailey, 1988; Bourne et al., 1991). In the present investigation, the digestive cells showed severe cytoplasmic vacuolization, after exposure of 3ppm fenvalerate for 96 hours. It is also noted that treatment with fenvalerate resulted in breakdown of digestive cells. Changes of digestive cells and decreased excretory functions due to intoxication were reported in hepatopancreatic digestive cells of *Mytillus galloprovincialis* exposed to heavy metals (Domouhsidou and Dimitriadis, 2000) and in the digestive cells of *Littorina littorea* after exposure to organic pollutants (Cajaraville et al., 1995). The effect of fenvalerate on the digestive gland of *B. bengalensis* resulted in extensive destruction and disorganisation of the intertubular connective tissue, folding of the basement membrane and detachment of tubular epithelium from basal lamina.

Antenna is a sensory organ of *B. bengalensis*. It is also affected by toxin's exposure. It's tip portion consisted with epithelium. Epithelium was damaged by toxin exposure as evident for its disrupted cytoarchitecture. In antenna of treated *B. bengalensis,* vacuolation and substantial disintegration of cells were noted. Histopathological damage of antennae was indicative to possible impairment of tactile sensitivity in *B. bengalensis*. This situation may lead to decrease in the ability of *B. bengalensis* specimens in food and mate recognition in its natural habitat.

ACKNOWLEDGMENTS

The authors thankfully acknowledge the grant support received from University Grants Commission (DRS- SAP) and the Department of Science and Technology (FIST), Government of India for development of instrument section of Zoology department of Calcutta University.

REFERENCES

Ahirrao, D.V. and Kulkarni, A.B. (2001). Sublethal stress of pyrethroids on biochemical contents in prostrate gland of freshwater snail, *Bellamya bengalensis*. *J.* of *Expt.Sc.*Vol-2, Issue 3, Pages 1-3.

Bayne, B.L., ed. (1976). Marine mussels: Their Ecology and Physiology. Cambridge, England: Cambridge University Press.506 p.

Boghen, A.and J. Farley. (1974). Phasic activity in the digestive gland cells of intertidal prosobranch, *Littorina saxatilis* (olivi) and its relations to the tidal cycle. *Proc. Malacol. Soc.Lond.* 41:41-56.

Bourne, N.B., Jones, G.W. and Bowen, I.D. (1991). Endocytosis in the crop of the slug *Deroceras reticulatum* (Muller) and the effects of the ingested molluscicides, metaldehyde and methiocarb. *J. Moll. Stud.*, 57:71-80.

Cajaraville, M.P., Robledo, Y., Etxeberria, M. and Marigomez, I. (1995).Cellular biomarkers as useful tools in the biological monitoring of environmental pollution: Molluscan digestive lysosomes. In: Cell Biology in Envirnmental Toxicology, Cajaraville, M.P. (Ed.). University of the Basque Country Press, Bilbo, pp: 29-55.

Chakraborty, S., Ray, M., Ray, S. (2010). Toxicity of sodium arsenite in the gill of an economically important mollusc of India. *Fish Shellfish Immunology*. 29, 136 -148.

Domouhsidou, G.P.and Dimitriadis, V.K. (2000). Ultrastructural localizationon of heavy metals (Hg, Ag, PbandCu) ingills and digestive gland of mussels, *Mytillus galloprovincialis* (L.) *Arch. Environ. Contam. Toxicol.* 38:472-478.

Fretter, V. and Graham, A. (1962). British prosobranch molluscs. Their functional anatomy and ecology. London. *The Royal Society*. 755pp.

Khan, R.A. and Chaudhuri, S. (1984). The population and production ecology of a freshwater snail *Bellamya bengalensis* (Lamarck) (Gastropoda: Viviparidae) in on artificial lake of Calcutta, India. *Bull. Zool. Surv. India*. 5 (2 and 3): 59-76.

Lopes, M., Rodrigues, A. and Marigomez, I. (2001). Morphology and histology of the digestive gland of *Oxychilus* (Drouetia) *atlanticus* (Morelet and Drouet) (Gastropoda: Pulmonata) Arquipelago, *Life and Marine Sc.* Supp 2 (Part-B): 71-76. Ponta Delgada.ISSN 0873-4704.

Lutfy, R.G. and Demian, E.S. (1967). The histology of the alimentary system of *Marisa cornuaretis* (Mesogastropoda: Ampullariidae). *Malacologia,* 5 (3): 375-422.

Marshall, D.J., Hodgson, A.N. and Trueman, E.R. (1989).The muscular hydrostat of limpet tentacle. *J.Molluscan Stud.*55: 421-422.

Martoja, M. (1964). Contribution a l'etude de l'appareil digestif et de la digestion chez les gasteropodes carnivores de la famille des nassarides (Prosobranches stenoglosses). *Cellule* 64: 235-334.

Merdsoy, B. and J. Farley. (1973). Phasic activity in the digestive gland cells of the marine prosobranch gastropod, *Littorina littorea* (L). *Proc. Malacol. Soc. Lond.*40: 474-482.

Neff, M.J., Hillman, R.E., Carr, R. C., Buhl, R. L. and Lahey, J.I. (1987). Histopathologic and biochemical responses in arctic marine bivalve molluscs exposed to experimentally spilled oil. *Arctic.* 40. (supp.1): 220-229.

Pal, S.G., Ghosh, B. and Modak, S. (1990). Fine structure of the digestive tubules of meretrix.In the bivalvia- proceedings of a memorial symposium in honour of Sir Charles Maurice Yonge, Edinburg, 1968. B. Morton (ed.): Hongkong University press, pp. 168-175.

Pattnaik, S., Chainy, G.B.N. andJena, J.K. (2007). Characterization of $Ca2^{+}$ ATPase activity in the gill microsomes of freshwater mussel, *Lamellidens marginalis* (Lamarck) and heavy metal modulations.*Agricult.*, 270 (1-4): 443-450.

Rahman, R.MD., Parvez, A.MD., Sarker, M. MD. and Jahan, S.MD. (1998). Histopathology of infected organs of *Bellamya bengalensis* (Lamarck) by larval Helminth. *Univ .j. zool. Rajshahi Univ*. Vol.17. pp.19-27.

Rao, N.V. and Rao, K.V.S. (1980). On a rare nudibranch, *Thordisa crosslandi* Eliot (Mollusca: Dorididae) from the west coast of India. *Bulletin Zoological Survey of India 2(2–3)*: 219, pl. IV.

Raut, S.K. (1991). Laboratory Rearing of Medically and Economically Important Molluscs.In "Snails, Flukes and Man, "(Jairajpuri, M.S.ed.). *Zoological Survey of Indian Pub*. pp 79-83.

Victor B. (1993). Responses of hemocytes and gill tissues to sublethal cadmium chloride poisoning in the crab *Paratelphusa hydrodromous* (Herbst). *Arch Environ Contam Toxicol* 24:432–439.

Wedgwood, M.A.and Bailey, S.E.R. (1988).The inhibitory effects of the molluscicide metaldehyde on feeding, locomotion and faecal elimination of three pest species of terrestrial slugs. *Ann. Applied Biol.*, 112: 439-457.

Wigham, G.D. (1976). Feeding and digestion in the marine prosobranch *Rissoa parva* (Dacosta). *J. Molluscan Stud*. 42:74-94.

In: Vertebrate and Invertebrate Animal Approaches
Editor: José Rosa Gomes
ISBN: 978-1-63117-102-4

Chapter 19

STRUCTURAL IMPAIRMENT OF GILL AND DIGESTIVE GLAND OF *LAMELLIDENS MARGINALIS* EXPOSED TO CYPERMETHRIN

Krishnendu Das, Mitali Ray and Sajal Ray*
Department of Zoology, Aquatic Toxicology Laboratory,
University of Calcutta, Kolkata, West Bengal, India

ABSTRACT

Lamellidens marginalis (Mollusca; Bivalvia ; Eulamellibranchiata) the freshwater bivalve of commercial, nutritional, medicinal and ecological importance is under threat due to cypermethrin, a pyrethroid pesticide. Cypermethrin is a newly introduced pesticide in the agricultural sector of India with insufficient information of its toxicity in aquatic organism. The present study describes the histopathological changes of gills and digestive glands of *L. marginalis* under the exposure of cypermethrin. Cypermethrin (0.09 ppm for 30 days) induced major alteration in gill and digestive gland tissue includes vacuolization, loosening of attachment with basement membrane, abnormal changes in lumen diameter, appearances of disintegrated dense fibrosis. Possible dysfunction of gills and digestive glands reduce the survival value of *L. marginalis* by lowering gas exchange, food transport and detoxification potential. Prolonged sublethal exposure of this neurotoxic pyrethroid may lead to gradual decline in the population of this economically important freshwater bivalve. Histopathological investigation on gill and digestive gland would provide an useful information of cypermethrin toxicity in aquatic bivalve.

Keywords: *Lamellidens marginalis*; Cypermethrin; Biomarker; Gills; Digestive glands

* Corresponding author: Fax: 913324614849. Telephone:0913324615445, E-mail: raysnailmail@ rediffmail.com.

INTRODUCTION

Lamellidens marginalis is distributed in India, Bangladesh, Srilanka and Myanmar (Rao, 1989) and are located in freshwater ditches, ponds, lake, streams and prefer stagnant shallow water bodies. This species is being used as an important indigenous source of dietary protein (Mishra et al., 2008).Being a filter feeding organism, they accumulate and concentrate pollutants in different organs. This property makes freshwater bivalve potentially useful as biomonitor species of water quality and its management. Habitat of this economically, nutritionally and medicinally important species (Prabhakar et al., 2009, Baby et al., 2010) is under threat due to different anthropogenic activities like habitat destruction and contamination (Chakraborty et al., 2008). Cypermethrin, the neurotoxic synthetic pyrethroid pesticide is now widely used to control pests of cotton, rice, fruits and vegetables (Tendulkar and Kulkarni, 2012). This is graded as the first toxicity category by Environmental Protection Agency (US) (Bouya et al., 2012). Cypermethrin often causes toxicity in non target organisms (Raksheskar, 2012 ; Pugazhendy et al., 2012) which often became more vulnerable than the target organism (Mulla et al., 1979; Smith and Stratton, 1986) even at very low concentration (Jahanbakhshi et al., 2012). This neurotoxin (Curtis et al., 1995) induced significant alteration in glycogen level (Tendulkar et al., 2012), total protein, free amino acid, nucleic acid, RNA (Tripathi et al., 1992;Schimmel et al., 1983;Tantarpale, 2011;Patil A.G., 2011) and in enzyme activity like AChE, Alkaline phosphatase, Acid phosphatase, Cytochrome oxidase, Protease etc. were reported by various authors (Tiwari, 2012). Cypermethrin is also responsible for chromosomal aberration in human lymphocyte (Suman et al., 2006; Chakravarthi et al., 2007), significant alteration in oxidative enzymes in muscle, kidney (Sekhar et al., 2010), increase in blood uric acid, urea, creatinine (Saxena et al., 2010) of mammals. During monsoon the agricultural runoff with cypermethrin contaminates the natural habitat of *L. marginalis*. Soil and sediments act as major reservoirs of cypermethrin (Bacci et al., 1987). Extended exposure of *L. marginalis* to cypermethrin triggers their morphological, physiological and biochemical alterations and leads toward decline in population. Information of cypermethrin induced organ damage in invertebrate is limited. Histological approach is used as an important tool to assess the impact of toxicant on tissue level (Sprague et al., 1971). Histological study conveys information related to tissue damage at structural and functional level. This study was made as a histopathological survey of gills and digestive glands of *L. marginalis* in freshwater habitat under the exposure of cypermethrin. The impact of pollutants on digestive gland and gills were used as biomarker by a number of authors (Etxeberria et al., 1995; Lowe et al., 1981;Tripp et al., 1984). However, limited information is available (Ahirrao, D.V. et al., 2011) regarding histopathological changes of gills and digestive glands for *L. marginalis* exposed to cypermethrin. Histopathological changes have ecological significance and the study can be used in a monitoring programme (Sunila et al., 1987) for ecosystem screening. This investigation will help to analyze comparative status of *L. marginalis* under cypermethrin exposure.

MATERIALS AND METHODS

Adult *L. marginalis* specimens of shell size (5-6 cm) were manually collected from selected unpolluted fresh water ponds of South 24 Parganas district of West Bengal.

Immediately after collection those specimens were transported in well aerated fiber-glass tanks for 5 days in room temperature and allow them to acclimatize. To ensure sufficient oxygen supply continuous flow of tap water was maintained. In this investigation we followed the procedure for rearing and maintenance according to Rout, 1991 with little modifications. An aqueous solution of cypermethrin (United Phosphorous Limited, India; 10% EC; CAS number 52315-07-8) was prepared in borosilicate glass containers at concentration of 0.09 ppm. Cypermethrin free analytical grade water was used to prepare desired solutions for the experiment. The experimental sets of animals were exposed to 0.09 ppm of cypermethrin for 30 days of time span. Along with the controls the animals were exposed to a volume of 5 L of cypermethrin solution. During experiment routine replenishment of water was carried out in every 12 hours to avoid residual toxicity. Aseptically dissected organs (Gill and digestive gland) of *L. marginalis* were collected with fine, sharp and sterile scalpel. Removal of mucus, blood from small pieces of collected tissue samples were carried out with chilled sterile snail saline (SSS, 5mM HEPES, 3.7mM NaOH, 36 mM NaCl, 2mM KCL, 2mM $MgCl_2.2H_2O$ and 4mM $CaCl_2.2H_2O$ at pH 7.8) (Chakraborty et al., 2008). Immediately thereafter the tissues were fixed in Bouin's solution (saturated picric acid: 75 parts, 37-40 % formaldehyde: 25 parts and glacial acetic acid: 5 parts). Then those fixed tissue samples were passed through standard alcoholic gradation. After 5 days of immersion in cedar wood oil the tissue samples were embedded in paraffin and blocks were prepared. Tissue sections with an average thickness of 5-6 μ were made by using a microtome. Harris haematoxylin and eosin stains were used to stain those tissue sections. Stained sections were studied under camera attached digital microscope (Olympus BX 51, Germany) (Chakraborty et al., 2010).

RESULTS

In control specimen anterior posteriorly extended two plate-like gill plates (demibranch) on each side of visceral mass were found as major component of gills in *L. marginalis.* Gill plates unite to form water tube and then open in to the mantle cavity. Vertically arranged filaments form dorsoventrally lengthened demibranch. In control specimens water channel were clear and guided by ciliated epithelial cells (Figure-1a, c). Gill has slender tubules, anchored to the body surface by the mantle border. Ciliated cuboidal cells (Figure-2a) in row run parallel to the longitudinal axis of the lamella to form the branchial filaments. These cells found closely associated with dark, oval shaped nuclei. Under sublethal concentration of cypermethrin for 30 days connective tissue and epithelial cells lost their normal structure (Figure-1b, d). Reduction of space between water channel and interlamellar junctions were noticed. Epithelial lining at the tip of gill filaments with swelled epithelium were found disintegrated. Hyperchomatic anaplastic cells (Figure-2f) infiltrated in the affected gill tissue. Cilia became clumped (Figure-2b).

Ruptured and swelled mucus secretory cells (Figure-2b) were studied with their fused form in gill filaments. Reduction in thickness is followed by detachment of epithelial cells from the muscular layer. Severe damage in the tip regions of lamellar epithelium were noticed. Dense fibrosis (Figure-2f) in the matrix of gill filaments was noticed. Water channels were found clogged with hyper chromatic neoplastic cells. Microscopic study shows that digestive gland is composed of numerous hepatic lobules. Each lobule is lined by columnar epithelial cells resting on basement membrane (Figure-3a).Hepatic lobules in its core region possess narrow lumen (Figure-3c, e). Interlobular space is filled with connective tissue layer. The histopathological details of digestive glands of *L. marginalis* appeared to be identical to observations made by previous workers on bivalves (Usheva et al., 2006; Agwuocha et al., 2011; Kumar et al., 2011).

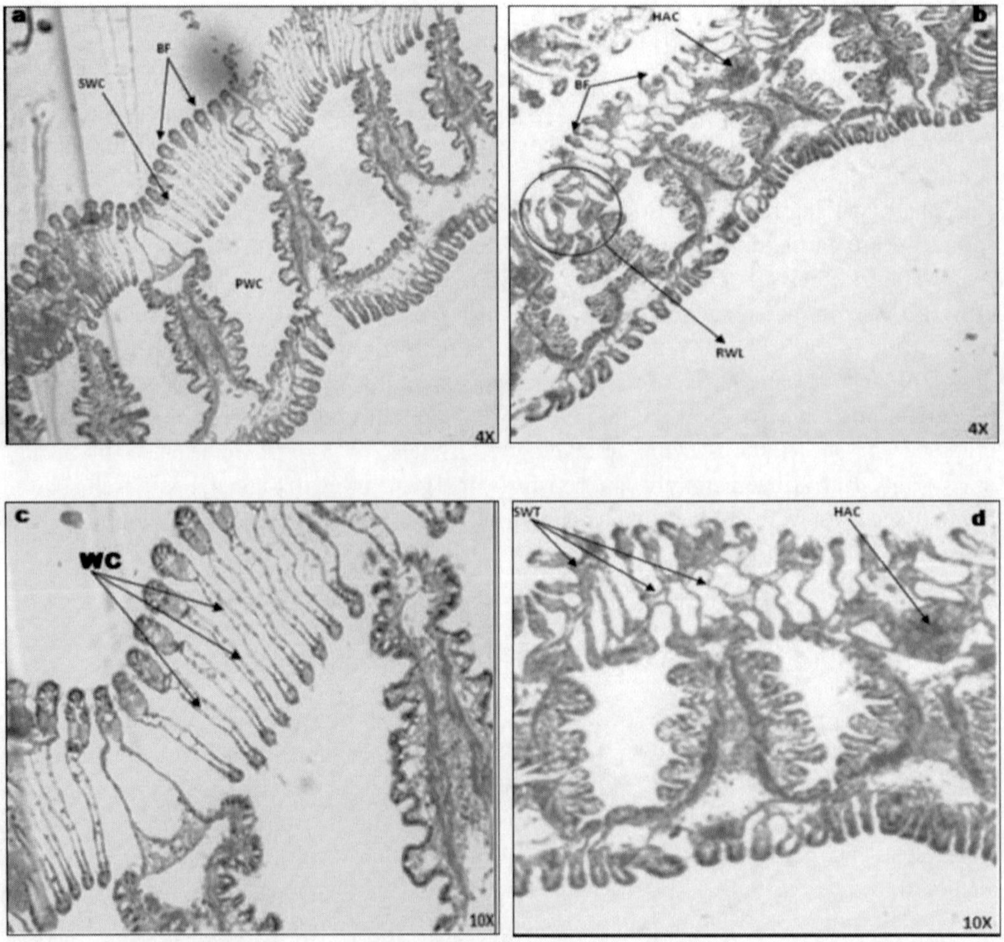

BF-branchial filaments; HAC- hyperchromatic anaplastic cells; PWC- primary water channel ; RWL- Reduction in space between water channel and inter lamellar junction; SWT-Shrunken water tube; SWC- secondary water channels; WC- Clear water channel.

Figure 1. Transverse section of gill of L.marginalis. (a), (c) normal gill with clear water channel, normal space between water channel and interlamellar junction; (b), (d) cypermethrin exposed (0.09 ppm /30days) gill exhibiting reduced space between water channel and interlamellar junction, intense infiltration of hyper chromatic anaplastic cells, tissue swelling, irregular shaped branchial filaments.

The digestive gland on exposure of cypermethrin exhibited atrophy, edema and lesion in epithelial cells of the organ (Figure-3d).In addition to vacuolization in the monolayer epithelium of digestive tubules (Figure-4b), many of the digestive tubules were found to be fused with each other (Figure-3f) under cypermethrin exposure. Some epithelial cells lost their attachment with the basement membrane (Figure-3d) and hemocytic infiltration (Figure-4b) in the disintegrated connective tissue were noticed in the cypermethrin exposed specimens.

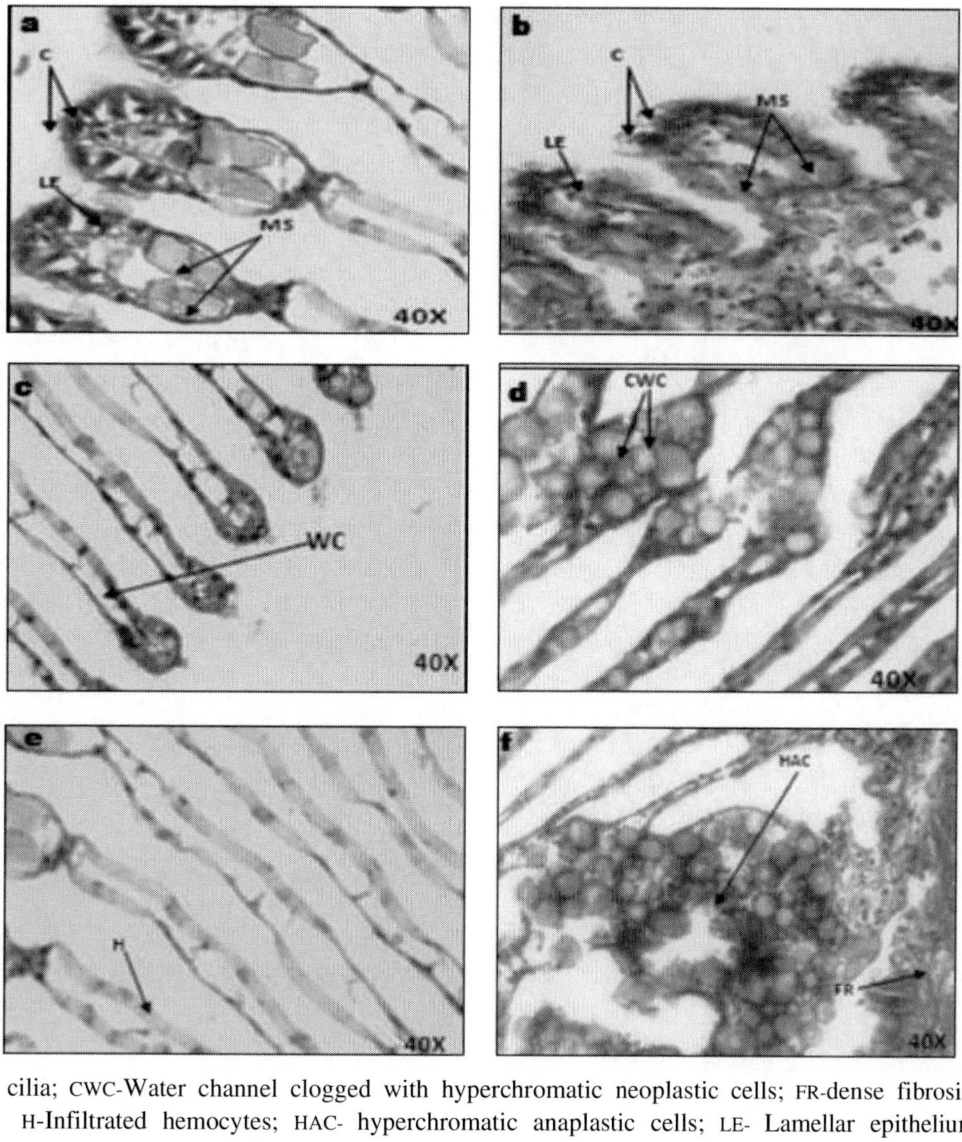

C- cilia; CWC-Water channel clogged with hyperchromatic neoplastic cells; FR-dense fibrosis;. H-Infiltrated hemocytes; HAC- hyperchromatic anaplastic cells; LE- Lamellar epithelium; MS- Mucus secretory cells; WC- Clear water channel.

Figure 2. Transverse section of the gill filaments of *L. marginalis.* (a), (c), (e) normal gill with intact lamellar epithelium, intact mucus secretory cells; (b), (d), (f) cypermethrin exposed (0.09ppm /30 days) gill exhibiting swelled, disintegrated lamellar epithelium, ruptured mucus secretory cells, clumped ciliated epithelium, dense fibrosis, water channel clogged with hyperchromatic anaplastic cells.

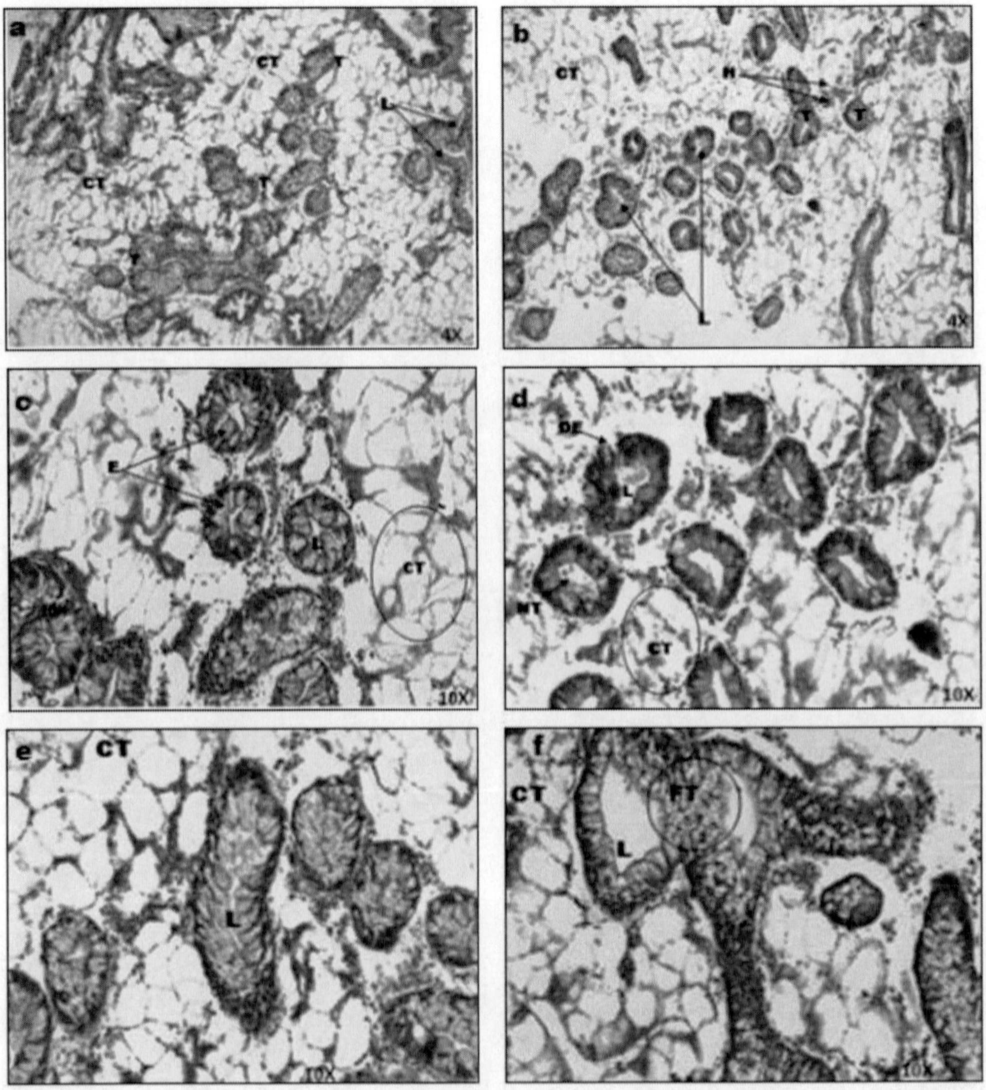

CT-Connective tissue; DE-Disruptive epithelial lining ; E-Mono layer of epithelial cells ; FT-Fused digestive tubule; H-Infiltrated hemocytes ; L- Lumen of digestive tubules ; NT-Necrotic tissue ; T-Hepatic lobule;

Figure 3. Transverse section of digestive gland of *L. marginalis.* (a), (c), (e) normal digestive gland exhibiting hepatic lobule lined with columnar epithelial cells and narrow lumen in the core region of hepatic lobule ; (b), (d), (f) cypermethrin exposed (0.09 ppm /30days) digestive gland exhibiting hepatic lobule, with wide lumen and disintegrated connective tissue.

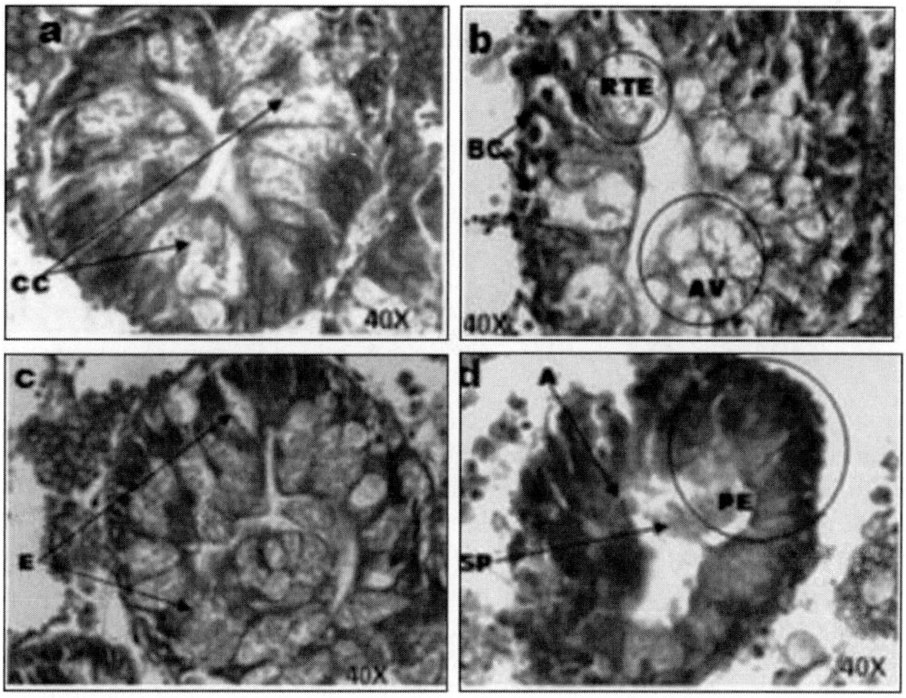

A-Acidophilic cells; AV- Cells with apical vacuole; BC-Basophilic cell; CC-Columnar epithelial cells; E-Mono layer of epithelial cells; PE-Pseudostratified epithelium. RTE-Reduced thickness in epithelium; SP-Secretory product.

Figure 4. Transverse section of hepatic tubule (T) of *L. marginalis* (a), (c) in normal specimen, hepatic tubule exhibiting intact columnar epithelial cells; (b), (d) cypermethrin exposed (0.09 ppm/30 days) digestive gland exhibiting hepatic tubules with disintegrated epithelial cells and infiltrated basophilic cells (BC).

DISCUSSION

L. maginalis being a filter feeder play major role as bioindicator of chemical contamination. In the present study of histopathology of the gills and digestive gland of cypermethrin exposed *L .marginalis* provides a vivid spectrum of damage of the organs. The gill of the control specimen exhibited clear water channel and normal lamellar epithelium. The gill of the cypermethrin exposed specimens exhibited a number of alarming features. Cypermethrin induced osmotic imbalance in ruptured mucus secretory cells (Hietanen et al, 1988; Siegfried et al., 1993) reduce the food transport system. Disintegrated gill lamellae reduces the filtration ingestion rate together (Payne et al., 1994; Mehedinti, 1999). Degeneration of ciliary surface area on gill epithelial cells reduces the rate of refreshing water penetration phase into lamellar network. Gill lesions affect gas exchange and food transport (Sunila, 1987). The occurrence of swollen membrane folds in epithelial cells of gill lamellar may reflect functional changes linked to increased pinocytic activity or intercellular transport (Smolarz, 2006). Presence of infiltrated hemocytes in the hemocelic areas of *L. marginalis* suggest immunological significance. Clogged water channels reduce the normal transport of food and in addition to that respiration level may get declined due to clogging. All these observations marked a possible dysfunction of the gill in *L. marginalis*.

Digestive gland in control specimens exhibited hepatic lobules with well structured columnar epithelial cells and narrow lumen in the core region. Digestive glands of cypermethrin exposed specimens exhibited a number of alterations in hepatic lobules, lumen and connective tissue. Reduced thickness of epithelial cells in cypermethrin treated digestive glands may be due to entire or partial release of calcium during the detoxification process (Marigomez, 1998).Sublethal dose of cypermethrin leads to increase in tubular lumen diameter (Andhale et al., 2011), irregular shape of digestive cells and presence of vacuoles. Histopathological changes in the digestive glands suggest the reduced activity in food absorption, digestion and detoxification (Agwuocha et al., 2011). Increase in basophilic cells is closely associated with degenerative changes in digestive cells (Thompson, 1974; Cajaraville, 1990; Marigomez, 1990). Increase in number of infiltrated hemocytes in the connective tissue region was correlated with environmental pollution and detoxification. Reduction in digestive gland epithelium thickness may be due to the loss of apical cytoplasm during the detoxification(Johnson et al, 1996; Snyman et al, 2003).Such adverse damages in digestive epithelium is considered as generalized stress response of bivalves under toxin exposure (Usheva et al., 2006; Agwuochal et al., 2010, 2011).

Degenerated digestive cells lead to insufficient nutrition and thus may affect the survival fitness of *L. marginalis* in contaminated habitat. Histopathological analysis confirmed alteration in the histology of gills and digestive gland consequent to cypermethrin exposure. Cypermethrin induced structural damage of gill and digestive gland was indicative to possible impairment of respiratory, digestive and detoxification efficacy of *L. marginalis* distributed in contaminated habitat. Present observation would provide useful information of toxicity of cypermethrin in relation to organ damage and subsequent damage in the functional attributes of gill and digestive gland. Chronic and uncontrolled exposure of cypermethrin may lead to develop a detrimental toxic effect in *L. marginalis* which might be irreversible and lethal for other invertebrate species.

ACKNOWLEDGMENTS

Grant support received from the Department of Science and Technology, Government of India, to MR is thankfully acknowledged.

REFERENCES

Chakraborty, S., Ray, M. and Ray, S. (2008).Sodium Arsenite Induced Alteration of Hemocyte Density of *Lamellidens marginalis* – An Edible Mollusk from India. *Clean*, *36 (2)*, 195 – 200.

Raksheskar, G.A. (2012).Influenced of cypermethrin on DNA, RNA and RNA/DNA ratio in gills of the freshwater fish *Channa striata*. *Bio. Sc. Dis,* 3(1), 17-19.

Agwuocha, S., Kulkarni, B.G. and Pandey, A.K. (2010).Toxicity of petroleum hydrocarbon and refined oil-water soluble fractions to *Gafrarium divaricatum*. *Fish. Chem,* 30 (1), 120-123.

Agwuocha, S., Kulkarni, B.G. and Pandey, A.K. (2011).Histopathological alterations in hepatopancreas of *Gafrarium divaricatum* exposed to xylene, benzene and gear oil.*Wsf. J. Environ. Biol,* 32(1), 35-38.

Ahirrao, D.V. and Kulkarni, A.B. (2011).Sublethal stress of pyrethroids on Biochemical contents in prostate gland of a freshwater snail, *Bellamya bengalensis. Journal of experimental sciences,* 2(3), 1-3.

Andhale, A.V., Bhosale, P.A. and Zambare, S.P. (2011).Histopathological study of nickel induced alterations in fresh water bivalve, *Lamellidens marginalis. J. exp. Sci,* 2(4):1-3.

Baby, R.L., Hasan, I., Kabir K. A. and Naser, M. N. (2010). Nutrient Analysis of Some Commercially Important Molluscs of Bangladesh. *Journal of Science Research, 2,* 390–396.

Bacci, E.D. Calamari, C. Gaggi and M. Vighi. (1987).An approach for the prediction of environmental distribution and fate of cypermethrin. *Chemosphere,* 16. (7), 1373-1380.

Bouya, H., Errami, M., Salghi, R., Bazzi, L h., Zarrouk, A., Al Deyab, S. S., Hammouti, B., Bazzi, L. Chakir.A. (2012). Elec-trochemical degradation of cypermethrin pesticide on a SnO_2 Anode. *Int. J. Electrochem. Sci,* 7, 3453-3465.

Cajaraville, M.P., Diez, G., Marigomez, I. A. and Angulo, E. (1990).Responses of basophilic cells of the digestive gland of mussels to petroleum hydrocarbon exposure.*Dis. Aqua. Org*, 9, 221-228.

Chakraborty, S., Ray, M. and Ray, S. (2010).Toxicity of sodium arsenite in the gill of an economically important mollusk of India.*Fish. Shelf. Immun*, 29, 136-148.

Chakravarthi, K., Naravaneni, B.R. and Philip, G.H. (2007). Study of Cypermethrin Cytogenesis effects on Human Lymphocytes Using In-Vitro Techniques *J. Appl. Sci. Environ. Manage,* 11 (2), 77 – 81.

Curtis, J. E. and Horne, P. A. (1995). Effect of Chlorpyrifos and Cypermethrin Applications on Non-Target Invertebrates in a Conservation-Tillage Crop. *J. Aust. Ent. Soc, 34,* 229-231.

Etxeberria, M., Cajaraville, M.P., Marigomez, I. (1995). Changes in digestive cell lysosomal structure in mussels as biomarkers of environmental stress in the Urdaibai Estuary (Biscay Coast, Iberian peninsula). *Mar. Pollut. Bull,* 30, 599-603.

Etxeberria, M., Sastre, I., Cajaraville, M.P. and Marigomez, I. (1994).Digestive lysosome enlargement induced by experimental exposure to metals (Cu, Cd and Zn) in mussels collected from a zinc-polluted site. *Arch.Environ.Contam.Toxicol*, 27, 338-345.

Hietanen, B., Sunila, I. and Kristoffersson, R. (1988). Toxic effect of zinc on the common mussel *Mytilus edulis* L. (Bivalvia) in brakish water. I. Physiological and histological studies. *Ann.Z ool.Fennici*, 25, 341-347.

Jahanbakshi, A., Shaluei, F. and Baghfalaki, M. (2012).Acute toxicity of cypermethrin on the Great Sturgeon (*Huso huso*) juveniles. *Worl. Jour. Fis. Mar. Sci,* 4(2), 170-174.

Johnson, M.A., Paulet, Y.M., Donval, A. and Pennee, M.L. (1996).Histology, Histochemistry, and enzyme biochemistry in the digestive system of the endosymbiont bearing bivalve (*Loripes lucinalis* Lamarck).*J. Exp.*197(1), 15-38.

Kumar, S., pandey, R.K., Das, S. and Das, V. K. (2011).Pathological changes in exposed to sublethal concentration of dimethoate.*Ger. Bull. Bio*, 2(2), 18-23.

Lowe, D.M., Moore, M.N., Clarke, K.R. (1981). Effects of oil in the digestive cells in mussels: quantitative alterations in cellular and lysosomal structure. *Aquat. Toxicol,* 1, 213-226.

Marigomez, I., Kortabitarte, M. and Dussart, G.B.J. (1998). Tissue- level biomarkers in sentinel stages as cost effective tools to assess metal pollution in soils.*Arch. Environ. Contam. Toxicol*, 34, 167-176.

Marigomez, I.A., Cajaraville, M.P. and Angulo, E. (1990).Histopathology of the digestive gland-gonad complex of the marine prosobranch *Littorina littorea* exposed to cadmium.*Dis. Aquat. Org*, 9, 229-238.

Mehedinti, R., Hancu, M., Coman, M. and Mehedinti, T. (1999).Histological and histochemical alteration in certain black sea bivalves submitted to accidental pollution. *Morphol. Embryol*, XLV, 143-152.

Misra, G. and Mukhapadhyay, P.K. (2008). Mussel farming: Alternate water monitoring practice. *Aquaculture Asia Magazine*, July–Sept, 32–34.

Patil, A.G. (2011).Protein changes in different tissues of freshwater bivalve *Parreysia cylindrical* after exposed to indoxacrab.*Rec.Res.Sci.Tech,* 3(3), 140-142.

Payne, B.S., Lel, J., Miller, A.C. and Hubertz, E.D. (1995). Adaptive variation in palps and gill size of the zebra mussel (*Dreissena polymorpha*) and Asian clam (*Corbicua fluminea*). *Can.J.Fish.Aquat.Sci*, 52, 1130-1134.

Prabhakar, A.K. and Roy, S.P. (2009). Ethno-medicinal Uses of Some Shell Fishes by People of Kosi River Basin of North-Bihar, India. *Ethno-Med, 3(1)*, 1-4.

Pugazhendy, K., Meenambal, M. and Vijayan, P. (2012).Chelating property of *Delonix elata* against toxic impact of cypermethrin on hematological changes in fresh water fish *Cyprinus carpio* (Linn). *Jour. Phar. Res.*5 (5), 2610-2614.

Rao, S. et al. *Handbook on freshwater molluscs of India*.Calcutta: *Zoological Survey of India*; 1989.

Regoli, F. (1998).Tracemetals and antioxidant enzymes in gills and digestive gland of the Mediterranean mussel *Mytilus galloprovincialis. Arch. Environ. Contam. Toxicol*, 34, 48-63.

Saxena, P. and Saxena, A.K. (2010). Cypermethrin Induced Biochemical Alterations in the Blood of Albino Rats. *Jordan Journal of Biological Sciences*, *3(3)*, 111-114.

Schimmel, S.C., Garnas, R.L., Patric, J.M. and Moore, J.C. (1983). Acute toxicity, bioconcentration and persistence of AC 222, 705, benthiocrab and C, Chloropyrifus, fenvalerate, methylparathion and permethin in the estuarine environment. *J. Agric. Food. Chem,* 31, 104-113.

Sekhar P.R., Savithri, Y., Rao, Y.P., Kishore, S. and Rao, K.J. (2010). Cypermethrin and sodium fluoridesynergistic effect on oxidative enzymes in muscle and kidney of albino mice.*Int.Jour.Pha.Bio.Sci,* 1(3), 1-9.

Siegfried, B. (1993).Comparative toxicity of Pyrethroid insecticides to terrestrial and Aquatic insects.*Env.Toxico.Chem,* 12, 1683-1689.

Smith, T.M. and Stratton, G.W. (1986).Effects of synthetic pyrethroid insecticides on nontarget organisms. *Residue. rev,* 97, 93-120.

Snyman, R.G., Reinecke, A.J. and Reinecke, S.A. (2005).Quantitative changes in the digestive gland cells of the snail *Helix aspersa* after exposure to the fungicide copper oxychloride. *Eco. Env. Saf.*, 60:47-52.

Sprague, J.B. (1971).The measurement of pollutant toxicity to fish III, sublethal effect and safe concentration. *Wat. Res,* 5, 245-266.

Suman, G., Naravaneni, R. and Jamil, K. (2006).Invitro cytogenetic studies of cypermethrin on human lymphocytes. *Ind. Jour. Exp. Bio,* 44, 233-239.

Sunila, I. (1987).Histopathology of mussels (*Mytilus edulis* L.) from the Tvarminne area, the Gulf of Finland (Baltic Sea). *Ann. Zool. Fennici*, 24, 55-69.

Tantarpale, S.A. (2011).Cypermethrin impact on total protein in muscle and liver of the freshwater fish *Channa striatus. Sci. Res. Rep,* 1(3), 155-158.

Tendulkar, M. and Kulkarni, A. (2012). cypermethrin –Induced Toxic effect of Glycogen metabolism in Estuarine clam, *Marcia opima* (Gmelin, 1791) of Ratnagiri Coast, Maharastra. *Journal of Toxicology,* 2012, 1-3.

Thompson, R.J., Ratcliffe, N.A. and Bayne, B.L. (1974).Effects on starvation on structure and function in the digestive gland of the mussel (*Mytilus edulis* L.). *J. Mar.Biol.Abs.U.K*, 54, 699-712.

Tiwari, S., Tiwari, R. and Singh, A. (2011). Impact of cypermethrin on fingerlings of common edible carp (*Labeo rohita*). *The. Sci. Worl. Jour.*, 2012, 1-7.

Tripathi, G. (1992).Relative toxicity of aldrin, fenvalerate, capton and diazinon to the freshwater food fish, *Clarias batrachus. Biomed. Environ. Sci,* 5, 33-38.

Tripp M.R., Fries, C.R., Craven, M.A., Grier, C.E. (1984). Histopathology of *Mercenaria mercenaria* as an indicator of pollutant stress. *Mar. Environ. Res.,* 14.521-524.

Usheva, L.N., Vaschenko, M.A. and Durkina, V.B. (2006).Histopathology of the digestive gland of bivalve mollusk *Crenomytilus grayanus* (Dunker 1853) from South Western Peter, The Great Bay, Sea of Japan. *Russ. J. Mar. Biol.,* 32(3), 166-172.

In: Vertebrate and Invertebrate Animal Approaches
Editor: José Rosa Gomes
ISBN: 978-1-63117-102-4

Chapter 20

MORPHOLOGICAL FEATURES OF THREE ENDEMIC PILL-MILLIPEDES OF THE GENUS *ARTHROSPHAERA* IN THE WESTERN GHATS OF INDIA

B. S. Kadamannaya[1], C. N. Ambarish[2] and K. R. Sridhar[2*]
[1]Department of Zoology,
Field Marshal K.M. Cariappa Mangalore University College, Karnataka, India
[2]Department of Biosciences, Mangalore University, Mangalagangotri, Karnataka, India

ABSTRACT

The giant pill-millipedes of the genus *Arthrosphaera* are endemic to the forests of Western Ghats of India and Sri Lanka. They are active during the wet period in monsoon and post-monsoon seasons. Being major saprophagous fauna, *Arthrosphaera* are of immense value in mineralizing the organic matter in soil and improve the fertility of forests and plantations of the Western Ghats. During regular surveys of forests and plantations of the Western Ghats, three species of *Arthrosphaera* were recovered in the mid- and high-altitude locations. Due to lack of systematic morphological account, several species of *Arthrosphaera* have not been assigned to species level. Based on the literature and recent developments, the present study attempts to provide precise information on the morphological status of three species of *Arthrosphaera* (*A. craspedota, A. lutescens* and *A. zebraica*). Such detailed morphological explanations help to distinguish *Arthrosphaera* from its sister genera *Sphaeromimus* and *Zoosphaerium*. Although the presence of two locking carinae in some *Arthrosphaera* species resembles the members of *Zoosphaerium*, presence of one locking carina relates with the members of *Sphaeromimus* of Madagaskar. *Arthrosphaera* species recovered in the present study showed bipartite vulvae, presence of two ridges on the first joint of anterior telopods and ridges on the wash board in females, which facilitates stridulation. Further detailed studies on the differences in the stridulatory organs (anterior and posterior telopods and washboard) by scanning electron microscopy facilitate to distinguish species of *Arthrosphaera* more precisely. Morphological details given in the present study include:

* Corresponding author: E-mail: kandikere@gmail.com.

colour, head, antennae, mouthparts, gnathochilarium, collum, thoracic shield, tergites, endotergum, anal shield, legs, female sexual features and male sexual features.

Keywords: *Arthrosphaera*, distribution, morphology, pill-millipedes, Western Ghats

INTRODUCTION

Although the estimated diversity of Diplopoda exceeds 80, 000, only about 12, 000 species are described (Golovatch et al., 1995; Hoffman et al., 1996, 2002). Based on these estimates, the Class Diplopoda constitutes third most diverse class of terrestrial arthropods after hexapods and arachnids (Pitz and Sierwald, 2010). Being major saprophagous fauna, millipedes are valuable in cycling of nutrients in a variety of habitats due to their ability of fragmentation of organic matter and transform into fecal pellets, which increases the surface area of the organic matter facilitating colonization of microorganisms for further decomposition (Hopkin and Read, 1992). Giant pill-millipedes belongs to the Order Sphaerotheriida are although most fascinating, they are the least studied group among Diplopoda (Wesener and Vanden Spiegel, 2009). Sphaerotheriida is a monophyletic order of the class Diplopoda (subclass, Pentazonia) dates back to Palaeozoic era (Jeekel, 1974; Enghoff, 1984; Sierwald et al, 2003; Regier et al., 2005). Three orders of Pentazonia (Glomeridesmida, Glomerida and Sphaerotheriida) share common characteristics especially the reduction of median sternites and possess condyles-like structures on the mandibles (Verhoeff, 1928; Enghoff, 1984). Recently Wesener and Vanden Spiegel (2009) performed phylogenetic analysis of Sphaerotheriida by considering morphological character matrix comprising 89 characters. This study speaks about monophyletic status, geographical distribution, island gigantism in Madagascar and suggests the break-up of Gondwana fragments about 160-90 Ma. Thus, being highly restricted dispersal, the giant pill-millipedes become a potential model taxon for future investigation.

In most millipedes the pleurites are fused to the tergites and form pleurotergal arch, but in Sphaerotheriida, tergites, pleurites and sternites are separate. Unlike other millipedes, many species of the order Sphaerotheriida are attractively marked by red, orange and yellow spots. The pill-millipedes are shorter than other millipedes and consist of 11-13 segments. Various authors addressed the morphological parameters differently for the purpose of classification of *Arthrosphaera* (Silvestri, 1917; Verhoeff, 1927, 1928; Jeekel, 1974). Wesener and Sierwald (2005a) based on earlier literature provided a standard systematic scheme of morphological characters to be employed for identification of species. Earlier investigators in the Western Ghats of India listed several identified species along with unidentified morphotypes of *Arthrosphaera* (e.g. Chowdaiah, 1966; Achar, 1986, 1987; Ashwini and Sridhar, 2008; Kadamannaya et al., 2010). Based on the descriptions of different species of *Arthrosphaera* by earlier investigators (Pocock, 1899; Attems, 1936), a systematic account has been given following the scheme proposed by Wesener and Sierwald (2005a) to fill the gaps in our knowledge on *Arthrosphaera*. In continuation of our earlier studies on four pill-millipedes belongs to *Arthrosphaera* in the Western Ghats of India (Kadamannaya and Sridhar, 2009), we describe general morphological characteristics of three more species (*Arthrosphaera craspedota, A.lutescens* and *A. zebraica*) harboring the forests of the Western Ghats.

Materials and Methods

Survey was carried out on the occurrence of *Arthrosphaera* in the litter strata of forests and plantations of high altitude and foothill locations of the Western Ghats during monsoon and post-monsoon season (2005-06). Representative individuals of pill-millipedes were immersed in ethylacetate and preserved in ethanol (70%). The dimensions of the individuals (length, breadth and diameter on conglobation) were recorded using vernier calipers. Morphological characteristics were evaluated based on the magnification under a low- high-power microscope. Details of various body structures (antennae, mandibles, gnathochilarium, section of endotergam, 2nd leg pair in females and males, 9th pair of legs in females and males, sub-anal plates with wash board in females, anterior and posterior pairs of telopods) were assessed on treating with 4% KOH.

Results

Color of the animal, different features of head, antennae, mandibles, gnathochilarium, collum, thoracic shield, tergites, anal shield, legs, female sexual characters and male sexual characters are given for each species. Details of dimensions for male and female individuals of three species are given in Table 1 and the distinct distinguishing features are given in Table 2. Distribution and ecology of each species based on the literature and the present survey are furnished.

Table 1. Weight (g) and dimensions (mm) of males and females of *Arthrosphaera* spp. (mean; n=10; range in parenthesis)

Weight and dimension	Male	Female
Arthrosphaera craspedota		
Weight	8.30 (6.7-9.85)	7.57 (5.9-9.2)
Length	37.76 (35.4-39.4)	35.62 (32.3-38.4)
Breadth	18.26 (15.93-20)	15.98 (13.5-17.4)
Conglobated diameter		
Widest axis	25.88 (23.4-28.3)	24.04 (20.9-28)
Narrow axis	20.96 (19.8-22.2)	9.58 (18.0-19.9)
Arthrosphaera lutescens		
Weight	10.44 (9.45-10.98)	13.96 (9.7-16.32)
Length	42.13 (34.7-47.4)	46.15 (33.5-52.4)
Breath	26.73 (22.6-29.3)	24.00 (17.2-27.7)
Conglobated diameter		
Widest axis	32.66 (28.3-35.2)	30.17 (22.9-33.4)
Narrow axis	29.03 (24.9-31.2)	25.60 (18.7-29.9)
Arthrosphaera zebraica		
Weight	9.60 (8.88-11.94)	10.67 (9.15-12.78)
length	38.86 (31.4-44.5)	44.58 (40.9-47.2)
Breadth	21.64 (19.2-23.0	30.72 (24.9-35.1)
Conglobated diameter		
Widest axis	25.84 (23.1-27.9)	30.52 (29.1-31.5)
Narrow axis	22.96 (20.9-24.3)	27.47 (26.4-28.6)

Table 2. Diagnostic characteristics of three *Arthrosphaera* spp. recovered from the Western Ghats os India

Diagnostic features	*Arthrosphaera craspedota*	*Arthrosphaera lutescens*	*Arthrosphaera zebraica*
Coloration	Black with a thin line of ivory color	Olivaceous with a thin band of orange brown	Jet black with white band along the sides
Length (mm)	36.7	44.1	41.7
Width (mm)	17.1	25.4	26.2
Molar plate	No step	1 step	2 steps
Rows of teeth in pectinate lamellae	5	7	6-7
Locking carinae	1	2	1
Suture between carinae	Absent	Present	Absent
Coaxal lobe of legs	Absent	Coxal tooth	Absent
Stridulation ridges in the wash board	4-5 irregular	3-4	3-4 irregular
Nature of tibia and tarsus of anterior telopod	Coalased	Separated	Coalased

Arthrosphaera craspedota Attems (Figure 1)

Arthrosphaera craspedota, Attems, Mem. *Indian Mus., 11*, p. 156-157 (1936).

Color - Body of mature animal is black, posterior margin of each tergite possess thin line of ivory color. Head and collum are black, while the antennae and legs are greyish.

Head - Setiferous pits and hairs present around the clypeus and lateral part of the eyes. The posterior margin of the head possesses some isolated hairs.

Antennae – Six jointed. Length of the antennomeres: 1>2>3=4=5>6. Sixth antennomere oval shaped with more than 50 sensorial cones.

Mouthparts - Mandible molar plate process devoid of step. Five pectinate lamellae. Number of teeth declines proximally.

Gnathochilarium - Ventral side with bristles, numerous hairs are present on the palpi. Only a few hairs are present on the lingual lamella. On the central pads, different kinds of sensory cones are seen and a few are with central pit and hairs. Sensory cones are also present on the palpi.

Collum - Anterior margin with a few isolated hairs and the posterior margin bald.

Thoracic shield - Black with yellow posterior border. Lateral extension of the thoracic shield has a few hairs and ridges - Anterior marginal brim broader than the rest of the brim.

Tergites - Anterior part of the terga densely punctured. Posterior half sparingly punctured. Paratergite depression has hairs and possesses a few ridges in the posterior tergites. Tips of posterior margins of paratergites project posteriorly.

Endotergum - One row of marginal bristles. Internal part with short spines and few isolated bristles.

Anal shield - Lower margin at posterior end tapered. Anal shield with one well developed black locking carina on each side.

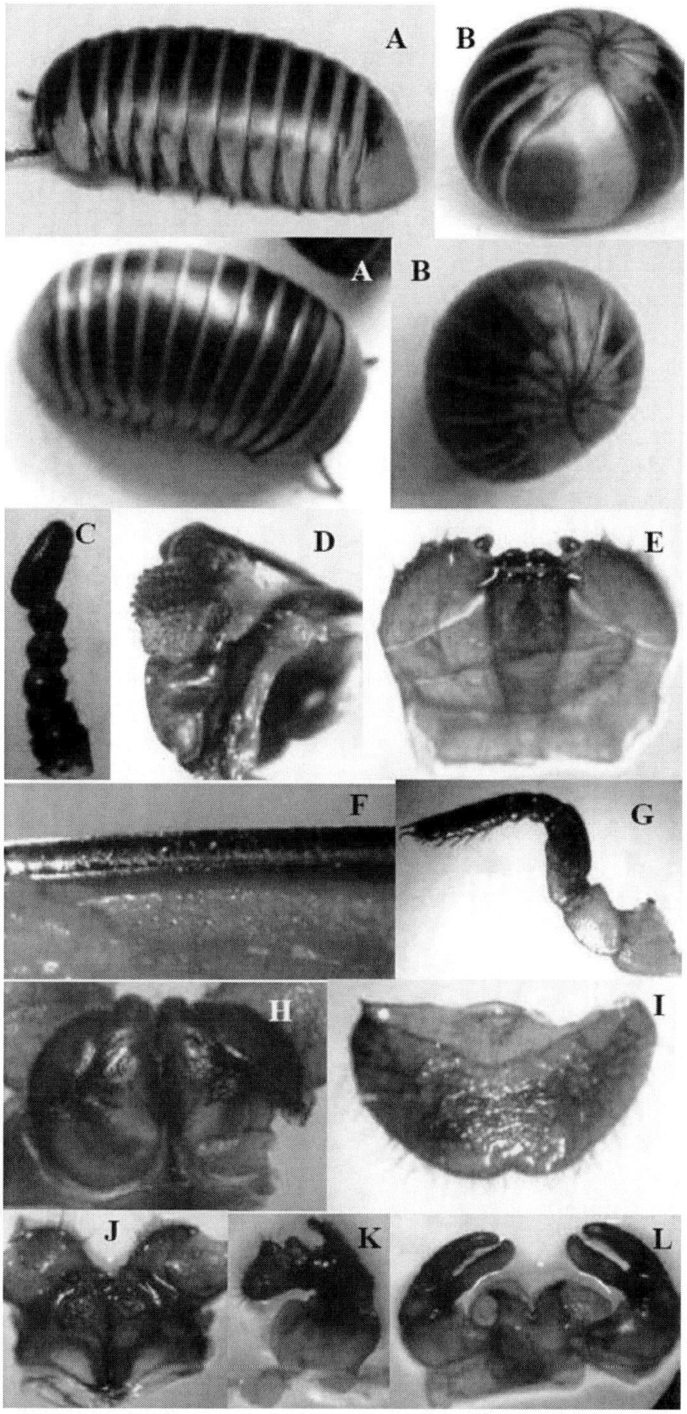

Figure 1. *Arthrosphaera craspedota* – whole animals (A), conglobated animals (B), antenna (C), mandible (D), gnathochilarium (E), endotergum (F), 9th leg (G), vulvae (H), washboard (I), male gonopore (J), anterior telopod (K) and posterior telopod (L).

Legs - The first and second leg pair with only 2-3 ventral spines and weakly curved claw. Remaining pairs of leg have 7-11 ventral spines with curved claw. Coxal lobe of ninth leg pair is not present. All the podomere possess isolated dense hairs.

Female sexual characters - Operculum of vulvae is small without any depression at the center, basal margin devoid of outgrowth and almost straight. The inner plate is long and broad reaching above the operculum. Exterior plate is not as long as the inner one reaching only the basal part of the operculum. Sub-anal plate with wash board shows a depression at the median line and has long hair at the periphery. Short hairs are present all over the surface. The wash board consists of irregular asymmetrical 1-5 stridulation ribs. Length of ribs is variable.

Male sexual characters - Male gonopores present on the 2nd leg pair covered with sclerotized, undivided rounded plate.

Anterior telopod - Syncoxite with a few hairs. First joint possesses a lobe and stimulatory ridge. Femoral process is broad with a lobe at the top. Tibia and tarsus coalesced, tibial part with a rounded lappet, the tassel part with a rounded lappet at the top and a blunt lappet at the lateral side.

Posterior telopod - The inner horns of the syncoxite has a cone shaped spine like structure. Movable finger possesses specific dentition and a row of crenulated teeth and two white lappets. Stout tips of the immovable finger is curved and hook-shaped.

Distribution - These animals are collected from the forests of Agumbe (657 m) and Ulvi (647 m) locations (Karnataka). They showed stridulation behavior in the field.

Arthrosphaera lutescens Butler (Figure 2)

Zephronia lutescens, Butler, *Ann. Mag. Nat. Hist., 10*, p. 356 (1872).

Zephronia lutescens, Butler, *Proc. Zool. Soc. London*, p. 179 (1873).

Arthrosphaera lutescens, Pocock, *J. Bombay Nat. Hist. Soc., 12*, p. 274 (1899).

Color - Tergites are olivaceous, posterior margin possesses a thin band of orange brown (sometimes grayish with a band of dull brown). Head is dark brown, collum is chestnut, antennae are dark green (black) legs are greenish, thoracic shield - anterior part is black whereas posterior zone is green.

Head - With numerous hairs and pits especially around the clypeus and eyes. Surface of the head wrinkled and bald for most of the area. A tuft of small hair was seen at the posterior end near the collum.

Antennae - Possess six antennomeres, the sixth one is the biggest with an oval terminal disc having many sensory papillae. Sclerotized teeth are visible at the base of the antennomere.

Mouthparts - Mandible with seven rows of pectinate lamellae, number of teeth decreased from apical to proximal rows. Molar plate consists of single strongly developed step.

Gnathochilarium – Possess many hairs and 6-7 sensorial cones on the palpi and 12-14 sensory pits on the central pads of lingual lamella.

Collum - Smooth with a very few hairs.

Thoracic shield - Smooth, lateral extension has a few thick hairs towards the marginal rim.

Tergites - Anterior zone is punctate with setaceous pits. Posterior portion is smooth without wrinkles and hairs. The anterior paratergite depression of the anterior tergites consists of ridges and anterior paratergite of posterior tergites densely covered with hairs. Tips of the posterior margins of paratergites project posteriorly.

Endotergum - The endotergum possesses a band of sclerotized nodules between the marginal ridge and the internal area. The marginal ridge is characterized by a single row of short hairs.

Figure 2. *Arthrosphaera lutescens*: whole animals (A), conglobated animal (B), antenna (C), mandible (D), gnathochilarium (E), endotergum (F), 9th leg (G), vulvae (H), washboard (I), male gonopore (J), anterior telopod (K) and posterior telopod (L).

Anal shield – Rounded, covered with short hair and punctate throughout. The anal shield carries two black locking carinae on both the sides. Anterior carina is similar to that of other tergites. Posterior ones are twice larger than the anterior ones. A clear suture separates two locking carinae. A small invagination is seen at the end of the suture, where it meets the margin, numerous hairs are seen throughout the ventral side.

Legs - First two pair are with few ventral spines (3-4) and weakly curved claw, remaining legs possess 6-7 ventral spines and a curved claw. Coxa of ninth leg pair bears coxal tooth. Coxae of all legs at the inside margin possess a few long hairs.

Female sexual characters - Second pair of legs possess coxal lobe. Operculum of the vulva is broad. Center of operculum is devoid of indentation. Exterior plate is large and broad surrounding the base of the operculum. Inner plate is comparatively small surpassing the apical end of the operculum. Outer margin of the operculum consists of many hairs. Wash board is sub-anal plate, semicircular with hairs on the margin and possesses 3-4 stridulating ridges.

Male sexual characters – Second leg pair possesses a coxal lobe. Male gonopore is situated on the coxae of 2nd leg pair, which covered with Selerotized rounded plate.

Anterior telopods - Syncoxite of anterior telopods on both sides devoid of hairs. Four joints are clearly separated. First joint with two ridges, lobe of the 2nd joint (femoral lobe) extends beyond the 3rd joint and situated at the side of the tibia (3rd joint). The 3rd joint (tibia) possesses a leaf like lobe (superior lappet), and tarsal part (4th joint) with a shallow superior lappet and an apical lobe with hairs.

Posterior telopods - Syncoxite and the subanal lobe with dense hairs. Inner horn of syncoxite with a blunt thorn. The movable finger has the characteristic white lappets (two) the 2nd joint (immovable finger) longer than the movable finger with specific dentition along the inner side.

Distribution - This species were collected from the hilly forests of Agumbe (657 m) and Nalur (626 m) and mixed plantations near Thirthalli (610 m) (Karnataka). These locations consisting of areca plantation and semi-evergreen to evergreen forests receive heavy rains. Usually this species are common adjacent to channels of roads with ample organic matter.

Arthrosphaera zebraica Butler (Figure 3)

Zephronia zebraica, Butler, *Ann. Nat. Hist, 10*, p. 356 (1872).

Color - Tergite is jet black with white/cream color along the sides of the tergites (tips of the tergite on either side), anterior zone of head is light brown, posterior part is dark brown. Eyes grayish with more than 75 ocelli and collum is greenish along the lateral sides. Legs are greenish, thoracic segment black with cream along the lateral sides and anal segment is black with cream color along the posterior end.

Head - Anterior portion is highly pubescent with setiferuos pits, clypeus with a single tooth and surrounded by hairs and a few isolated hairs present around the eyes. Posterior zone is punctured with a few rows of hairs.

Antennae - With six jointed, 6th joint is broad, oval in shape and bears numerous sensory cones with hairs.

Figure 3. *Arthrosphaera zebraica*: whole animals (A), conglobated animals (B), antenna (C), mandible (D), gnathochilarium (E), endotergum (F), 9th leg (G), vulvae (H), washboard (I), male gonopore (J), anterior telopod (K) and posterior telopod (L).

Mouthparts - Mandible is with 6-7 rows of pectinate lamellae. The teeth in each row are comparatively longer and their number decreases proximally. Condyles possess two steps.

Gnathochilarium - Possesses many hairs on the lamellae, linguale and lateral palpi. Pits with long hairs are seen on the central pads as well as on the lateral palpi, which may be sensory in nature.

Collum - Punctured throughout with short hairs.

Thoracic shield - More or less polished, a few setaceous pits with hair throughout. Anterior marginal brim slightly broader than the rest.

Tergites - Anterior end with many pits joined together (beaded) bearing hairs. Posterior end is polished. Anterior paratergites depression possesses numerous short hairs. Ventral side consists of hair and a single locking carina. Tips of the paratergite project posteriorly. Sternite with many hairs and upper margin rounded.

Anal shield - Smooth on the surface and rounded. Ventral side is with a distinct broad, short single locking carina. There is no suture and no invagination along the periphery. Ventral surface covered with many hairs.

Legs - Tarsi of first ten pairs of legs with 3-4 ventral spines and an apical spine. Tarsi of next pairs of legs have 11-12 ventral spines with a claw and an apical spine. Ninth pairs of legs are without coaxal lobe. Hairs present all over the margins of podomeres.

Female sexual characters - Coaxe of the 2nd leg with a small coaxal tooth. Operculum of the vulvae is broad without any depression at the middle reaching the basal part of the 2nd joint. Inner plate is short stout with a hook-shaped broad apical end, which surpasses the tip of the operculum above. Exterior plate is stout surrounds the base of the operculum and inner plate. Margin of the operculum consists of a few short hairs. Sub-anal plate rounded with an invagination at the center wash board possesses 3-4 small irregular stridulating ribs on either side. Hairs present all over the surface.

Male sexual characters - Anterior telopod - Syncoxite with a few hairs. First joint possesses a harp like process with 2 ridges and both ridges possess sharp border. They form a rim around the 2nd joint. The 2nd joint is curved with a superior slender lobe. The 3rd and 4th joints are coalased with 3 lobes, ten rounded superior lappets and the 3rd blunt apical lobe consist of bristles all over the surface.

Posterior telopod - Syncoxite with dense hairs. The coxal horn (inner horn) is slender bent in an angle of 45 degrees with hook-like thorn. The sub-anal lobe with numerous hairs. The 2nd joint (femoral process) is equal to the 3rd joint (tibia or movable finger) in length. The movable finger has 2 white lappets and other crenulations. A wide gap is visible between these two fingers.

Distribution - Collected from the forests of Varadalli (607 m) and Ulvi (647 m) (Karnataka).

DISCUSSION

Wesener and Sierwald (2005a) revised the genus *Sphaeromimus* based on the electron microscopic investigations. Considering 89 morphological characters, the genus *Sphaeromimus* resembles with *Arthrosphaera* (Wesener and VandenSpiegel, 2009). It is evident that the genus *Sphaeromimus* shares many characteristics with the Indian genus *Arthrosphaera* (four jointed anterior telopod, six jointed antennae and, flat and broad sixth antennomere). But, the presence of harp on the anterior telopod connects *Arthrosphaera* with the genus *Zoosphaerium*. Based on the common characteristics, it is possible to predict that the Sphaerotheriidae were inhabitants of Indo-Australia and Madagascar-South Africa, but later they were disappeared in the intermediate geographical region possibly due to unfavorable climatic conditions.

Arthrosphaera craspedota showed distribution in Southern India (Chowdaiah and Kanaka, 1974) especially in Kerala (Parambikulam: 520-980 m) (Attems, 1936). In the present study, this species was found in the forest locations of Agumbe (657 m) and Ulvi (647 m) of the Western Ghats of Karnataka. *Arthrosphaera lutescens* were recovered from Maddathoray (Tamil Nadu) and Kulattupuzha (Kerala) (Attems, 1936). This species was also recovered from the forests of Thirthalli (610 m) (Achar, 1986) of Karnataka. In our study, it was common in mixed plantations of Thirthalli and hilly forests of Agumbe (657 m) and Nalur (626 m). Based on the reports of Pocock (1899) and Attems (1936), *Arthrosphaera zebraica* showed distribution in Bombay (Maharashtra). They were also recovered from the forests of Soraba (610 m) (Chowdaiah, 1966), Varadalli (607 m) and Ulvi (647 m) (Kadamannaya et al., 2010).

Pocock (1899) characterized *Arthrosphaera* based on the presence of broad and truncate apex of the legs and upper angle possess a long spine above the claw and thus more space exists between the claw and the spine. The vulva in females composed of three distinct segments assembled together in to a cone. In the males, the movable digit of the hinder pair of copulatory leg composed of a single segment. These characteristics were summarized by Attems (1936) in his key to the Indian genera of Sphaerotheriida and also provided additional characteristics (coxae of the 2nd leg of female separated, tarsal part of the tibio-tarsus of the anterior telopod possess broad rounded lobe, the margin of the lobe beset with knobs and the tibia of the posterior telopods possess two white lappets). Attems (1936) has documented the characteristics of bipartite basal part of the vulva and presence of two white lappets in the tibia of the posterior telopods, which serve as an important diagnostic feature of *Arthrosphaera*. Such bipartite vulva and presence of ridge on the first joint of anterior telopods seems to facilitate the stridulation in *Arthrosphaera*. In the present study, all the three species of *Arthrosphaera* exhibited stridulatory behavior and there seems to be a considerable variation in the position of the ridges in the anterior telopod. These species are distinguishable from each other based on the characteristic features summarized in Table 2. According to Attems (1936), the stridulatory organ is absent in the *Arthrosphaera*. However, in other species of *Arthrosphaera* (*A. dalyi, A. davsoni, A. fumosa* and *A. magna*), the first joint of anterior telopods endowed with two sharp ridges (Kadamannaya and Sridhar, 2009). These ridges were also prominent in *A. craspedota, A. lutescens* and *A. zebraica* in the present study.

At the outset, *A. craspedota* and *A. zebraica* show some striking similarities as the tibial and tarsal part of the anterior telopod is not separated and they possess only single locking carina in the anal shield (Figure 1, 3). They also differ in body length and breadth, the rows of teeth on the pectinate lamellae and presence of steps on the molar plate process (*A. zebraica* has two steps, while *A. craspedota* devoid of such steps) (Table 2). Presence of two locking carinae in *A. lutescens* in the present study resembles the members of *Zoosphaerium*, while one locking carina in *A. craspedota* and *A. zebraica* resembles the members of *Sphaeromimus* of Madagaskar (Wesener and Sierwald 2005a, 2005b). Sub-anal plates and wash board of *Arthrosphaera* needs further detailed description. Future studies on the sensory cones on the 6th joint of antennae, central pads of lamellae lingual and palpi of gnathochilarium with details of mandibles provide precise insight of the genus *Arthrosphaera*. The homology between the stridulatory oranges and the ridge on the anterior telopods of *Arthrosphaera* based on scanning electron microscopy is under investigation in our laboratory.

ACKNOWLEDGMENTS

Authors are grateful to Mangalore University for permission to carry out the study in the Department of Biosciences. BSK acknowledges the award of fellowship under Faculty Improvement Program (University Grants Commission, New Delhi, India). CNA acknowledges the award of Inspire Fellowship, Department of Science and Technology, New Delhi, Government of India [DST/INSPIRE Fellowship/2011/(294): Award # IF110540].

REFERENCES

Achar, K.P. (1986). Analysis of male meiosis in seven species of Indian pill-millipede (Diplopoda: Myriapoda). *Caryologia 39*, 89–101.

Achar, K.P. (1987). Chromosomal evolution in Diplopoda (Myriapoda: Arthropoda). *Caryologia 40*, 145–155.

Ashwini, K.M. and Sridhar, K.R. (2008). Distribution of pill millipedes (*Arthrosphaera*) and associated soil fauna in the Western Ghats and west coast of India. *Pedosphere 18,* 749–757.

Attems, C. (1936). Diplopods of India. *Memoirs of the Indian Museum 11*, 133–167.

Chowdaiah, B.N. (1966). Chromosome studies in two species of pill-millipedes (Diplopoda–Myriapoda). *Caryologia 19*, 135–141.

Chowdaiah, B.N. and Kanaka, R. (1974). Cytological studies in six species of pill-millipedes (Diplopoda-Myriapoda). *Caryologia 27*, 55–64.

Enghoff, H. (1984). Phylogeny of millipedes – a cladistic analysis. *Z. Zool. Syst. Evolutionsforsch.* 22, 8–26.

Golovatch, S.I., Hoffman, R.L., Adis, J. and de Morais, J.W. (1995). Identification plate for the millipede orders populating the Neotropical region south of central Mexico (Myriapoda: Diplopoda). *Stud. Neotrop. Fauna Environ. 30*, 159–164.

Hoffman, R.L., Golovatch, S.I., Adis, J., de Morais, J.W., 1996. Practical keys to the orders and families of millipedes of the Neotropical region (Myriapoda: Diplopoda). *Amazoniana 14*, 1–35.

Hoffman, R.L., Golovatch, S.I., Adis, J. and de Morais, J.W. (2002). Diplopoda. In: *Arachnida and Myriapoda* (Ed. Adis, J.), Pensoft, Sofia-Moscow, 501–533.

Hopkin, S.P. and Read, H.J. (1992). *The Biology of Millipedes.* Oxford University Press, Oxford.

Jeekel, C.A.W. (1974). The taxonomy and geography of the Sphaerotheriida (Diplopoda). *Symposium of the Zool. Soc. London 32*, 41–52.

Kadamannaya, B.S. and Sridhar, K.R. (2009). Morphology and distribution of four pill millipedes (*Arthrosphaera*) of the Western Ghats. *Animal biology Journal 1*, 81–98.

Kadamannaya, B.S., Sridhar, K.R. and Sreepada, K.S. (2010). Assemblage and distribution of pill millipedes and earthworms in relation to soil edaphic features in the Western Ghats and the west coast of India. *Frontiers of Agriculture in China 4*, 243–250.

Pitz, K.M. and Sierwald, P. (2010). Phylogeny of the millipede order Spirobolida (arthropoda: Diplopoda: Helmenthomorpha). *Cladistics 26*, 497–525.

Pocock, R.I. (1899). A monograph of the pill-millipedes (Zephroniidae) inhabiting India, Cylon and Burma. *Journal of Bombay Natural History Society 12*, 269–285.

Regier, J.C., Wilson, H.M. and Shultz, J.W. (2005). Phylogenetic analysis of Myriapoda using three nuclear protein-coding genes. *Molecular Physiology and Evolution 34*, 147–158.

Sierwald, P., Shear, W.A., Shelley, R.M. and Bond, J.E. (2003). Millipede phylogeny revisited in the light of the enigmatic order Siphoniulida. *J. Zool. Syst. Evol. Res. 41*, 87–99.

Silverstri, F. (1917). Specie di Sphaeroteridae delle regioni australiana e neozelandse a me note. *Bullettina del Laboratario di Zoologia Generale e Agraria della R. Scuola Superiore d'Agricoltura in Portici 12*, 61–85.

Verhoeff, K.W. (1927). Myriapoda: Diplopoda. Results of Mjöberg's Swedish scientific expeditions to Australia 1910-1913. *Arkiv für Zoologi 16*, 40–69.

Verhoeff, K.W. (1928). Klasse Diplopoda I. In: *HG Bronn's Klasseu und Ordnungen des Tierrichs* 5.2.II Acad. Verlag, Leipzig, Germany.

Wesener, T. and Sierwald, P. (2005a). The giant pill-millipedes of Madagascar. *Proc. California Acad. of Sci. 56*, 557–599.

Wesener, T. and Sierwald, P. (2005b). New giant pill-millipede species from the littoral forest of Madagascar (Diplopoda, Sphaerotheriida, *Zoosphaerium*). *Zootaxa 1097*, 1–60.

Wesener, T. and VandenSpiegel, D. (2009). A first phylogenetic analysis of giant pill-millipedes (Diplopoda: Sphaerotheriida), a new model Gondwanan taxon, with special emphasis on island gigantism. *Cladistics 25*, 545–573.

In: Vertebrate and Invertebrate Animal Approaches
Editor: José Rosa Gomes
ISBN: 978-1-63117-102-4

Chapter 21

IMPACTS OF AMMONIA TOXICITY ON SOME ENZYMES OF A FRESHWATER FISH AT DIFFERENT PH AND TEMPERATURES

A. Shiwanand and G. Tripathi*
Department of Zoology, J.N.V. University, Jodhpur, India

ABSTRACT

Acute effects of ammonium chloride on the activities of glutamate dehydrogenase (GDH), catalase (CAT) and acetylcholinesterase (AChE) of liver, gill and skeletal muscle of the freshwater fish, *Clarias batrachus* were studied at different pH and temperatures. In control the activities of GDH, CAT and AChE declined significantly at 5.5 and 8.5 pH as compared to pH 7 of the water within a period of 96 hr. In contrast, the GDH and CAT activities increased gradually, whereas AChE activity decreased with the enhancement of pH from 5.5 to 8.5 in ammonia exposed group. However, GDH, CAT and AChE activities in different tissues declined in both control and ammonia treated groups as a function of increase in temperatures from 15 to 35 °C within 96 hr. Maximum effect of pH and temperatures on GDH and CAT was in liver followed by gill and skeletal muscle. But, in case of AChE, the maximum effect was on gill followed by liver and skeletal muscle at different pH and temperatures. It demonstrated that neutral pH and 25 °C temperature decrease ammonia toxicity. Therefore, adverse effects of ammonia on fish can be minimized by altering pH and temperature of water in an aquaculture.

INTRODUCTION

Ammonia toxicity is a severe problem in fish farming. Fish growth yield may be depressed by usual ambient ammonia concentrations, under intensive farming conditions (Person-Le Ruyet et al., 1997). Suboptimal water quality or detrimental ecological conditions can limit fish production by creating stress to the resident fish (Sinha, 1997; Das, 2002). As dietary protein is broken down in the body, some of the nitrogen is excreted through the gills

* E-mail: drgst@rediffmail.com.

as ammonia. The decomposition of organic matter also produces ammonia, which diffuses from the sediment into the water column (Floyd et al., 1990). Ammonia accumulates easily in aquatic systems because it is a natural byproduct of fish metabolism. High level of ammonia in water which results from agricultural or mining operation runoff, excessive biological waste accumulation, insufficient water aeration or inadequate tank conditioning (Noga, 1996; Randall and Tsui, 2002; Spencer et al., 2008). In general, the toxicity of chemicals varied with respect to species and size of fish and duration of exposure (Oh et al., 1991; Dutta et al., 1995). Chronic stress in fishes occurs due to variations in pH, dissolved oxygen and unionized ammonia in water (Das et al., 1994). Ammonia causes stress and damages gills and other tissues, even in small amounts. Rodrigues et al., (2007) investigated acute toxicity and sublethal effects of ammonia and nitrite for juvenile cobia *Rachycentron canadum*. Oliveira et al. (2008) found that cardinal tetra (*Paracheirodon axelrodi*) is tolerant to high temperature (33.3 °C), to a wide pH range (acid pH=2.9 and alkaline pH=8.8) and to high total ammonia concentration (23.7 mg/L). Wagner et al. (1997) studied the effects of temperature and high pH on mortality and the stress response of rainbow trout after stocking. Effect of malathion and temperature combination on blood glucose and urea levels of gobiid fish *Glossogobius giuris* (Venkataramana et al., 2005). Kumar and Gupta (2006) concluded acute toxicity of mercury to the fingerlings of Indian major carps (catla, rohu and mrigal) in relation to water hardness and temperature. Ciardiello et al. (1997) reported enzymes of antarctic fishes: effect of temperature on catalysis. Ecological conditions *viz.* water temperature, pH, alkalinity, stocking density etc. of aquatic environment play a significant role on the colonization of bacterial flora and incidence of diseases in culturable and wild species of fish (Snieszko, 2008; Prasad, 2009).

Sun et al. (2012) studied combined effects of ammonia and microcystin on survival, growth, antioxidant responses and lipid peroxidation of bighead carp *Hypophthalmythys nobilis* larvae. Chen et al. (2011) described oxidative stress responses of grass carp *Ctenopharyngodon idella* larvae exposed to purified microcystin under different ammonia concentrations. Yang et al. (2011) investigated the effect of long term ammonia exposure on feeding, growth and some antioxidant parameters of liver, gill and blood of juvenile crucian carp (*Carassius auratus*). Oxidative stress and antioxidant enzyme activities in liver and white muscle of Nile tilapia juveniles in chronic exposure to sublethal total ammonia nitrogen (TAN) were studied (Hegazi et al., 2010). Yang et al. (2010) reported toxicity of ammonia and its effects on oxidative stress mechanisms in juvenile crucian carp (*Carassius auratus*). Ching et al. (2009) found that environmental ammonia exposure induces oxidative stress in gills and brain of *Boleophthalmus boddarti* (mudskipper). Kaizer et al. (2009) reported the effects of ammonia concentration on acetylcholinesterase activities in brain of teleost fish (*Rhamdia quelen*).

AChE is a tissue enzyme found principally in the nervous tissue (gray matter), red cells and muscle. GDH, CAT and AChE enzymes are normally found within the cells of liver, heart, gills and kidneys (Shalaby, 2009) but their increase in plasma indicates the tissue injury or organ dysfunction (Wells et al., 1996).

The differences in toxic potential of the pollutants may be attributed to the susceptibility of the test animals and factors like pH and hardness of the water (Ahmad, 2012). The present paper examines the impact of ammonia toxicity on nitrogen metabolism, antioxidative and neurotoxic responses at different pH and temperatures of the freshwater air-breathing catfish, *Clarias batrachus*.

MATERIALS AND METHODS

Fish and Chemicals

Adult and healthy specimens of the air-breathing walking catfish, *Clarias batrachus* weighing 184±6 g having body length 23-29 cm were collected from local fish market of Jodhpur, India. This freshwater catfish is used for human nutrition and available in fish markets of India. Collections were done from November to February. This is the resting phase of the fish. The fish were maintained under laboratory conditions one week prior to experimentation. Control groups were without ammonia treatment. However, experimental groups were treated with sublethal concentration of ammonium chloride (100 mg L^{-1}) for 96 hr. Water was changed on alternate day and a fresh concentration of ammonium chloride was maintained during ammonia exposure period. Fish were starved during experimentation. All chemicals used in this study were of analytical grade and procured from Sisco Research Laboratory (SRL), Mumbai, India. Double glass-distilled water was used for preparation of chemicals. Different pH values were adjusted by dilute sulphuric acid and sodium hydroxide solution and fish were maintained at 25°C. There were three control and three ammonia treated groups for three different pH i.e., 5.5, 7 and 8.5 of the medium. Likewise, in case of temperature, there were three control and three ammonia treated groups for three different temperatures i.e., 15, 25 and 35°C of the water. Different water temperatures at pH 7 were maintained with a thermostate control system. The pH values were measured with a pH meter and temperatures were measured with Multi-thermometer. There were control and ammonia treated groups as follows:

(A) Groups for ammonia treatment at 25 °C temperature and different pH

Group 1: Normal individuals were kept untreated for 96 hr at pH 5.5 of the water (Control 1)
Group 2: Normal individuals were kept untreated for 96 hr at pH 7 of the water (Control 2)
Group 3: Normal individuals were kept untreated for 96 hr at pH 8.5 of the water (Control 3)
Group 4: Individuals were exposed to sublethal concentration of ammonium chloride (100 mg L^{-1}) for 96 hr at pH 5.5 of the medium (Treated 1)
Group 5: Individuals were exposed to sublethal concentration of ammonium chloride (100 mg L^{-1}) for 96 hr at pH 7 of the medium (Treated 2)
Group 6: Individuals were exposed to sublethal concentration of ammonium chloride (100 mg L^{-1}) for 96 hr at pH 8.5 of the medium (Treated 3)

(B) Groups for ammonia treatment at 7 pH and different temperatures

Group 1: Normal individuals were kept untreated for 96 hr at temperature 15 °C of the water (Control 1)
Group 2: Normal individuals were kept untreated for 96 hr at temperature 25 °C of the water (Control 2)
Group 3: Normal individuals were kept untreated for 96 hr at temperature 35 °C of the water (Control 3)

Group 4: Individuals were exposed to sublethal concentration of ammonium chloride (100 mg L^{-1}) for 96 hr at temperature 15 °C of the medium (Treated 1).
Group 5: Individuals were exposed to sublethal concentration of ammonium chloride (100 mg L^{-1}) for 96 hr at temperature 25 °C of the medium (Treated 2).
Group 6: Individuals were exposed to sublethal concentration of ammonium chloride (100 mg L^{-1}) for 96 hr at temperature 35 °C of the medium (Treated 3).

Tissue Processing and Enzyme Preparation

Fish were sacrificed by decapitation and tissues (liver, gill, skeletal muscle) of *C. batrachus* were removed immediately and washed in 0.6% saline. They were properly cleaned and weighed quickly. Tissues were stored at –40 ºC in Remi Deep Freezer (BDI-8158) until used for analysis. Glutamate dehydrogenase (GDH, EC 1.4.1.3), acetylcholinesterase (AChE, EC 3.1.1.7) and catalase (CAT, EC 1.11.1.6) activities remained almost unaltered during period of storage for several weeks. Tissues were homogenised (10% w/v) in potassium phosphate buffer (50mM, pH 7.5), containing 0.32M sucrose and 1 mM EDTA for GDH and for CAT and AChE sodium phosphate buffer (0.1M, pH 7.4) was used. Homogenization was done with the help of Potter-Elvehjem homogenizer fitted with a Teflon Pestle (Remi RB1278). Homogenization was done at about 4 ºC. For GDH, the suspension was treated with Triton X-100 (0.5%) for 1 hr. Homogenates were centrifuged at 12000g for 15 min in a high speed refrigerated centrifuge (Remi Cooling Compufuge, CPR 24). The supernatant was collected and used for the assay of GDH, CAT and AChE activities.

Enzyme Assay

GDH activity was assayed as described by Das et al. (1991) with some modifications. The assay mixture for reductive amination reaction contained 150mM of potassium phosphate buffer (pH 7.4), 500mM ammonium chloride, 15mM α-ketoglutarate, 0.8mM of NADPH and the tissue extract as enzyme source in a total volume of 3.0 ml. The oxidation of NADPH were recorded at 340nm using Cintra 5 UV-VIS spectrophotometer (GBC, Germany). Activity of the enzyme was expressed in μmol/min/g wet mass. The procedure adopted for the assay of catalase was that of Aebi (1984). Catalase was assayed in a medium containing 50-150mM H_2O_2 and 300mM phosphate buffer (pH 7). Total volume of reaction mixture was 3 ml.

The decrease in absorbance was noted at 240nm. The enzyme activity was expressed in mmol/min/g wet mass. AChE activity was determined spectrophotometrically following the method of Ellman et al. (1961) using acetylthiocholine iodide as the substrate. Volume of reaction mixture was 3ml.

Absorbance was recorded at 412nm using Cintra 5 UV-VIS spectrophotometer. Reading of blank without enzyme was substrated to yield the enzymatic activity. Reaction mixture contained 2.7ml phosphate buffer (0.1M, pH 7.4), 0.05ml DTNB (5, 5'- Dithiobis (2-Nitrobenzoic acid) (0.18mM), 0.1ml enzyme sample and 0.15ml acetylthiocholine iodide. Enzyme activity was expressed in μmol/min/g wet mass.

Statistical Analysis

A one way analysis of variance (ANOVA) followed by DMRT was performed to determine the level of significance. Computations were done with the help of statistical package.

RESULTS

Effects of acute ammonia toxicity at different pH and temperatures on the activities of glutamate dehydrogenase, catalase and acetylcholinesterase of liver, gill and skeletal muscle of *Clarias batrachus* are described as follows:

A) Effects of Ammonia Toxicity on Enzymes at Different pH and 25°C Temperature

Activity of GDH of liver, gill and skeletal muscle of control fish decreased significantly at 5.5 and 8.5 pH of water as compared to 7 pH. However, the GDH activity increased gradually in these tissues from 5.5 to 8.5 pH of the medium in ammonia-exposed group. The increases were significantly higher in liver (1.33-2.74 fold) followed by gill (1.45-2.64 fold) and skeletal muscle (1.28-2.62 fold) in ammonia exposed fish as compared to control fish at all pH levels. Maximum elevation in the enzyme activity was at pH 8.5. CAT showed decrement in its activity in liver, gill and skeletal muscle at pH 5.5 and 8.5 as compared to 7 pH of water in control fish within 96 hr. While CAT activity enhanced gradually and it was maximum in liver (1.27-2.72 fold) followed by gill (1.26-2.64 fold) and skeletal muscle (1.24-2.53 fold) from 5.5 to pH 8.5 in ammonia treated fish as compared to control. Maximum induction was found at 8.5 pH. Activity of AChE declined at 5.5 and 8.5 pH in liver, gill and skeletal muscle of control fish as compared to the activity at 7 pH within 96 hr. Likewise, the AChE activity maximally declined in gill (22.10-45.09%) followed by liver (21.09-36.20%) and skeletal muscle (19.26-31.05%) in ammonia exposed fish as compared to control group. The maximum reduction was at pH 8.5.

(B) Effects of Ammonia Toxicity on Enzymes at Different Temperatures and 7 pH

Increase in temperatures from 15 to 35 °C gradually decreased the GDH activity in liver, gill and skeletal muscle of both control and ammonia treated fish within 96 hr. Maximum reduction in GDH activity was in gill (22.44-38.93%) followed by skeletal muscle (22.18-35.84%) and liver (12.59-33.05%) in control fish as compared to ammonia treated fish. The maximum reduction was at 35 °C. CAT activity decreased gradually in liver, gill and skeletal muscle of control and ammonia exposed catfish for 96 hr with the elevation of temperatures from 15 to 35°C. CAT activity was maximally declined in skeletal muscle (21.52-26.92%) followed by gill (17.45-23.69%) and liver (16.78-32.77%) in control fish as compared to ammonia exposed group. Maximum declinement was at 35°C of the medium. The activity of

AChE declined gradually in liver, gill and skeletal muscle of control and ammonia treated fish for acute toxicity as a function of increase in temperatures 15 to 35 °C. Maximum reduction of AChE activity was in skeletal muscle (27.23-37.34%) followed by liver (26.24-37.50%) and gill (12.70-29.34%) in ammonia treated fish as compared to control. Increase of temperatures and ammonium chloride declined the AChE activity in these tissues of fish as compared to control group. Maximum declinement was found at 35°C of water.

DISCUSSION

GDH activity of liver, gill and skeletal muscle of *Clarias batrachus* increased gradually in ammonia treated group as compared to control with elevation of 5.5 to 8.5 pH of the medium for 96 hr (Figure 1).

The important function of GDH is that the amino group of most amino acids is transferred to α-ketoglutarate to produce glutamate. The increased GDH activity may indicate increased rapid utilization of amino acids (Nelson and Cox, 2005; Sathyanarayana, 2005). Ammonia significantly increases body weight, specific growth rate and affect antioxidant system in bighead carp larvae (Sun et al., 2012). The oxidation of glutamate in the Kreb's cycle leads to energy production (Narasimha and Ramana, 1985). Sousa and Meade (1977) suggested that the mechanism of ammonia toxicity involves stimulation of glycolysis by the ammonium ions and the simultaneous suppression of the Krebs cycle due to the depletion of α-ketoglutarate as it removes ammonia by amination to form glutamine. Maximum increment in GDH activity was found in liver followed by gill and skeletal muscle. This may be assigned to the tissue-specific differences in ammonia toxicity.

Activity of CAT in different tissues of *C. batrachus* enhanced in ammonia-exposed group as compared to control with respect to increasing pH from 5.5 to 8.5 of the water (Figure 2). CAT is an adaptation and antioxidant enzyme in fish (Rudnick, 1967; Orbea et al., 2000; Sole et al., 2004; Camargo et al., 2005; Fernandez-Diaz et al., 2006). Maximum activity of CAT was in liver followed by gill and skeletal muscle. It reflects that increase in pH induces different effects of ammonia in fish. Hegazi et al. (2010) also described increase in catalase activity in liver and white muscle of Nile tilapia (*Oreochromis niloticus*) exposed to ammonical nitrogen. Chen et al. (2011) documented catalase as a biomarker for measuring oxidative stress response in grass carp larvae. Like present observations ammonia has been reported to significantly affect CAT which show its protective function when oxidative stress occurs, but specific results are variable depending on time, ammonia concentration and the tissue (Yang et al., 2011). In contrast, sublethal concentration of ammonium chloride (at pH 7) decreases the catalase activity in brain and gills of *Boleophthalmus boddarti* (Ching et al., 2009). AChE activity showed decrement in different tissues of fish from 5.5 to 8.5 pH of water in ammonia treated group with compared to control within 96 hr (Figure 3). Ammonia-associated decrease in AChE activity at lower and higher pH may be due to enzyme modification or decrease enzyme synthesis.

It is well known that environmental factors such as temperature, have profound effects on the catalytic activity and on enzyme regulation and structure. Temperature can influence fish in multiple ways, affecting biochemical and physiological activities and can act as a lethal factor when its effect destroys the integrity of the organism (Currie et al., 1998). Activity of GDH decreased as a function of increase in temperature from 15 to 35 °C of water in control group as compared to ammonia exposed group of liver, gill and skeletal muscle of *C. batrachus* within 96 hr (Figure 4).

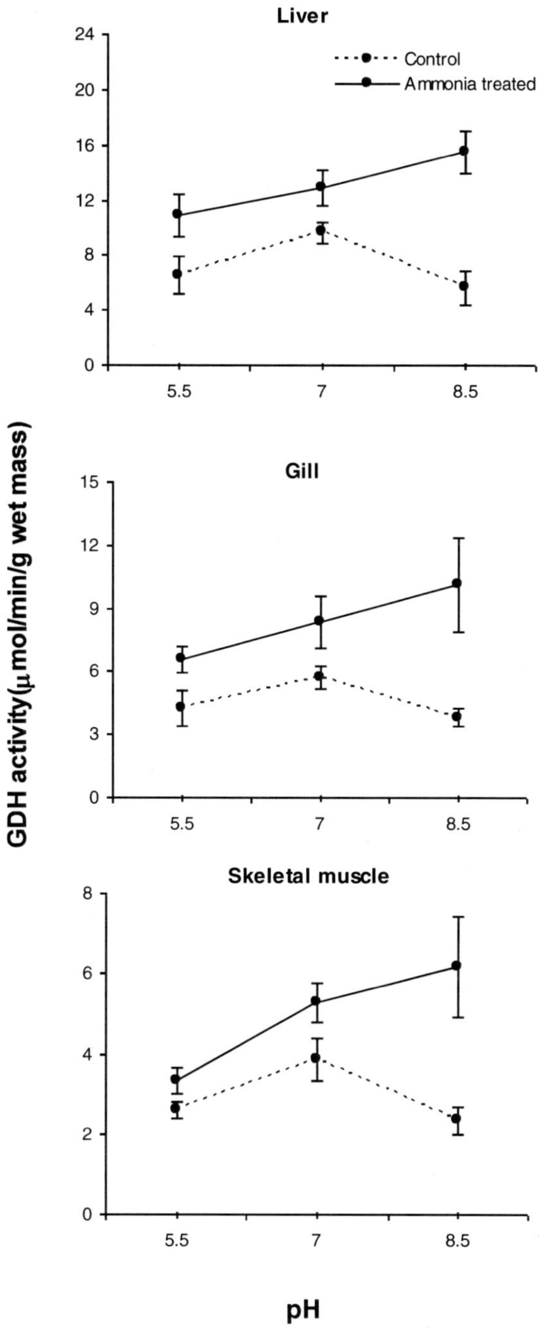

Figure 1. Effects of ammonia toxicity on glutamate dehydrogenase (GDH) activity (µmol/min/g wet mass) at different pH in liver, gill and skeletal muscle of cat fish, *C. batrachus*.

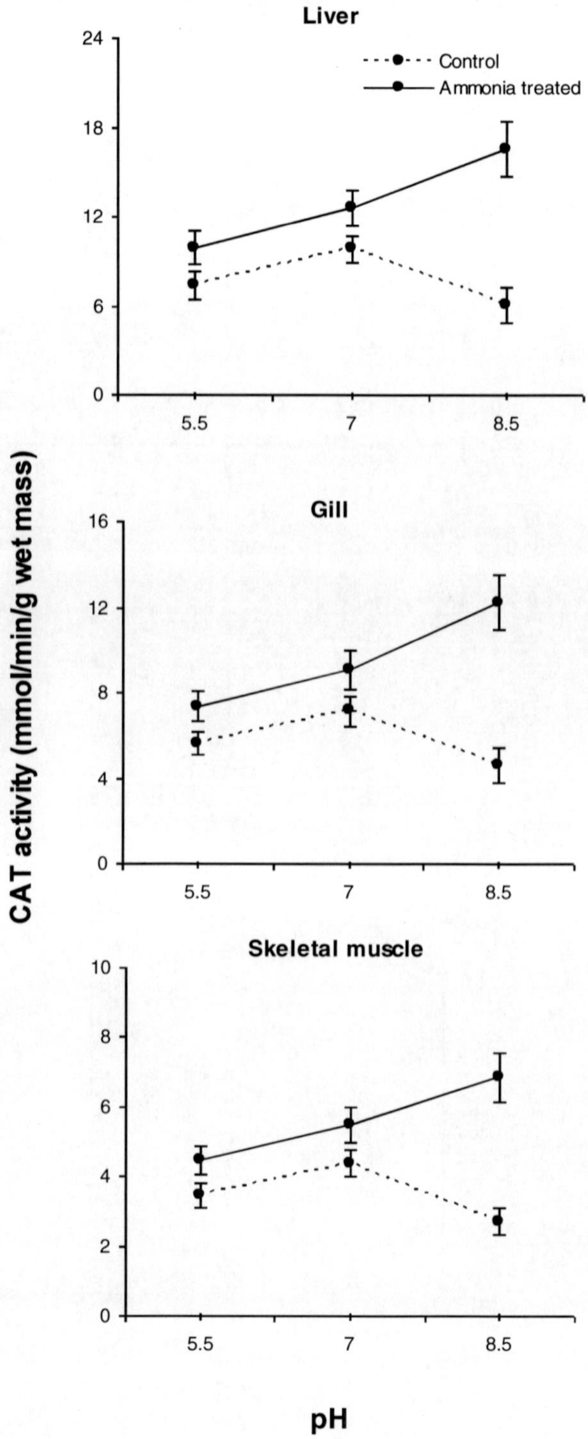

Figure 2. Effects of ammonia toxicity on catalase (CAT) activity (mmol/min/g wet mass) at different pH in liver, gill and skeletal muscle of catfish, *C. batrachus.*

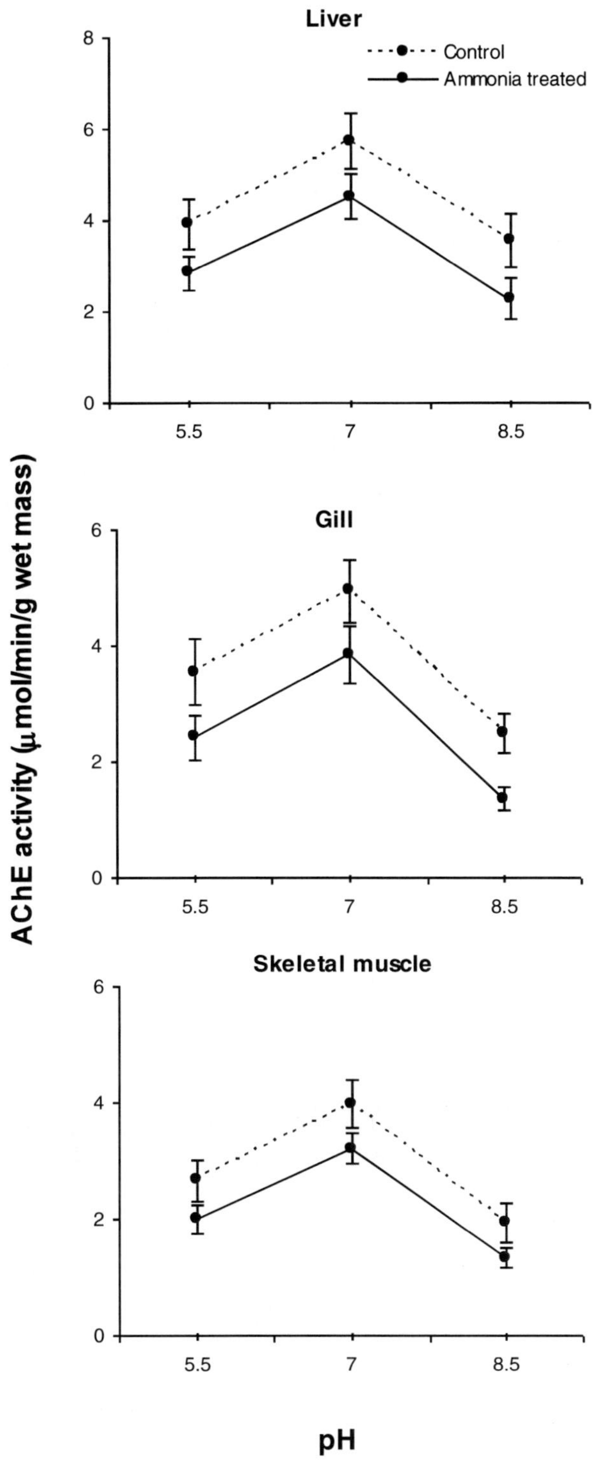

Figure 3. Effects of ammonia toxicity on acetylcholinesterase (AChE) activity (μmol/min/g wet mass) at different pH in liver, gill and skeletal muscle of catfish, *C. batrachus.*

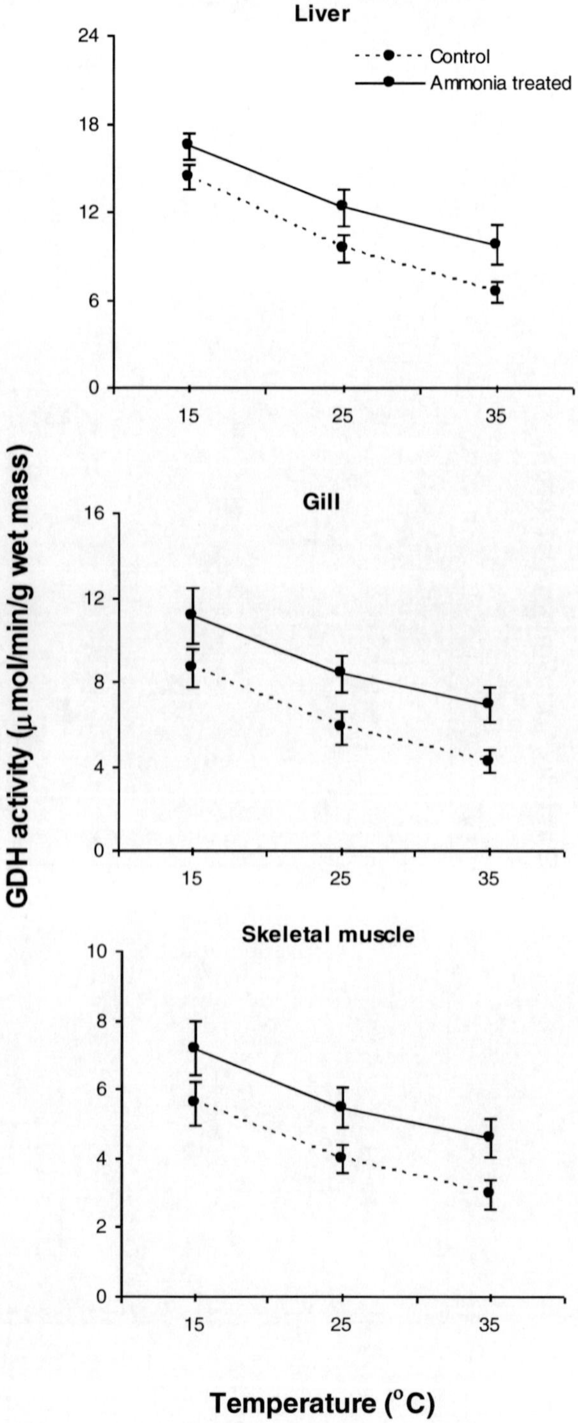

Figure 4. Effects of ammonia toxicity on glutamate dehydrogenase (GDH) activity (μmol/min/g wet mass) at different temperatures (°C) in liver, gill and skeletal muscle of catfish, *C. batrachus.*

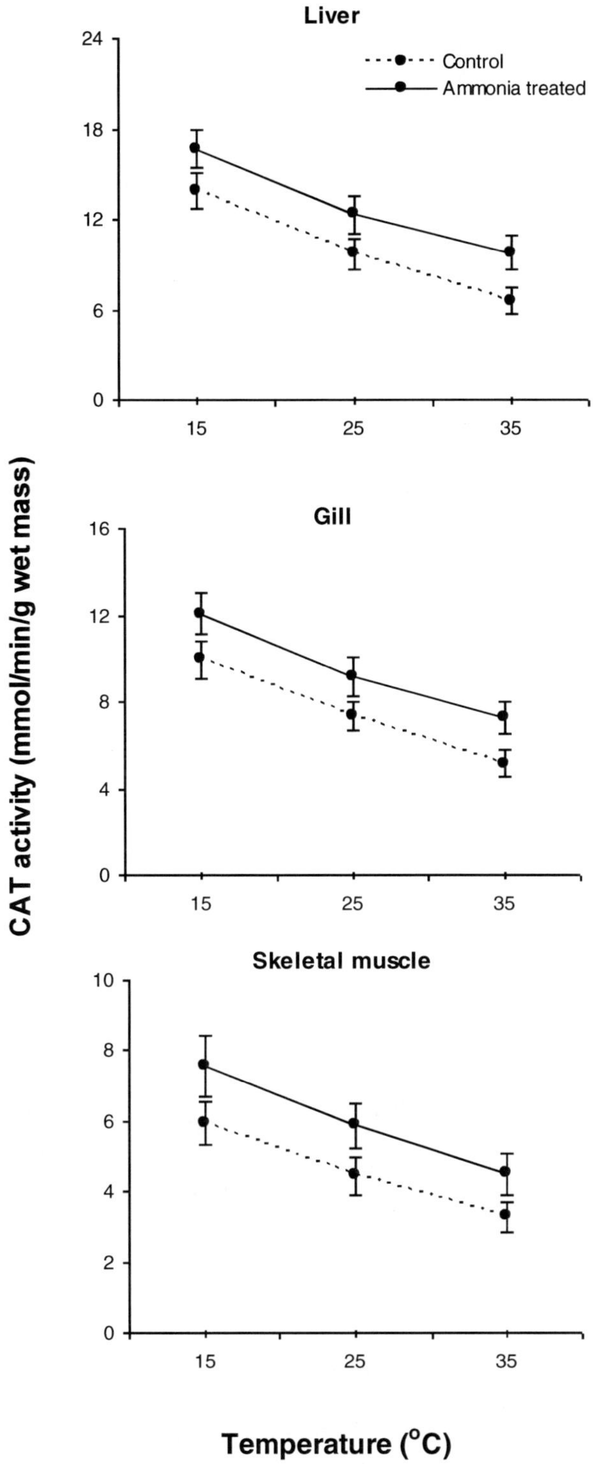

Figure 5. Effects of ammonia toxicity on catalase (CAT) activity (mmol/min/g wet mass) at different temperatures (oC) in liver, gill and skeletal muscle of catfish, *C. batrachus*.

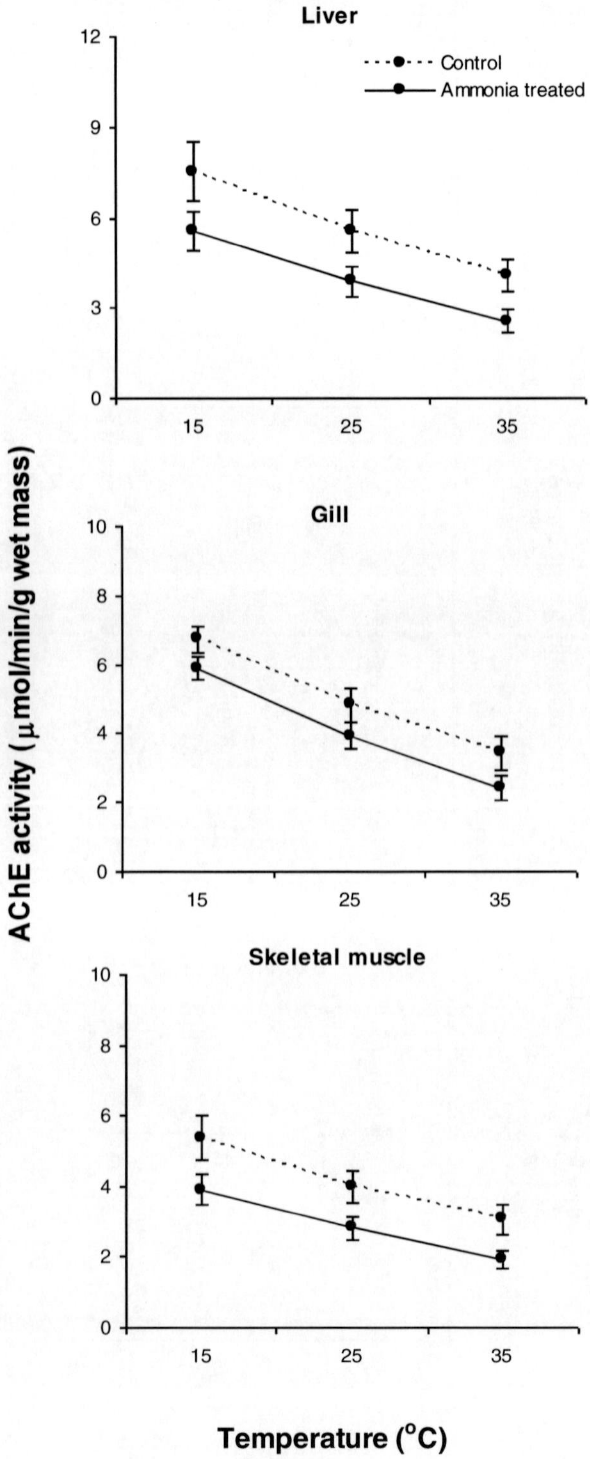

Figure 6. Effects of ammonia toxicity on acetylcholinesterase (AChE) activity (µmol/min/g wet mass) at different temperatures (°C) in liver, gill and skeletal muscle of catfish, *C. batrachus*.

Ammonia causes cellular destruction or necrosis in hepatocytes (Hedayati et al., 2010). Ciardiello et al. (1997) found that temperature effect the catalytic activity of GDH from liver of the icefish, *Chaenocephalus aceratus*. GDH activity maximally reduced in gill, followed by skeletal muscle and liver. High levels of ammonia cause stress and produce harmful response including retarded growth, inefficient immune function (Cheng et al., 2004; Pinto et al., 2007) and reduced survival (Jobling, 1994). The presence of ammonia suppresses feed utilization and increases water content of the body and urea excretion rate in turbot juveniles (Person-Le Ruyet et al., 1997).

Ammonia damages the physiological functions, growth and reproduction. Initially the fish may be seen gasping at the surface for air. The fish being to lose its appetite and become lethargic. Internal damage occurs to the brain and central nervous system (Kroupova et al., 2005). Increase in temperatures from 15 to 35°C of the medium gradually decreased the CAT activity in different tissues of both control and ammonia treated fish within 96 hr (Figure 5). CAT activity significantly declined in control fish at different temperatures as compared to ammonia treated groups. Ammonia induced changes in CAT activity might be due to allosteric alteration in the enzyme molecule. Kannadi et al. (2008) reported the effects of temperature on the behavioural and respiratory responses of catfish *Mystus gulio* (Hamilton). Nasr et al. (1998) reported that exposure to unionized ammonia induced gill damage and subsequent liver tissue hypoxia associated with blood anaemia. Antioxidant defenses (CAT) in the liver of juvenile crucian carp fluctuated with time and ammonia concentration (Yang et al., 2010).

AChE activity of different tissues of fish declined significantly as a function of enhancement of temperatures from 15 to 35 °C of water in ammonia treated group as compared to control (Figure 6). It is utilized for the hydrolysis of the acetylcholine to choline and acetate in cholinergic synapses. AChE can be accepted as a potent indicator of ammonia toxicity. Modesto and Martinez (2010) studied AChE inhibition by pollutant with anti-AChE action. Maximum reduction of AChE activity was in skeletal muscle, followed by liver and gill. Both pH and temperature of water affect the quantity of the toxic form of ammonia present in water. Ammonia is responsible for more unexplained losses in aquaculture than any other water quality parameter. Therefore, it can be suggested that the toxicity of ammonia can be minimized by altering physicochemical condition of water. The present findings may have some help to aquaculture.

ACKNOWLEDGMENTS

Authors gratefully acknowledge the University Grants Commission (UGC), New Delhi for providing financial support in the form of UGC major research project (No. 36-104/2008(SR). AS is thankful to UGC for giving Project Fellowship.

REFERENCES

Aebi, H. (1984) Catalase *in vitro*. *Methods in Enzymology,* 108. Academic Press, New York. pp. 121-126.

Ahmad, Z. (2012) Toxicity bioassay and effects of sub-lethal exposure of malathion on biochemical composition and naematological parameters of *Clarias gariepinus*. *Afr. J. Biotechnol.,* 11: 8578-8585.

Camargo, J.A., Alonso, A. and Salamanca, A. (2005) Nitrate toxicity to aquatic animals: a review with new data for freshwater invertebrates. *Chemosphere*, 58: 1255-1267.

Chen, Y., Sun, H., Yang, W. and Yang, Z. (2011) Oxidative stress responses of grass carp *Ctenopharyngodon idella* larvae exposed of purified microcystin under different ammonia concentrations. *Fresen. Environ. Bull.*, 20: 2869-2874.

Cheng, W., Hsiao, I.S. and Chen, J.C. (2004) Effect of ammonia on the immune response of Taiwan abalone *Haliotis diversicolo*r supertexta and its susceptibility to *Vibrio parahaemolyticus. Fish Shellfish Immunol.,* 17: 193-202.

Ching, B., Chew, S.F., Wong, W.P. and Ip, Y.K. (2009) Environmental ammonia exposure induces oxidative stress in gills and brain of *Boleophthalmus boddarti* (mudskipper). *Aquat. Toxicol.,* 95: 203-212.

Ciardiello, M.A., Camardella, L. and Di Prisco, G. (1997) Enzymes of antarctic fishes: effect of temperature on catalysis. *Cybium.*, 21: 443-450.

Currie, R.J., Bennett, W.A. and Beitinger, T.L. (1998) Critical thermal minima and maxima of three freshwatern game-fish species acclimated to constant temperatures. *Env. Biol. Fish.,* 51: 187-200.

Das, J.R., Saha, N. and Ratha, B.K. (1991) Tissue distribution and subcellular localization of glutamate dehydrogenase in a freshwater air-breathing teleost, *Heteropneustes fossilis. Biochem. System. Ecol.,* 19: 207-212.

Das, M.K., Das, R.K., Ghosh, S.P. and Bhowmick, S. (1994) Fish diseases, its relation with environmental factors in a sewage fed wetland. *J. Inland Fish.Soc. India*, 26: 100-105.

Das, M.K. (2002) Social and economic impacts of disease in Inland open water cultuered based fisheries in India. In: primary aquatic animal health care in rural, small scale, aquaculture development. (Eds: J.R. Arthur, M.J. Philips, R.P. Subasinghe, M.B. Reantaso and I.H.M. Rae). FAO, Fish. Tech. Pap. No. 406 pp 333-344.

Dutta, H.M., Munshi, J.S.D., Dutta, G.R., Singh, N.K., Adhikari, S. and Richmonds, C.R. (1995) Age related differences in the inhibition of brain acetylcholinesterase activity of *Heteropneustes fossilis* (Bloch) by malathion. *Comp. Biochem. Physiol.,* 111: 331-334.

Ellman, G.L., Courtney, D.K. and Andres, R.M. (1961) A new and rapid colorimetric determination of acetylcholinesterase activity. *Biochem. Pharmacol.* 7: 88-95.

Fernandez-Diaz, C., Kopecka, J., Canavate, J.P., Sarasquete, C. and Sole, M. (2006) Variations on development and stress defences in *Solea senegalensis* larvae fed on live and microencapsulated diets. *Aquacult.*, 251: 573-584.

Floyd, R.F., Watson, C., Petty, D. and Pouder, D.B. (1990) Ammonia in aquatic systems. Series FA 16. Fisheries and Aquatic Science Department, Florida Cooperative Extension Service, Institute of Food and Agricultural Sciences, University of Florida.

Hedayati, A., Safahieh, A., Savari, A. and Marammazi, J.G. (2010) Assessment of aminotransferase enzymes in yellowfin sea bream (*Acanthopagrus latus*) under experimental condition as biomarkers of mercury pollution. *World J. Fish Mar. Sci.,* 2: 186-192.

Hegazi, M.M., Attia, Z.I. and Ashour, O.A. (2010) Oxidative stress and antioxidant enzymes in liver and white muscle of Nile tilapia juveniles in chronic ammonia exposure. *Aquat. Toxicol.,* 99: 118-125.

Jobling, M. (1994) *Fish bioenergetics*. Chapman &Hall, London. 294 pp.

Kaizer, R.R., Loro, V.L., Schetinger, M.R.C., Morsch, V.M., Tabaldi, L.A., Rosa, C.S., Garcia, L.O., Becker, A.G. and Baldisserotto, B. (2009) NTPDase and acetylcholinesterase activities in silver catfish, *Rhamdia quelen* (Quoy & Gaimard, 1824) (Heptapteridae) exposed to interaction of oxygen and ammonia levels. *Neotrop. Ichthyol.,* 7: 635-640.

Kannadi, P., Kannan, K. and Raveendran, S. (2008) Effects of temperature on the behavioural and respiratory responses of catfish *Mystus gulio. Indian. J. Environ. Ecoplan.*, 14: 750-754.

Kroupova, H., Machova, J. and Svobodova, Z. (2005) Nitrite influence on fish: A review. *Vet. Medic.,* 50: 461-471.

Kumar, A. and Gupta, A.K. (2006) Acute toxicity of mercury to the fingerlings of Indian major carps (catla, rohu and mrigal) in relation to water hardness and temperature. *J. Environ. Biol.,* 27: 89-92.

Modesto, K.A. and Martinez, C.B.R. (2010) Roundup® causes oxidative stress in liver and inhibits acetylcholinesterase in muscle and brain of the fish *Prochilodus lineatus. Chemosphere.* 78: 294-299.

Narasimha, M.B. and Ramana Rao, K.V. (1985) Alteration in the nitrogen metabolism in liver of fish *Tilapia mossambica* under lindane toxicity. *J. Pids 6th Nat. Sym. Life Sci.* 17-19 Jan. 13-22.

Nasr, F.A., El-Shafai, S.A. and Abo-Hegab, S. (1998) Suitability of treated domestic wastewater for raising *Oreochromis niloticus. Egypt. J. Zool.,* 31: 81-94.

Nelson, D.L. and Cox, M.M. (2005) Lehninger principles of Biochemistry. 3rd edition Macmillan worth Publishers.

Noga, E.J. (1996) *Fish Disease: Diagnosis and Treatment*. St. Louis: Mosby.

Oh, H.S., Lee, S.K., Kim, Y.H. and Roh, J.K. (1991) Mechanism of selective toxicity of diazinon to killfish (*Oryzias latipes*) and loach (*Misgumus anguillicaudatus*). *Aquat. Toxicol. Risk Assess.* 14: 343-353.

Oleveira, S.R., Souza, R.T.Y.B., Nunes, Ė.S.S., Carvalho, C.S.M., Menezes, G.C., Marcon, J.L., Roubach, R., Akifumi Ono, E. and Affonso, E.G. (2008) Tolerance to temperature, pH, ammonia and nitrite in cardinal tetra, *Paracheirodon axelrodi*, an amazonian ornamental fish. *Acta Amazonica*, 38: 773-780.

Orbea, A., Fahini, D.H. and Cajaraville, P.M. (2000) Immunolocalization of four antioxidant enzymes in digestive glands of mollusks and crustaceans and fish liver. *Histochem. Cell. Biol.,* 114: 393-404.

Person-Le Ruyet, J., Delbard, C., Chartois, H. and Le Dellion, H. (1997) Toxicity of ammonia to turbot juveniles: 1. Effects on survival, growth and food utilization. *Aquat. Living Resour.,* 10: 307-314.

Pinto, C., Aragão, C., Soares, F., Dinis, M.T. and Conceição, L.E.C. (2007) Growth, stress response and free amino acid levels in Senegalese sole (*Solea senegalensis* Kaup 1858) chronically exposed to exogenous ammonia. *Aquac. Res.,* 38: 1198-1204.

Prasad, Y. (2009) Prevalence, surveillance and virulence characterization of *Flavobacterium columnare* and *Flavobacterium psycrophilum* in Indian catfish of sub-Himalayan region. *Biochem. Cell. Arch.,* 9: 71-82.

Randall, D.J. and Tsui, T.K. (2002) Ammonia toxicity in fish. *Mar. Pollut. Bull.,* 45: 17-23.

Rodrigues, R.V., Schwarz, M.H., Delbos, B.C. and Sampaio, L.A. (2007) Acute toxicity and sublethal effects of ammonia and nitrite for juvenile cobia *Rachycentron canadum. Aquaculture*, 271: 553-557.

Rudnick, Von Manfred, H. (1967) Vergleichende untersuchungen über die einwirkung hōherer und tieferer temperaturen auf die katalase aktivitāt. *Zool. J. Physiol.,* 1: 227-250.

Sathyanarayana, U. (2005) Biochemistry book and allied (P) Ltd. 8/1 Chintamani Das Lane Kolkata 700009 India.

Shalaby, A.M.E. (2009) The opposing effects of ascorbic acid (Vitamin C) on ochratoxin toxicity in Nile tilapia (*Oreochromis niloticus*). http://www.ag.arizona.edu/ista/ista6web/pdf/209.pdf. Rterieved: 05-04-09.

Sinha, M. (1997) Fisheries enhancement of small reservoirs and floodplain lakes in India. Sugunan. *Bull. Cent. Inland Capture Fish. Res. Inst.,* 75: 1-11.

Sneiszko, S.F. (2008) The effects of environmental stress on outbreaks of infectious diseases of fishes. *J. Fish. Biol.,* 6: 197-208.

Sole, M., Potrykus, J., Fernanadez-Diaz, C. and Blasco, J. (2004) Variations on stress defences and metallothionein levels in the senegal sole, *Solea senegalensis*, during early larval stages. *Fish. Physiol. Biochem.,* 30: 57-66.

Sousa, R.J. and Meade, T.L. (1977) The influence of ammonia on the oxygen delivery system of coho salmon haemoglobin. *Comp. Biochem. Physiol.,* 58: 23-58.

Spencer, P. Pollock, R. and Dube, M. (2008) Effects of un-ionized ammonia on histological, endocrine and whole organism endpoints in slimy sculpin (*Cottus cognatus*). *Aquat. Toxicol.,* 90: 300-309.

Sun, H., Lü, K., Minter, E.J.A., Chen, Y., Yang, Z. and Montagnes, D.J.S. (2012) Combined effects of ammonia and microcystin on survival, growth, antioxidant responses and lipid peroxidation of bighead carp *Hypophthalmythys nobilis* larvae. *J. Hazard. Mater.,* 221-222: 213-219.

Venkataramana, G.V., Sandhya Rani, P.N., Nadoni, M.B. and Murty, P.S. (2005) Effect of malathion and temperature combination on blood glucose and urea levels of gobiid fish *Glossogobius giuris* (Ham.). *J. Indl. Polln. Contl.,* 21: 257-258.

Wagner, E.J., Bosakowski, T. and Intelmann, S. (1997) Effects of temperature and high pH on mortality and the stress response of rainbow trout after stocking. *Trans. Am. Fish. Soc.,* 126: 985-998.

Wells, R.M. Mclntyre, R.H., Morgan, A.K. and Davis, P.S. (1996) Physiological stress responses in big gamefish after exposure: Observations on plasma chemistry and blood factors. *Comp. Biochem. Physiol.,* 84: 565-571.

Yang, W., Xiang, F., Liang, L. and Yang, Z. (2010) Toxicity of ammonia and its effects on oxidative stress mechanisms of juvenile crucian carp (*Carassius auratus*). *J. Freshwat. Ecol.,* 25: 297-302.

Yang, W., Sun, H., Xiang, F., Yang, Z. and Chen, Y. (2011) Responses of juvenile crucian carp (*Carassius auratus*) to long-term ammonia exposure: Feeding, growth and antioxidant defenses. *J. Freshwat. Ecol.,* 26: 563-570.

SECTION II: STUDIES ON MORPHOLOGY AND REPRODUCTION IN VERTEBRATE AND INVERTEBRATE ANIMALS

In: Vertebrate and Invertebrate Animal Approaches
Editor: José Rosa Gomes ISBN: 978-1-63117-102-4

Chapter 22

ULTRASTRUCTURE AND DISTRIBUTION OF IMMUNOPOTENT LEUCOCYTES IN AN AIR-BREATHING FISH, *HETEROPNEUSTES FOSSILIS* (BLOCH)

Rajarshi Ghosh and Sumit Homechaudhuri*
Aquatic Bioresource Research Laboratory,
Department of Zoology,
University of Calcutta, Kolkata, India

ABSTRACT

The present study analyzed the ultrastructural details of the immunopotent leucocytes from haemopoietic tissues of head kidney and relative distribution of leucocytes from peripheral circulation of adult *Heteropneustes fossilis* (Bloch) to provide a useful database for understanding its immunoefficiency. Leucocytes in the head kidney of eight *H. fossilis* from wetlands near Kolkata, India, were studied through fixed thin sections for Transmission Electron Microscopy (TEM). Neutrophils, eosinophils and basophils were identified among granulocytes based on their granular morphology and shape of the nucleus. Lymphocytes of the lymphoid series were also identifiable in *H. fossilis*. Macrophages were monocyte-derived cells having well developed pseudopodia in its surface and contained free ribosomes, lysosomes, RER, mitochondria, intracytoplasmic vacuoles and many cytoplasmic vesicles of variable size. The relative distribution of leucocytes was also enumerated from differential counts.

Keywords: Heteropneustes fossilis, Transmission Electron Microscopy, leucocytes

---

* Corresponding Author: Prof. Sumit Homechaudhuri, Department of Zoology, University college of Science, University of Calcutta, 35 Ballygunge Circular Road, Kolkata 700019, India. Email Address: sumithomec@ yahoo.com Telephone: +91-9830392284.

INTRODUCTION

An increasing interest in the phylogenetic development of cellular immune mechanisms in fish has underlined the importance of a clear understanding of the identity of leucocytes of individual species prior to any significant advance in classifying their immunological role. Much knowledge on the ultrastructure of the immunopotent leucocytes of different species of fish has been accumulating in the literature (Lester and Desser, 1975; Blaxhall, 1983; Cenini, 1984; Hine and Wain, 1987; Modra et al., 1998, Ghosh and Homechaudhuri, 2007). The functions of the leucocyte populations in fish are also being documented (Praveen et al., 2000; Neumann et al., 2001). However, no significant information is available on the siluroid cat fish *Heteropneustes fossilis* (Bloch) which is highly tolerant to oxygen tension having bimodal gas exchange. *H. fossilis*, an air-breathing siluroid cat fish shows adaptive response to hypoxia and various toxicities in their natural environment but are highly susceptible to infections. It is now well accepted that teleostean fish have leucocyte subpopulation consisting of granulocytes (neutrophils, eosinophils and basophils) and agranulocytes (lymphocytes, monocytes and macrophages). However, their identification, functions and relative contributions to specific and non-specific immune responses are still a matter of debate (Rowley et al., 1988; Hine, 1992). Therefore, this study attempts to define the immunopotent leucocytes of *H. fossilis* and is concerned primarily with their morphology at the ultrastructural level. The differential count of the leucocytes in circulating blood was also determined.

MATERIALS AND METHODS

Live specimens of *H. fossilis* were obtained from the local point sources in and around the wetland areas of Kolkata, WB, India. The healthy and disease-free fish were isolated and acclimatized to laboratory conditions for 10 days in glass aquaria (size: 0.6m x 0.3m x 0.3m; 4 fish per aquarium). The eight specimens selected for study were 35.73 ± 0.4 g in weight and 18.93 ± 0.12 cm in length. The fish samples were euthanized by anaesthetic overdose using a 200 mg/l solution of MS-222. The fish were then bled by severing the tail, and the head kidney was aseptically removed. For TEM analysis, small pieces (1 mm^3) of head kidney tissue were washed in 0.1 M phosphate buffer solution (pH: 7.2) and fixed in freshly prepared solution of fixative for 12 h at 4°C, followed by washing with freshly prepared buffer (Karnovsky, 1965). The composition of fixative (100ml) was as follows: 4 % Paraformaldehyde solution - 45 ml; 0.1 M phosphate buffer solutions (pH: 7.2) - 45 ml and 25 % Glutaraldehyde (stock solution) - 10 ml. The tissues were post-fixed in 1 % aqueous osmium tetroxide (OsO_4) solution for 2 h at 4°C. After repeated washing in 0.1 M phosphate buffer, tissues were dehydrated in ascending grades of ethanol upto 100 %, infiltrated and embedded in Araldite CY 212 (TAAB, UK). Ultrathin sections (60-80 nm) were cut on an ultramicrotome (ultracut E, Reichert Jung). The sections were mounted on meshed copper grids and double-stained with aqueous uranyl acetate (10 min) and lead citrate (10 min) respectively (Reynolds, 1963). The sections were examined on a Morgagni 268 D Transmission Electron Microscope (Fei Electron Optics, The Netherlands) at accelerating voltage of 80 KV, in All India Institute of Medical Sciences, AIIMS, New Delhi. The blood smear was examined for differential count of WBC after staining with freshly prepared

Leishman stain. Blood cells were observed under oil immersion magnification in a research microscope. The differential count of leucocytes was done by counting 150 cells of uniform blood smears avoiding extreme edges of the slides. The percentage of leucocytes was calculated from such observations.

RESULTS

A. Granulocytes (Figure 1)

Neutrophils

The neutrophils showed an eccentric, bi-lobed nucleus and were common in the circulating blood (Figure 2).

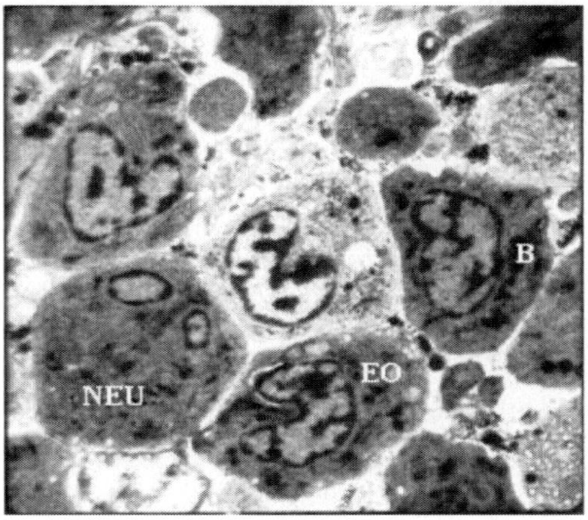

Figure 1. Transmission electron micrograph of granulocytes of *H. fossilis*. NEU-Neutrophil, EO-Eosinophil and B-Basophil (x 9, 600).

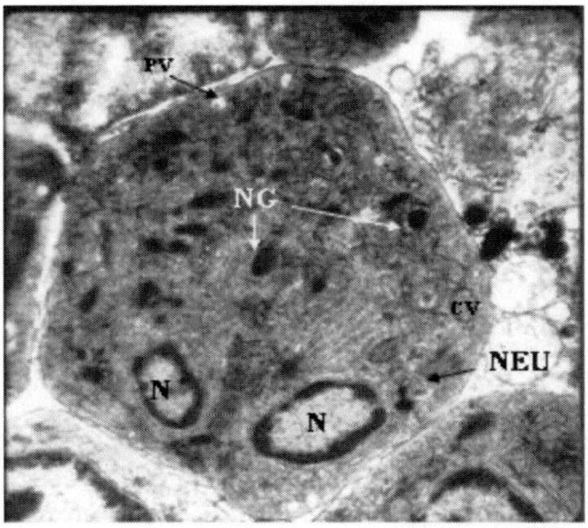

Figure 2. Transmission electron micrograph of neutrophil (NEU) of *H. fossilis*. N-Nucleus, NG-Neutrophilic granules, CV-Cytoplasmic vesicles and PV-Phagocytic vesicles (x 14, 400).

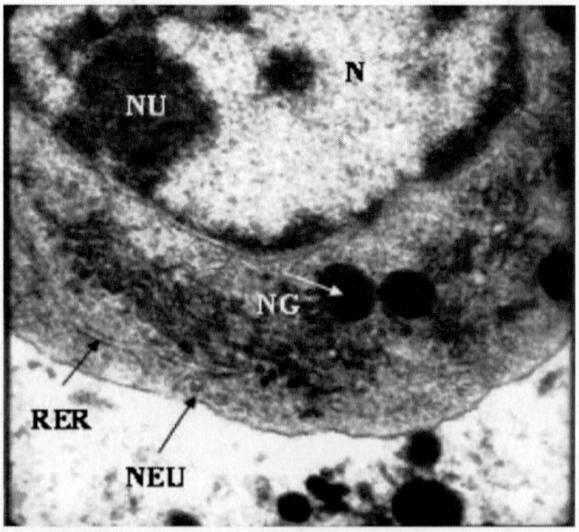

Figure 3. Neutrophil (NEU) in higher magnification showing abundant granules (NG) and RER in the cytoplasm. N- Nucleus and NU-Nucleolus (x 29, 100).

The cytoplasm also contained small vesicles (CV), Golgi apparatus, mitochondria. The RER was located near the nucleus or at the periphery of the cytoplasm. The nuclei of neutrophils showed a denser peripheral margin of chromatin. A few phagocytic vesicles (PV) were found at the periphery of the cytoplasm.

Eosinophils

These were polygonal cells with an irregular outline having a central, irregular-to-bilobed nucleus, the main characteristic being the abundant, round-to-elongated eosinophilic granules (Figure 4). The nucleolus was very rare. The eosinophilic granules (EG) were membrane-bound, uniformly distributed in the cytoplasm, and had a dark content surrounded by a light zones which were not separated by a membrane. A considerable number of Golgi apparatus, mitochondria and RER were also found in the cytoplasm (Figure 5).

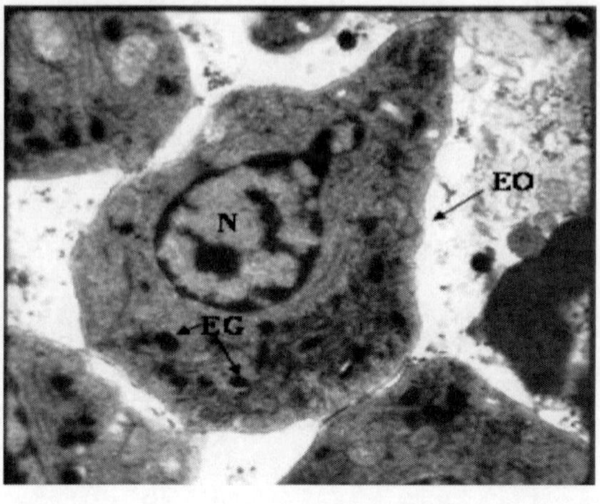

Figure 4. Transmission electron micrograph of eosinophil (EO) of *H. fossilis*. N-Nucleus and EG-Eosinophilic granules (x 11, 400).

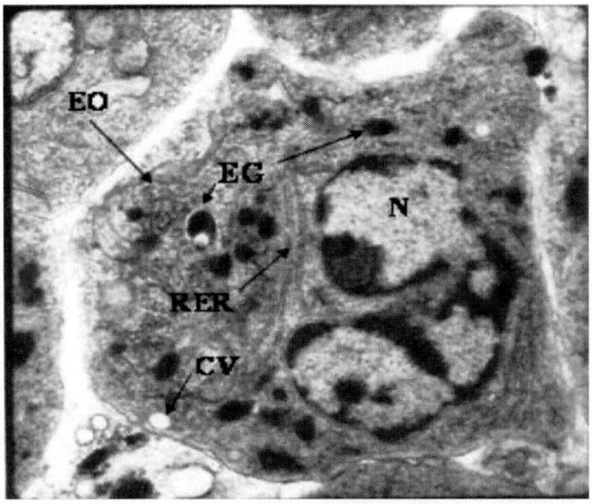

Figure 5. Eosinophil (EO) showing characteristic bi-lobed nucleus (N), eosinophilic granules (EG), few cytoplasmic vesicles (CV) and abundant RER in the cytoplasm (x 14, 400).

Cytoplasmic vesicles (CV) and filopodia were well developed at the periphery of the cytoplasm.

Basophils

These cells were similar to the eosinophils, but they possessed somewhat larger, round, membrane-bound granules (BG) which showed varying proportions of dark and light zones of different shapes and had an internal membrane separating the dark from the light zones (Figure 6). Golgi bodies (G) were abundant and few big vacuoles (V) were found near the nucleus.

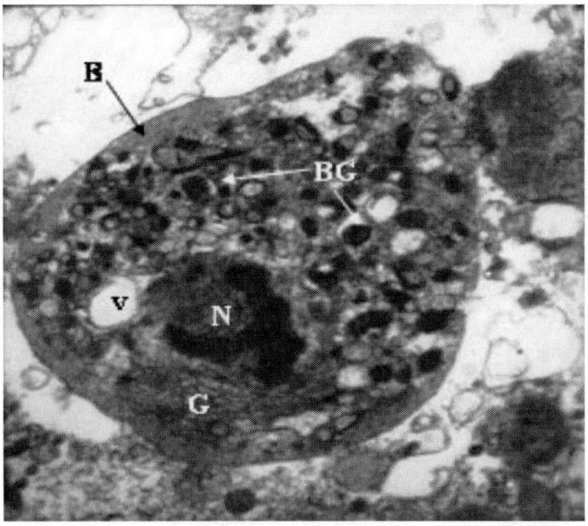

Figure 6. Transmission electron micrograph of basophil (B) of *H. fossilis*. BG-Basophilic granules, N-Nucleus, G-Golgi bodies, V-Vacuoles (x 11, 400).

B. Agranulocytes

Lymphocytes

Lymphocytes (Figure 7), the smallest of leucocytes types were characterized by a large, approximately circular, sometimes indented nucleus, surrounded by a narrow cytoplasmic margin (Figure 8). A dense peripheral margin of heterochromatin was observed in the dark nucleus (Figure 9). There may be a large nucleolus. The plasma membrane was folded in places and in some cases showed microvilli (MV). The narrow cytoplasm contained few mitochondria and numerous scattered ribosomes. Golgi apparatus was poorly developed. Some intracytoplasmic vacuoles were found in the periphery of the cytoplasm.

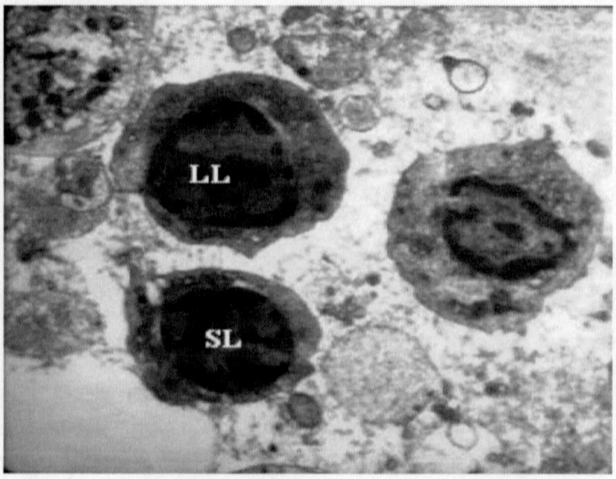

Figure 7. Transmission electron micrograph of lymphocytes among agranulocytes from head kidney of *H. fossilis*. LL-Large lymphocyte and SL- Small lymphocyte (x 10, 260).

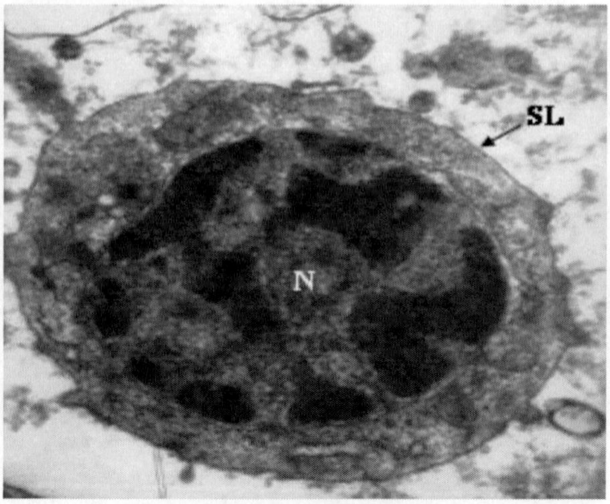

Figure 8. Small lymphocyte (SL) showing large round central nucleus (N) covering most of the cell area (x 19, 760).

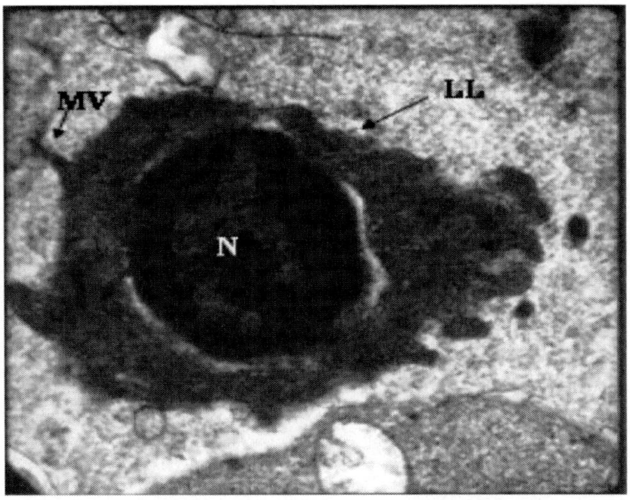

Figure 9. Large lymphocyte (LL) showing round but somewhat indented nucleus (N) and microvilli (MV) in the cell margin (x 8, 100).

Monocytes and Macrophages

Monocytes were the largest cells in the circulating blood and their shape was round to oval (Figure 10). Macrophages may be highly variable in size and shape (Figure 11). The monocytes and macrophages of *H. fossilis* were the main phagocytic leucocytes which contained a largely euchromatic nucleus, without lobes, generally in a rather eccentric position. The nucleus of the macrophage was rimmed by patchy heterochromatin which also appeared clumped throughout, it may contain a nucleolus.

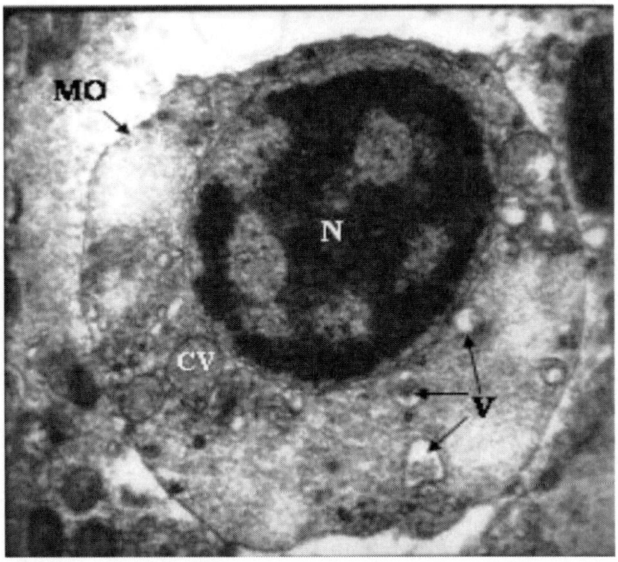

Figure 10. Transmission electron micrograph of monocyte of *H. fossilis* showing eccentric nucleus (N), many vacuoles (V) and cytoplasmic vesicles (CV). (x 23, 520).

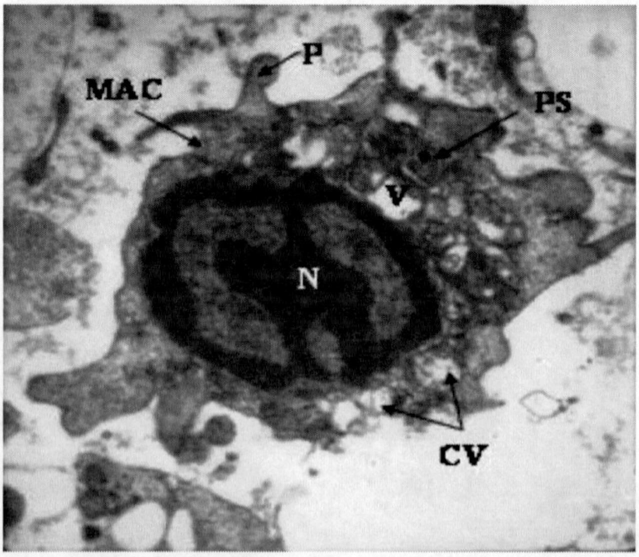

Figure 11. Transmission electron micrograph of macrophage (MAC) of *H. fossilis* showing large central nucleus (N) and characteristic numerous pseudopodia (P) at the cell surface, many cytoplasmic vesicles (CV), vacuoles (V) and phagosomes (PS). (x 14, 400).

A large cytoplasm was a prominent characteristic of these cells which was rich in free ribosomes, lysosomes, RER, mitochondria, intracytoplasmic vacuoles (V) and contained many clear cytoplasmic vesicles (CV) of variable size. Occasional small, round-to-oval, membrane-bound granules were seen in the cytoplasm. The filopodia were developed at the cell surface of monocytes whereas pseudopodia (P) were well developed in macrophages. Phagosomes (PS) were found in the cytoplasm of macrophages.

Neutrophils, eosinophils, basophils, large lymphocytes, small lymphocytes and monocytes in the peripheral circulation were 38-42, 4-6, 0.5-1, 17-21, 25-29 and 9-11 %, respectively as obtained from the differential count of leucocytes.

DISCUSSION

Each type of leucocytes analogous to mammals is evident in the head kidney and circulation of *H. fossilis* although the structure and distribution varying widely from the mammalian system. It is apparent from the observations that the fish neutrophil, eosinophil and basophil structures are similar to the respective myelocytes of mammals. Electron micrographs of neutrophils showed prominent round-to-oval cytoplasmic granules. Ferguson (1976) studied the ultrastructure of the plaice granulocytes and described the granules as having a fibrillar structure with bands or striations. Cannon et al. (1980) observed that the ultrastructure of the granules of channel catfish neutrophil were similar to granules in developing netrophilic myelocytes of mammals. Praveen et al. (2000) confirmed the presence of myeloperoxidase (MPO) in azurophilic granules of fish neutrophils. Interestingly, the fish neutrophil and eosinophil possess fewer lobations (not more than 2 lobes) in their nuclei than in mammals which are equipped with more lobes in the nuclei rendering the cells to be more efficient than the fish blood cells. This fact is in agreement with the observation of Fujimaki

(1986) that polymorphonucleus was extremely rare in the neutrophils of gold fish, *Carassius auratus*. Therefore, the ultrastructural study of fish leucocytes provides an insight into the probable evolutionary significance with regard to functional efficiency of the vertebrate immunopotent cells. The monocyte is a partly differentiated cell with moderate phagocytic properties and is generally found in the blood. Under appropriate conditions it leaves the vascular system and completes its differentiation becoming a mature macrophage within the tissues. Monocytes have been reported in various investigations on fish leucocytes. They have been reported in plaice (Ellis, 1976; Ferguson, 1976) and in *Scyliorhinus cancicula* (Morrow and Pulsford, 1980). The macrophage populations were frequently seen in *H. fossilis* but compared to other agranulocytes, their distribution is much less in circulation. Non-specific defences involving tissue macrophages and circulating neutrophils are extremely important in immune protection in fish (Anderson and Siwicki, 1994). If however, invading pathogens survive the non-specific barrier, specific defence systems involving lymphocytes respond to the challenge. The lymphocytes took the same general form as those in other fish species and mammalian lymphocytes. From differential counts, neutrophils and lymphocytes were the commonest cell types. The other leucocytes are highly variable from one species to another. Morrow and Pulsford (1980) reported that monocytes represented 13% of the leucocytes in *S. cancicula* which was highly comparable with the present study.

CONCLUSION

In this study, the ultrastructural details of different leucocytes of *H. fossilis* have been described. This study will aid in ultrastructural pathology of this fish whose population decline can be attributed to proneness to infections in their shallow natural habitat. Thus this study has direct implication for conservation of this important food fish in their natural environment.

ACKNOWLEDGMENTS

The authors acknowledge University Grant Commission (UGC), New Delhi, India for financial assistance and the Head, Department of Zoology, University of Calcutta for providing infrastructural facilities. Special thanks are due to Dr. T. C. Nag, Electron Microscopy Unit, AIIMS, New Delhi for technical help.

REFERENCES

Anderson, D.P. and Siwicki, A.K. (1994). Duration of protection against *Aeromonas salmonicida* in brook trout immunostimulated with glucan or chitosan by injection or immersion. *Prog. Fish Cult.*, 6: 258-261.

Blaxhall, P.C. (1983). Electron microscopic studies of fish lymphocytes and thrombocytes. *J. Fish Biol.*, 22: 223-229.

Cannon, M.S., Mollenhauer, H.H., Cannon, A.M., Eurell, T.E. and Lewis, D.H. (1980). Ultrastructural localization of peroxidase activity in neutrophil leucocytes of *Ictalurus punctatus*. *Can. J. Zool.*, 58: 1139-1143.

Cenini, P. (1984). The ultrastructure of leukocytes in carp (*Cyprinus carpio*). *J. Zool. Lond.*, 204: 509-520.

Ellis, A.E. (1976). Leukocytes and related cells in the plaice *Pleuronectes platessa*. *J. Fish Biol.*, 8: 143-156.

Ferguson, H.W. (1976). The ultrastructure of plaice (*Pleuronectes platessa*) leucocytes. *J. Fish Biol.*, 8: 139-142.

Fujimaki, Y. (1986). Classification of goldfish leukocytes observed in the circulating blood of normal goldfish, *Carassius auratus*. *Bull. Nippon Vet. Zootech. Coll.*, 35: 158-165.

Ghosh, R. and Homechaudhuri, S. (2007). Micromorphometry, Cell distribution and Ultrastructure of some Immunopotent Leucocytes from Pronephric Kidney of *Channa punctatus* (Bloch). *Asian Fish. Sci.*, 20: 95-107.

Hine, P.M. (1992). The granulocytes of fish. *Fish and Shellfish Immunol.*, 2: 79-98.

Hine, P.M. and Wain, J.M. (1987). Composition and ultrastructure of elasmobranch granulocytes. I. Dogfishes (Squaliformes). *J. Fish Biol.*, 30: 547-556.

Karnovsky, M.J. (1965). A formaldehyde-glutaraldehyde fixative of high osmolality for use in electron microscopy. *J. Cell Biol.*, 27: 137A.

Lester, R.J.G. and Desser, S.S. (1975). Ultrastructural observations on the granulocytic leukocytes of the teleost, *Catostomus commersoni*. *Can. J. Zool.*, 53: 1648-1657.

Modra, H., Svobodova, Z. and Kolarova, J. (1998). Comparison of differential leukocyte counts in Fish of Economic and Indicator importance. *Acta Vet. Brno*, 67: 215-226.

Morrow, W.J.W. and Pulsford, A. (1980). Identification of peripheral blood leucocytes of the dogfish (*Scyliorhinus cancicula* L.) by electron microscopy. *J. Fish Biol.*, 17: 461-475.

Neumann, N.F., Barreda, D.R., Stafford, J.L., Ainsworth, A.J. and Belosevic, M. (2001). Anti microbial mechanisms of Fish phagocytes and their role in host defense. *Dev.Comp. Immunol.*, 25: 807-825.

Praveen, K.N., Namseer, W. and Goodwin, A. (2000). Evidence for a Novel Vertebrate peroxidase in channel cat-fish (*Ictalurus punctatus*). *Proc. 25th Annual Eastern Fish Health Workshop*, March 10-13.

Reynolds, E.S. (1963). The use of lead citrate at high pH as an electron opaque stain in electron microscopy. *Stain Technol.*, 35: 313-323.

Rowley, A.F., Hunt, T.C., Page, M. and Mainwaring, G. (1988). Fish. In: *Vertebrate Blood Cells* (A.F. Rowley and N.A. Ratcliffe, eds), pp. 257-334. Cambridge: Cambridge University Press.

In: Vertebrate and Invertebrate Animal Approaches
Editor: José Rosa Gomes
ISBN: 978-1-63117-102-4

Chapter 23

ARE SEX AND REPRODUCTIVE PHASE-SPECIFIC CHANGES IN CATALASE AND PROTEIN PROFILE OF A FRESHWATER CATFISH SEASON-DEPENDENT?

G. Tripathi* and N. Bandooni
Department of Zoology, J.N.V. University, Jodhpur, India

ABSTRACT

The sex specific seasonal changes in profiles of catalase (CAT) and total protein were studied to analyze the role of antioxidative defenses and protein in annual reproductive cycle of the freshwater fish, *Clarias batrachus*. The CAT activity and specific activity in liver, heart, kidney and intestine were maximum in prespawning and minimum in resting phase showing a sharp decline in oxidative defenses after spawning. Subsequently an increase was found in the activity and specific activity of CAT in different tissues of both sexes. The changes in catalase profile were positively correlated with the ambient water temperatures showing an inducing effect on oxidative defense. Protein content showed an increase during preparatory and resting phase which indicated an enhancement in protein synthesizing capacity. In contrast to CAT, the protein content declined in prespawning and spawning, which might be due to its greater utilization to meet the demand of reproductive activities. Protein content was higher in liver, heart, kidney and intestine of female than male which showed a greater biosynthetic requirement for reproductive preparation in female. Protein content in liver, heart, kidney and intestine of both male and female showed negative correlation with ambient water temperature showing inhibitory effect of temperature during seasonal changes. It might be due to greater utilization of protein during resting phase to enter into the next reproductive cycle. The present findings suggest temperature associated sex-specific variation in antioxidative defense and reproductive phase-specific changes in protein content. The degree of alterations in both parameters (catalase and protein) was sex-specific. This information may be useful for a strategic planning in fish breeding and aquaculture practices.

* E-mail: drgst@rediffmail.com; Tel. +9102912722759.

Keywords: Catalase, seasonal, freshwater fish

INTRODUCTION

Seasonal fluctuations in profile of few enzymes of fish have been documented (Marquez, 1976; Mc Corkle et al., 1979; Shehepkin and Emeretli, 1986; Singh and Singh, 1988; Sole et al., 2006). The effects of various seasons on the rates of oxidative phosphorylation in the liver and skeletal muscle of the *Tinca tinca* have been demonstrated (Pye et al., 1976). Seasonal variation in the activity of cytoplasmic malate dehydrogenase (cMDH) has been shown in the fish, *Astyanax fasciatus* (De Luca et al., 1983). Similarly, Seasonal changes in the rate of protein synthesis and the quality of different proteins in fish have been worked out (Krauskopf et al., 1981; Fauconneau 1985). Season-dependent changes occur in expression of B-isoform of liver cMDH of a catfish (Tripathi and Shukla, 1991). During seasonal acclimatization of the rainbow trout, quantitative and qualitative changes in mitochondria occur for metabolic compensation (St-Pierre et al., 1998). Cold-induced increase in mitochondrial density or aerobic capacity and its reversal during seasonal warming does not ignore the importance of integrated seasonal modifications in lipid saturation, kinetic properties of metabolic enzymes, contractile proteins and transmembrance transporters (Hazel, 1995; Storelli et al., 1998; Pörtner et al., 1998, 2000). Seasonal variations in the activities of superoxide dismutase, catalase, glutathione peroxidase, glutathione reductase, glucose-6-phosphate dehydrogenase, glutathione S-transferase and lipid peroxidation have been reported in *Salmo trutta caspius, Salmo trutta labrax* and *Salmo trutta macrostigma* (Aras et al., 2009).

Although the seasonal interference with the antioxidant defense system has not been well established in fish, the data available in the literature suggest a general antioxidant defense enhancement during spring and summer as compared to low temperature periods (Härdig and Höglund 1983; Gabryelak et al., 1983; Palace and Klaverkamp 1993). The research in fish revealed that environmental factors like water, temperature, salinity, season and feeding habits exert changes in peroxisomal enzyme activities that vary greatly among species (Fahimi; and Cajarville, 1995). The walking catfish, *Clarias batrachus* inhabits muddy or swampy water of high turbidity (Courtenay et al., 1974; Hensley and Courtenay 1980). This is an economically important and native species of tropical southeastern Asia. Since enzymes and other proteins participate in the intricately balanced subcellular functions (Sakeri et al., 1968), the seasonal changes in catalase and protein profile can be measured to monitor variation in nutritional value of *C. batrachus.* The antioxidant enzyme catalase catalyzes the transformation of hydrogen peroxide into water and oxygen. This enzyme has been reported to increase under oxygen stress conditions, and it is also mediated by the biotransformation of organic chemicals (Livingstone, 2001).

The climate change will affect the seasonality and make the conditions worse, placing fish at greater risk, due to temperature rise, pollution and disturbance in metabolism. This will hamper fish growth and reproduction severely leading to food crisis. Hence, there is an urgent need to understand the impact of season on metabolism of fish so that adequate measures may be taken in prior to meet the challenge. Though attempts have been made to analyze season-dependent changes in metabolism of fish, no unifying concept could be developed due to species-and place-specific variation and complexity of metabolic system. Therefore, it was decided to study the influence of seasonal climatic variation on the profile of catalase and protein in the liver, kidney, heart and intestine of the freshwater catfish *C. batrachus*. Season related sex-specific variations were also analysed to reflect differences in antioxidative defense and protein synthesizing capacity of the fish.

MATERIALS AND METHODS

Fish and Chemicals

The freshwater catfish *C. batrachus* is a seasonal breeder with five phases in the annual reproductive cycle (Singh and Singh, 1981). These phases are preparatory (March-April), prespawning (May-June), spawning (July-August), post spawning (September-October) and resting (November-February). Adult and healthy specimens of *C. batrachus*, weighing 75-85g, having body length of 18-19 cm, were collected from the fish market between the 10th and 15th day of each month. They were acclimatized under laboratory conditions at least one week prior to experimentation. Water temperature was monitored which varied according to the temperature of surrounding environment (Figure 1). The study was done with respect to sex from January to December 2008. Catalase activity, specific activity and protein content in liver, heart, kidney and intestine of the fish were estimated in each month. All chemicals were of analytical grade and procured from Sisco Research Laboratory, Mumbai. The glass distilled water was used for reagent preparation.

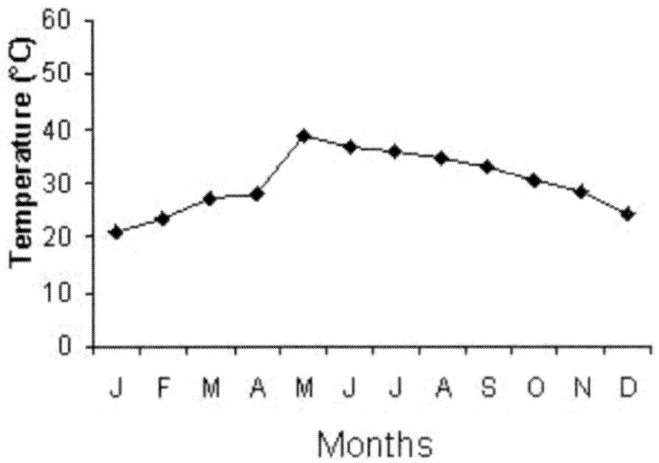

Figure 1. Seasonal variation in water temperature of fish holding tanks.

Enzyme Preparation

The specimens of *C. batrachus* were sacrificed by decapitation. The liver heart, kidney and intestine were removed and washed in 0.6% saline. They were properly cleaned and weighed rapidly. A 10% homogenate (w/v) was prepared in sodium phosphate buffer (0.1 M, pH 7) using Potter-Elvehjem homogenizer fitted with a Teflon pestle (Remi, RB 1278).

Homogenization was done at about 4^0C. Homogenate was first centrifused at 700g to remove cell debris. Then the supernatant was centrifuged at 12000g for 15 min in a high speed refrigerated centrifuge (Remi Cooling Compufuge, CPR 24) and the resulting supernatant was taken as the cytoplasmic fraction for assay of catalase.

Assay Method

The procedure adopted for the assay of catalase (CAT; EC 1.11.1.6) was that of Aebi (1984). Assay was done at temperatures which fluctuated with water temperatures. Catalase was assayed in a medium containing sodium phosphate buffer (0.1M, pH 7). Optimum concentration of substrate was used and the total volume of reaction mixture was 3ml. The decrease in absorbance was recorded at 240 nm using Cintra5 UV-VIS Spectrophotometer (GBC, Germany). Activity has been given in mmol/min/g wet mass of tissue. Specific activity was expressed as mmol/min/mg protein.

Protein Estimation

The protein content was determined by Folin-Ciocalteau method of Lowry et al. (1951). Standard plot was used for estimation of protein content in cytoplasmic fraction of tissue homogenate.

Statistical Analysis

Data on seasonal changes in enzyme and protein profiles were analyzed using SPSS statistical package. Since observations were taken in each month in different organs of both the sexes, the data were subjected to repeated measure analysis of variance (ANOVA). Months were tests within subject effects, whereas organs and sex were tests between subject effects. Pearson correlation coefficient was calculated to know the relationship of catalase activity, specific activity and protein content with the ambient water temperature. The level of significance was set at 0.05.

RESULTS

Catalase

Activity: The activity of catalase (CAT) in liver, kidney and intestine of *C. batrachus* increased gradually from January to May and in heart from January to June in both sexes. Thereafter, the enzyme activity in these tissues declined sharply in July and August and the decrease was continued from September to December. The maximum CAT activity was obtained in May / June and minimum activity was found in December / January in both males and females. Liver of male showed maximum CAT activity, while heart of female exhibited minimum activity. The CAT activities were about 2.74, 1.31, and 1.61 fold greater in liver, kidney and intestine respectively as compared to the activity in heart during regressed phase (Nov-Feb). Higher activity was observed in all tissues of males as compared to females.

The repeated measure ANOVA indicated significant (P<0.001) variation in CAT activity with respect to months. Month x sex interaction was non-significant. However, month x organ and month x sex x organ interactions were highly significant (P<0.001). Catalase activity varied significantly between sex (P<0.001). The variation in activity was also significant with respect to organ and sex x organ interaction. The catalase activity in different organs of both the sexes showed a significant positive correlation with ambient water temperature.

Specific activity: The specific activity of catalase in liver, kidney and intestine increased gradually from January to May and in heart from January to June in males and females both. The specific activity of catalase in these tissues declined sharply in July and August. Thereafter, CAT specific activity decreased slightly from September to December. The maximum CAT specific activity was obtained in May or June and minimum specific activity was found in December or January in both sexes. Liver of male showed maximum CAT specific activity, while heart of female indicated minimum specific activity.

The CAT specific activities were about 1.95, 1.53 and 1.61 fold greater in liver, kidney and intestine respectively as compared to the specific activity in heart during regressed phase (Nov – Feb). Catalase specific activity was always higher in all the tissues of males than the females. The repeated measure ANOVA indicated significant (P<0.001) variation in CAT specific activity with respect to months. Month x sex interaction was non-significant. Month x organ interaction was highly significant (P<0.001). Month x sex x organ interaction was insignificant. However, CAT specific activity was significant with respect to sex (P<0.001). Organ and sex x organ interactions were also significant (P<0.001). The specific activity of catalase showed a significant positive correction with water temperature in all tissues of both sexes.

Protein

Protein content of liver, heart, kidney and intestine showed slight decrease from January to February in both sexes. Thereafter, it increased upto April in these tissues of male. In female, liver indicated an increase in protein profile upto April. However, heart, kidney and intestine showed slight decrease in protein content in March followed by an increase in April. Subsequently in May and June a decline in protein content was found. But a sharp decline was observed in July and August which followed a slight increase upto December in all tissues of both sexes.

The protein content was maximum in April in liver, heart, kidney and intestine and minimum in the month of July. The protein contents were about 1.67, 1.17 and 1.15 fold greater in liver, heart and intestine respectively as compared to the protein content of kidney in resting phase (Nov-Feb). The repeated measure ANOVA indicated significant (P<0.001) variation in protein content with respect to month and month x sex interaction. Month x organ and month x sex x organ interactions were non-significant. Protein content varied significantly with respect to sex (P<0.001). Organ and sex x organ interactions were also significant (P<0.001). The protein content showed a significant negative correction with water temperature in all tissues of both sexes.

Discussion

The activity and specific activity of catalase in liver, heart, kidney and intestine of *C. batrachus* varied significantly with respect to month, sex and organ (Figures 2, 3). Likewise sex x organ interaction was also significant. The activity and specific activity of catalase in both males and females showed a distinct seasonal variation. The catalase profile also different between the sex throughout the year. This supports the report of Ronisz et al. (1999) who documented a much clear seasonal variation in catalase activity of females as compared to male eelpout. The CAT activity and specific activity were higher in different tissues (liver, heart, kidney, intestine) of male as compared to female *C. batrachus.*

The trend of seasonal changes in activity and specific activity of catalase was almost same. The increase in activity and specific activity of catalase of liver, heart, kidney and intestine during preparatory (March-April) and pre-spawning (May-June) indicated an enhancement in defensive efficiency to protect from the oxidative toxicity. Perhaps protection from oxidative damages was a pre-requisite for preparation of the fish to enter into the next reproductive cycle. The present observations on seasonal changes in activity and specific activity are in agreement to some extent with the report on catalase showing peak activity during autumn in *Salmo trutta caspius, Salmo trutta labrax* and *Salmo trutta macrostigma* (Aras et al., 2009). Filho et al. (2001) also described similar seasonal changes in liver of *Geophagus brasiliensis.*

Other studies using immature fish erythrocytes (Gabryelack et al., 1983) and adult eelpout liver (Ronisz et al., 1999) demonstrated higher activity of catalase in spring than in autumn. The decrease in CAT activity and specific activity during spawning (July-August) and post spawning (September-October) suggested lower oxidative stress during these periods. This showed the possibility of reduced oxidative damages in spawning related processes. In contrast to the present observations, a small and random seasonal fluctuation was found in catalase activity which declined in July in liver of *Pomatoschistus minutus* (Sole et al., 2006). The CAT activity and specific activity decreased during resting phase (November-February) in *C. batrachus.* It is in again contrast to the report of Ronisz et al. (1999) who showed a small variation in the activity of catalase in liver of *Zoarces viviparus.* Sole et al. (2006) suggested that the random fluctuations in catalase activity were not season-dependent.

Liver of male and female *C. batrachus* showed highest CAT activity and specific activity as compared to other tissues revealing maximum defensive capacity of hepatic cells from oxidative toxicity. In contrast, heart of both sexes showed minimum CAT activity reflecting least protective capacity from oxidative damages.

Catalase activity and specific activity in female and male both increased with the increase in ambient water temperature throughout the year. Livingstone (2001) also showed similar observations between increased reactive oxygen species (ROS) formation and temperature as a consequence of increased metabolism.

Seasonal changes in metabolic activities may not be necessarily related to the trend of seasonal fluctuations in surrounding temperatures, rather it is dependent on a number of factors including food availability reproductive status and profile of metabolic hormones (Reshkin and Ahearn, 1987; Collie and Ferraris, 1995; Mommsen et al., 1999; Tripathi and Verma, 2004). Whereas an enhancement in antioxidant activity coinciding with high water temperatures have been reported for other fishes (Härding and Höglund, 1983; Gabryelak et al., 1983; Palace and Klaverkamp, 1993).

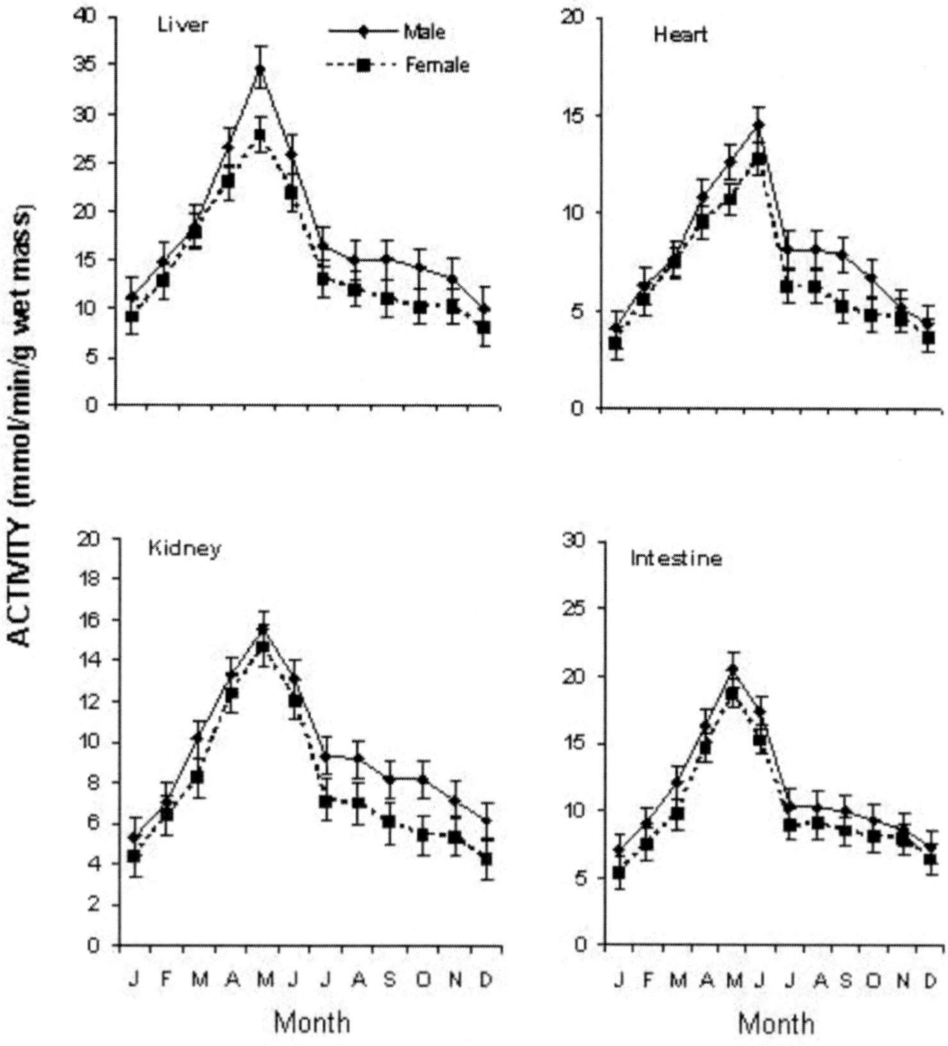

Figure 2. Seasonal changes in activity(mmol/min/g wet mass) of catalase from liver, heart, kidney and intestine of the freshwater catfish. C. *batrachus*. Each datum represents the mean ±SEM of six individuals (n=6). Results of repeated measure ANOVA: Month (M), P<0.001; Sex (S), P<0.001; Organ (O), P<0.001; Mxs, P>0.05 (NS), Mxo, P<0.001; Sxo, P<0.001; Mxsxo, P<0.05 (NS), NS, Non-significant.

Protein content of liver, heart, kidney and intestine of *C. batrachus* varied significantly with respect to month, sex and organ (Figure 4). Similarly, month x sex and sex x organ interactions were also significant. The increase in protein content of tissues in both sexes during preparatory as compared to resting phase indicated an enhancement in protein synthesizing capacity of the fish for reproductive preparation. Protein after being synthesized is usually deposited in cytoplasm. This is rapidly used to replace the loss of protein that occurs during physiological stress.

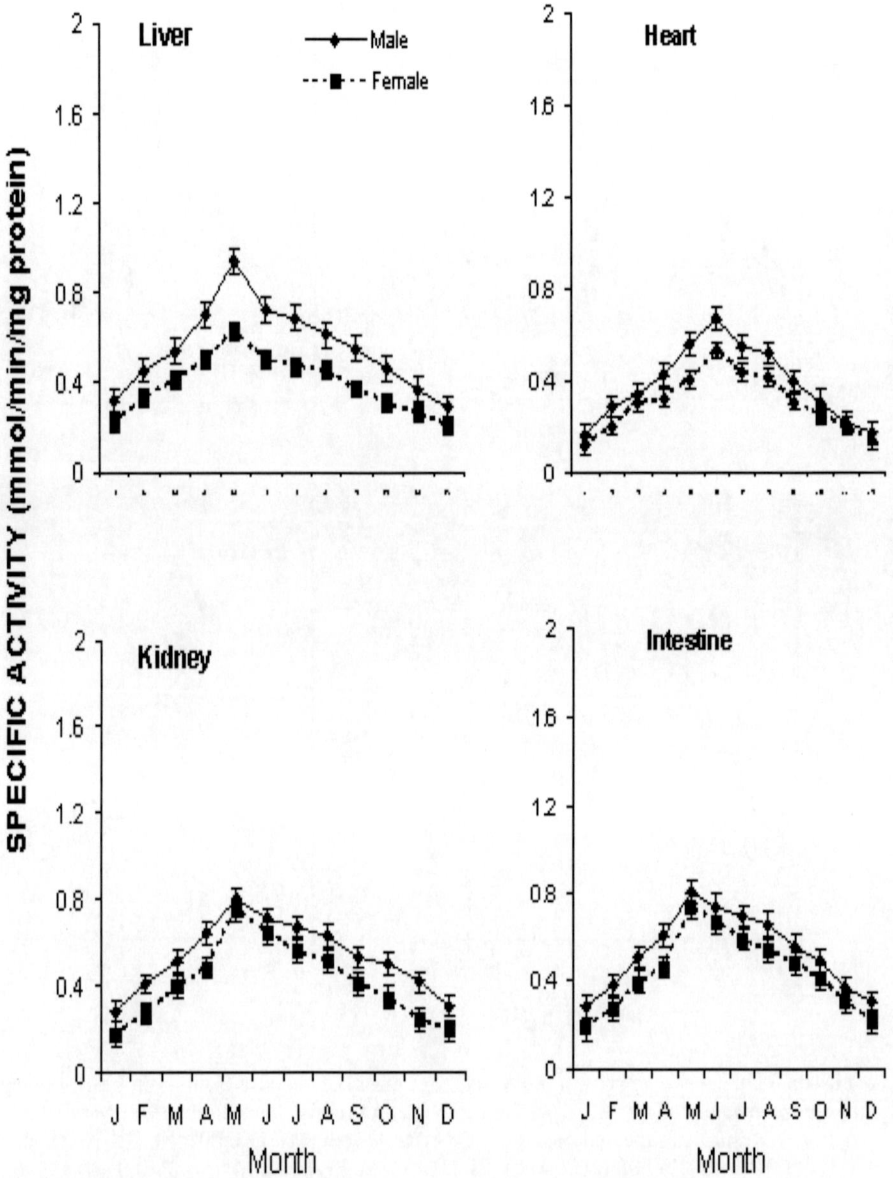

Figure 3. Seasonal changes in specific activity(mmol/min/protein) of catalase from liver, heart, kidney and intestine of the freshwater catfish. C. *batrachus*. Each datum represents the mean ±SEM of six individuals (n=6). Results of repeated measure ANOVA: Month (M), P<0.001; Sex (S), P<0.001; Organ (O), P<0.001; Mxs, P>0.05 (NS), Mxo, P<0.001; Sxo, P<0.001; Mxsxo, P>0.05 (NS), NS, Non-significant.

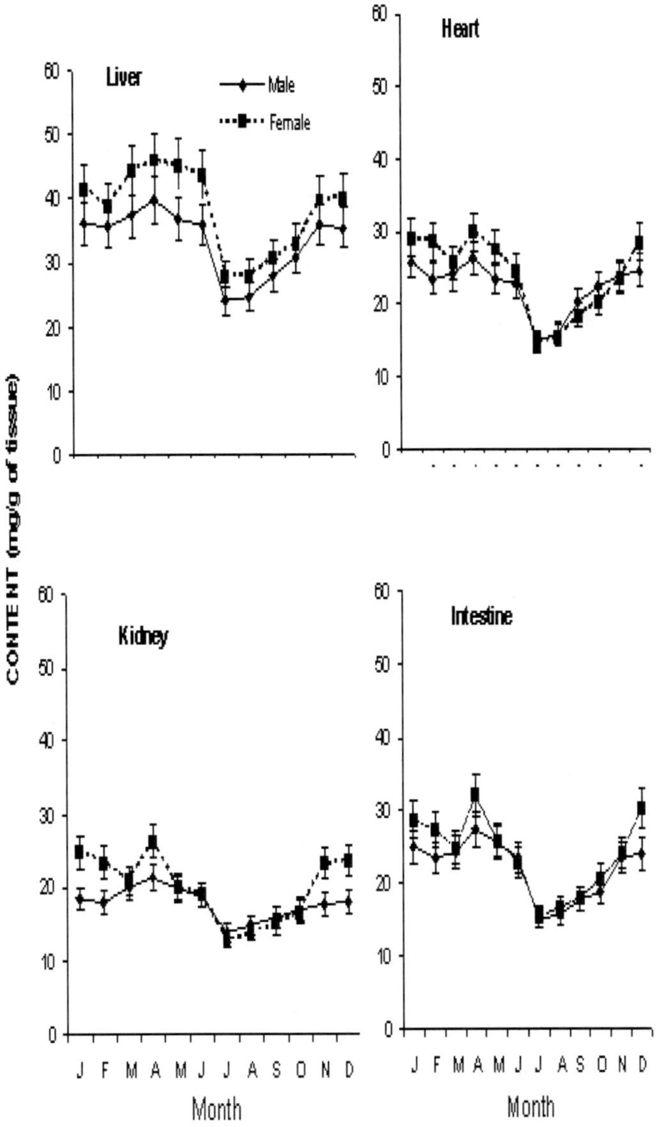

Figure 4. Seasonal changes in protein (mg/g of tissues) from liver, heart, kidney and intestine of the freshwater catfish. C. *batrachus*. Each datum represents the mean ±SEM of six individuals (n=6). Results of repeated measure ANOVA: Month (M), $P<0.001$; Sex (S), $P<0.001$; Organ (O), $P<0.001$; Mxs, $P>0.05$ (NS), Mxo, $P<0.001$; Sxo, $P<0.001$; Mxsxo, $P>0.05$ (NS), NS, Non-significant.

The reduction in total protein content of fish tissues in both sexes during pre-spawning and spawning might be due to greater utilization of protein for reproductive activities. The subsequent increase in protein content until the onset of resting phase showed reduced utilization of protein in non-breeding season. These observations support the seasonal dynamics of protein synthesis in other fish (Fauconneau, 1985; Mandal and Das, 1991). An increase in total protein synthesis in liver of *Pomatoschistus minutus* during late spring and early summer coincides with the reproductive period (Sole et al., 2006).

All tissues of female showed higher protein content as compared to male which reflected a greater biosynthetic need of proteins for reproductive preparation in female. This sex-specific difference may be assigned to the differences in synthesis of various precursor proteins and vitellogenin in liver of female fish (Singh and Singh, 1981). Protein content decreased with an increase in ambient water temperature in liver, heart, kidney and intestine of male and female. Temperature certainly plays an important role, but cannot be accepted as a single reason for seasonal changes in metabolic constituents. Perhaps seasonal influences need to be better understood in order to make further inferences regarding enzymatic and other biochemical changes in metabolism of freshwater fish. Thus the present studies provide baseline information for season-related sex-specific alterations in antioxidative defense and reproductive phase-specific changes in protein content of a freshwater catfish.

REFERENCES

Aebi, H. (1984). Catalase *in vitro*. *Methods in Enzymology*. Academic Press, New York. 105: 121-127.

Aras, N.M., Bayir, A., Sirkecioglu, A.N., Bayir, M., Aksakal, E. and Haliloglu, H.I. (2009). Seasonal changes in antioxidant defense system of liver and gills of *Salmo trutta caspius*, *Salmo trutta labrax* and *Salmo trutta macrostigma*. *J. Fish. Biol.* 74: 842-856.

Collie, N.L. and Ferraris, R.P. (1995). Nutrient fluxes and regulation in fish intestine. In: Hochachka, P.W., Mommsen, T.P. (Eds.), *Biochemistry and Molecular Biology of Fishes, Metabolic Biochemistry,* Vol. 4. Elsevier Science, Amsterdam, pp. 221-239.

Courtenay, W.R., Jr., Sahlman, H.F., Miley, W.W., II, and Herrema, D.J. (1974). Exotic fishes in fresh and brackish waters of Florida. *Biol. Conserv*. 6: 292-302.

De Luca, P.H., Schwantes, M.L.B., and Schwantes, A.R. (1983). Adaptative features of ectothermic enzymes-IV Studies on malate dehydrogenase of *Astyanax fasciatus* (Characidae) from Lobo reservoir (Sao Carlos, Sao paulo, Brasil). *Comp. Biochem. Physiol.* 74B: 315-324.

Fahimi, H.D., and Cajaraville, M.P. (1995). Induction of peroxisome proliferation by some environmental pollutants and chemicals in animal tissues. In: Cajaraville, M.P. (Ed.), *Cell Biology in Environmental Toxicology*, Basque Country Press Service, Bilbo, 221-255.

Fauconneau, B. (1985). Protein synthesis and protein deposition in fish. In: Cowey, C.B., Mackie, A.M., Bell, J.G. (Eds.), *Nutrition and Feeding in Fish*. Academic Press, London, pp. 17-46.

Gabryelak, T., Piatkowska, M. Leyko, W. and Peres, G. (1983). Seasonal variations in the activities of peroxide metabolism enzymes in erythrocytes of freshwater fish species. *Compa. Biochem. Physiol.* 75C: 383-385.

Härdig, J. and Höglund, L.B. (1983). Seasonal and ontogenetic effects on methaemoglobin and reduced gluathione content in the blood of reared baltic salmon. *Comp. Biochem. Physiol.* 75A: 27-34.

Hazel, J.R. (1995). Thermal adaptation in biological membranes: Is homeoviscous adaptation the explanation ? *Annu. Rev. Physiol.* 57: 19-42.

Hensley, D.A. and Courtenay, W.R. Jr. (1980). *Clarias batrachus* (Linnaeus) Walking catfish. pp 475 In: Lee, D.S., Gilbert, C.R., Hocutt, C.H., Jenkins, R.E., McAllister, D.E. and Stauffer, J.R. Jr. (Eds.) *Atlas of North American Freshwater Fishes. North Carolina*

Biological Survey Publication # 1980-12. North Carolina State Museum of Natural History, p. 854.

Krauskopf, M., Amthauer, R., Saez, L. and Zuvic, I. (1981). On the role of protein synthesis in the strategies of adaptation to environmental changes. In: Siddiqui, M.A.Q., Krauskopf, M. and Weissbach, H. (Eds.), *Molecular Approaches to Gene Expression and Protein Structure,* Academic Press, New York, pp. 197-214.

Livingstone, D.R. (2001). Contaminant – stimulated reactive oxygen species production and oxidative damage in aquatic organisms. *Mar. Pollut. Bull.* 42: 656-666.

Lowry, O.H., Rosebrough, N.J., Farr, A.L. and Randall, R.J. (1951). Protein measurement with Folin phenol reagent. *J. Biol. Chem.* 193: 265-275.

Mandal, G. and Das, A.B. (1991). Seasonal dynamics of protein synthesis in the tissues of *Channa punctatus* (Bloch), an Indian air-breathing teleost. *J. Aquac.* 1: 1-10.

Marquez, E.D. (1976). A comparison of oxalo-acetate transaminase, lactate dehydrogenase, α-hydroxy butyrate dehydrogenase and creatine phosphokinase in non-spawning, pre-spawning and spawning salmon. *Comp. Biochem. Physiol.* 54: 121-127.

McCorkle, E.M., Chambers, J.E. and Yarbrough, J.D. (1979). Seasonal effects on selected tissue enzymes in channel catfish, *Ictalurus punctatus. Comp. Biochem. Physiol.* 62B: 151-153.

Mommsen, T.P., Vijayan, M.M. and Moon, T.W. (1999). Cortisol in teleosts: dynamics, mechanisms of action and metabolic regualtion. *Rev. Fish Biol. Fish.* 9: 211-268.

Palace, U.P. and Klaverkamp, J.F. (1993). Variation of hepatic enzymes in three species of freshwater fish from precambrian shield lakes and the effect of cadmium exposure. *Comp. Biochem. Physiol.* 104C: S147-S154.

Pörtner, H.O., Hardewig, I., Sartoris, F.J. and Van Dijk, P.L.M. (1998). Energetic aspects of cold adaptation: critical temperatures in metabolic, ionic and acid-base regulation ? In: Protner, H.O. and Playle, R. (Eds.), *Cold Ocean Physiology.* Cambridge University Press, Cambridge, pp. 88-120.

Pörtner, H.O., Van Dijk, P.L.M. and Hardewig, I. and Sommer, A. (2000). Levels of metabolic cold adaptation: tradeoffs in eurythermal and stenothermal ectotherms. In: Davison, W. and Howard Williams, C. (Eds.), *Antarctic Ecosystems: Models for Wider Ecological Understanding.* Caxton Press, Christ Church, New Zealand, pp. 109-122.

Pye, V.I., Wieser, W. and Zech, M. (1976). The effect of season and experimental temperature on the rates of oxidation of liver and skeletal muscle mitochondria from the tench, *Tinca tinca. Comp. Biochem. Physiol.* 54B: 13-20.

Reshkin, S.J. and Ahearn, G.A. (1987). Basolateral glucose transport by intestine of teleost, *Oreochromis mossambicus. Am. J. Physiol.* 252: R579-R586.

Ronisz, D., Larsson, D.G.J. and Forlin, L. (1999). Seasonal variations in the activities of selected hepatic biotransformation and antioxidant enzymes in eelpout (*Zoarces viviparus*). *Comp. Biochem. Physiol.* 124 C: 271-279.

Sakeri, K.E., Sakeria, C.E. and Karlson, P. (1968). Protein synthesis in subcellular fractions of the browfly during different develpment stages. *J. Insect. Physiol.* 43: 425.

Shehepkin, V.Y. and Emeretli, I.V. (1986). Seasonal variation in lactate dehydrogenase and succinate dehydrogenase activity in tissue of fish with different ecological and physiological characteristic. *J. Evol. Biochem. Physiol. Engl. Tr.* 22: 335-358.

Singh, R.A. and Singh, S.N. (1988). Tissue distribution, effect of starvation and seasonal variation of arginase in the fresh water teleost, *Clarias batrachus* (L). *Biochem. Arch*. 4: 329-334.

Singh, S.P. and Singh, T.P. (1981). Relative concentration of prolactin, gonadotropin and thyrotropin in the pituitary gland and blood serum during different phases of annual reproductive cycle in the freshwater catfish, *Clarias batrachus. Endokrinologie*. 77: 39-50.

Sole, M., Kopecka, J. and Garcia de la Parra, L.M. (2006). Seasonal variations of selected biomarkers in sand gobies *Pomatoschistus minutus* from the Guadalquivir estuary, Southwest Spain. *Arch. Environ. Contam. Toxicol*. 50: 249-255.

St. Pierre, J., Charest, P.M. and Guderley, H. (1998). Relative contribution of quantitative and qualitative changes in mitochondria to metabolic compensation during sesonal acclimaisation of rainbow trout *Oncorhynchus mykiss. J. Exp. Biol*. 201: 2961-2970.

Storelli, C., Acierno, R. and Maffia, M. (1998). Membrane lipids and protein adaptations in Antarctic fish. In: Pörtner, H.O. and Playle, R. (Eds.), *Cold Ocean Physiology*. Cambridge University Press, Cambridge, pp. 166-189.

Tripathi, G. and Shukla, S.P. (1991). Seasonal effects on malate and lactate dehydrogenase of the freshwater catfish, *Clarias batrachus. Zool. Jb. Physiol.* 95: 1-11.

Tripathi, G. and Verma, P. (2004). Sex-specific metabolic changes in the annual reproductive cycle of a freshwater catfish. *Comp. Biochem. Physiol*. 137B: 101-106.

In: Vertebrate and Invertebrate Animal Approaches
Editor: José Rosa Gomes
ISBN: 978-1-63117-102-4

Chapter 24

FECUNDITY, SEX RATIO AND GONADO-SOMATIC INDEX OF *EPINEPHELUS CHLOROSTIGMA* AND *EPINEPHELUS AREOLATUS* SAMPLED FROM ARABIAN GULF

Zubair Ahmad**, *Hmoud Fares Alkahem and Amir A. A. Aldhahi
Department of Zoology, College of Science, King Saud University,
Riyadh, Kingdom of Saudi Arabia

ABSTRACT

Data registered for the absolute and relative fecundity in the present study revealed that *E. chlorostigma* is more fecund than *E. areolatus*. It was found that absolute fecundity has a direct relationship with the body length and weight. The relative fecundity for both the species does not delineate any particular pattern of variation. The sex ratio for both the species was heterogeneous showing the domination of male over female in certain months while females outnumber males in other months. The gradual increase in the value of gonado-somatic index from March to September (*E. areolatus*) and June to September (*E. chlorostima*) observed in the present study shows the maturation / development of ovum in the ovaries. The highest value of it for both the species was registered in the month of September. A sharp decline in October indicates the spawning period (releasing of eggs from the ovary) of the fish.

INTRODUCTION

Studies on reproduction in fishes are salutary to make a better understanding of the annual regeneration of the stocks. Fecundity with other reproductive parameters is of a great value in fishery production and formulation of management measures. It is most common measure for studying the reproductive potential in fishes because it is relatively easy measure

* E-mail: zahmed@ksu.edu.sa; zubair.ahmad.ah@gmail.com.

to accomplish, and is generally given by the total number of yoked eggs produced by a single female in one spawning season or in a year. The fecundity and development of eggs in the ovaries of fishes have been studied in the past by various fishery biologists. Some recent and important published work on this aspect are those of Bal and Rao (1984), Daoulas and Kattoulas, (1985), Hussein (1986), Pusey et al. (2001), Aday et al. (2002), Geertjes and Videler (2002), Heibo and Vollestad (2002), Machado et al. (2002), Ganias et al. (2004), Orlando et al. (2007), Azaza et al. (2008) and Dominguez-Petit et al. (2008). Spawning of chub, *Leuciscus cephalus* (L.) in an artificial lowland canal was investigated by Arlinghaus and Wolter (2003). Recently, Zhang et al. (2009) and Al-Akel et al. (2010) have studied maturity stages, age at first maturity and fecundity of *Ilisha elongata* from Ariake Sound, Japan and *Poicilia latipinna* from Saudi Arabia, respectively.

The majority of fishes are oviparous, and only a few like guppies (live bearer) are ovoviviparous. Parental cares are apparent in ovoviviparous fishes hence they produce very little eggs. On the other hand oviparous fishes are more fecund (produce more eggs) probably because of less survival chance due to devastating environmental conditions.

The gonads, particularly ovary, undergo regular seasonal changes in weight and such changes are indicative of spawning season (Qasim, 1973). The gonado-somatic index is also used for studying the spawning season following the seasonal changes in gonad's weight in relation to body weight (Bal and Rao, 1984).

Other parameter studied in the present study like sex ratio indicates the proportion of males with females in the population which expected to be 1:1 in nature. This type of studies may indicate segregation or aggregation of sexes according to feeding, breeding and migratory behaviour of the fish. Some of the contributors in this field are Bhatt (1972), Thompson and Munro (1978), Coleman et al. (1996), Tamaru et al. (1996), Zabala et al. (1997) and Al-Akel et al. (2010).

In the present study an attempt was made to register the absolute and relative fecundity of E. *chlorostigma* and *E. areolatus* fished from Arabian Gulf. To find the sex ratio the number of male and female in the population were enumerated and to calculate the gonado-somatic index the weights of ovaries and fish of both species for a complete year beginning from January to December were registered.

MATERIALS AND METHODS

Fresh and healthy specimens of *E. chlorostigma* and *E. areolatus* of different sizes were collected from the local market on monthly basis for one complete year. These fishes were weighed for total weight nearest to 0.1g and measured for the total length nearest to 0.1cm. After morphometric measurements, they were dissected, sexed (male or female) and their gonad (ovaries) taken out and weighed, separately. The ovaries were preserved in Gilson solution for further studies. The gonado-somatic index was calculated according to the formula given below:

$$GSI = \frac{\text{Weight of gonads (g)}}{\text{Total weight of the fish (g)}} \times 100$$

GSI= Gonado-somatic index

The sex ratio (SR) was calculated from the formula given below:

$$SR\ (M:F) = \frac{\text{Number of Females}}{\text{Number of Males}}$$

Fecundity (number of eggs produced by a single female in one spawning season) was estimated from the ovaries having eggs with yolk. Three sub-samples from each ovary were taken and weighed separately. The eggs in these sub-samples were counted and a mean was calculated. Total number of eggs in the ovary was calculated from the weight and number of eggs in the sub-samples and total weight of ovary according to the method outlined by Pusey et al. (2001).

$$\text{Fecundity} = \frac{\text{Mean number of eggs in sub} - \text{sample}}{\text{Mean weight of sub} - \text{sample}} \times \text{total weight of ovary}$$

The relationship between the fecundity (F) and total weight (TW) (Log F= log a + b log TW) as well as total length (TL) (Log F = log a + b log TL) of the fish was established after transforming the values to logarithm.

RESULTS AND DISCUSSION

The absolute fecundity according to size and weight groups (Number of eggs produced by single female) and relative fecundity (number of eggs per unit length or weight) of the grouper, *Epinephelus chlorostigma* and *Epinephelus areolatus* are presented in table 1 and 2, respectively. Data embodied in these tables indicate that smaller fishes produce fewer eggs than the larger specimens. The absolute fecundity of *E. chlorostigma* varied from 235740 to 1847121 for the fish of total length 30.1 and 51.45 cm, respectively. Similarly, the fecundity of *E. areolatus* showed variations from 163815 for a fish of 25.9 cm total length to 421271 for the fish of 39.7 cm total length. Since the larger fish spend little or zero amount of energy on growth, hence their energetic investment in reproduction tends to be higher to produce more and larger eggs, whereas smaller fish spend more energy on growth hence produces less and smaller eggs (Moyle and Cech, 1982).

Ghorab et al. (1986) while working on the fecundity of *E. chlorostigma* from Red Sea reported that egg production varied from 42718 to 1242724 for the females ranging from 30.5 to 47.0 cm total length. The same species showed variations in the fecundity in the two different environments being more fecund in Arabian Gulf than the Red Sea. Mikhail (1980) presented the fecundity of *E. aeneus* from the Alexandrian water as 783988 for fish of 66.0 cm total length and 3873271 for the fish Of 93.5 cm which is greater than the one reported in the present study. Abu-Hakima et al. (1982) while studying the fecundity of different fish species from the Kuwaiti water reported that fecundity varies from 310000 to 1690000 (sobaity), 1360000 to 2150000 (sheim), 850000 to 2900000 (hamour), 50000 to 71400000 (nubaiby), 77000 to 254000 (zubaidy) and finally 63000 to 214000 (nagroor). Their findings indicate that the fecundity of these fishes is more than *E. chlorostigms* and *E. areolatus* registered in the present study.

Table 1. Absolute and relative fecundity (in relation to length) of groupers, *Epinephelus chlorostigma* and *Epinephelus areolatus* (in parentheses) of Arabian Gulf

Length groups (cm)	Mean T.L. (cm)	No. of fish	Absolute fecundity	Relative fecundity (No./cm)
30---31.9 (25---26.9)	30.10 (25.90)	1 (5)	235740.74 (163815.08)	7831.92 (6324.91)
34---35.9 (27---28.9)	34.90 (28.24)	4 (5)	425913.21 (160493.75)	12203.82 (5683.21)
36---37.9 (29---30.9)	37.25 (29.89)	2 (14)	401916.63 (167297.93)	10789.71 (5597.12)
38---39.9 (31---32.9)	38.60 (31.76)	2 (13)	311598.48 (262949.93)	8072.50 (8279.28)
40---41.9 (33---34.9)	41.00 (33.90)	5 (9)	412312.62 (222063.04)	10056.41 (6550.53)
42---43.9 (35---36.9)	42.59 (35.95)	7 (2)	547398.03 (309105.04)	12852.74 (8598.19)
44---45.9 (37---38.9)	44.60 (38.50)	2 (1)	938473.74 (321320.23)	21042.01 (8345.98)
46---47.9 (39---40.9)	46.94 (39.70)	5 (2)	1098087.71 (421271.38)	23393.43 (10611.37)
48---49.9	48.65	8	1083669.54	22274.81
50---51.9	51.45	2	1847121.88	35901.30

Table 2. Absolute and relative fecundity (in relation to weight, in gram) of groupers, *Epinephelus chlorostigma* and *Epinephelus areolatus* (in parentheses) of Arabian Gulf

Weight groups (g)	Mean weight (g)	No. of fish	Absolute fecundity	Relative fecundity (No./g)
207--- 406.9 (191---390.9)	269.15 (250.7)	2 (13)	255269.11 (145023.18)	948.42 (578.47)
407---606.9 (291---390.9)	507.73 (349.1)	4 (15)	393892.07 (212820.95)	775.79 (609.63)
607---806.9 (391---490.9)	716.20 (436.7)	5 (14)	455564.66 (238985.04)	636.09 (547.25)
807---1006.9 (491---590.9)	913.33 (525.9)	9 (6)	508478.69 (252572.14)	556.73 (480.27)
1007---1206.9 (591---690.9)	1104.30 (651.1)	4 (1)	760907.11 (413346.77)	689.04 (634.84)
1207---1406.9 (691---790.9)	1306.39 (760.1)	8 (2)	973341.33 (430664.11)	745.06 (566.59)
1407---1606.9	1504.16	5	1264959.73	840.97
1407---1806.9	1750.00	1	1787577.09	1021.47

Generally the fecundity of a fish species vary considerably with other species but even the individuals of same species produce varying number of eggs depending upon the age, weight, gonad's length and weight and environmental conditions (Bal and Rao, 1984). Daoulas and Kattoulas (1985) and Hussein (1986) have related the changes in fecundity to the feeding intensity of fish and environmental conditions. The fish produces more and frequently larger eggs in food rich environment (Wooten, 1973 and Moyle and Cech, 1982).

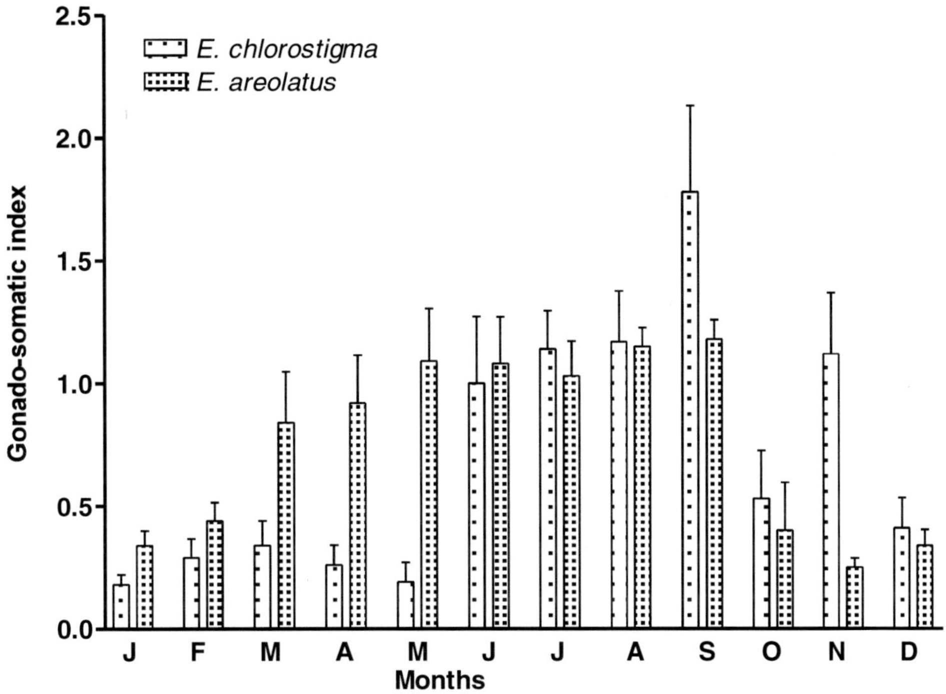

Figure 1. Monthly changes in the gonado-somatic index of *E. chlorostigma* and *E. areolatus* from the Arabian Gulf. Vertical lines indicate the standard error of the mean.

The relationship between absolute fecundity (F) and total weight (TW) of the fish were presented in Figures 2 and 3, and the regression equations registered were: Log F = 2.787 + 0.987 log TW (*E. chlorostigma*) and Log F = 2.904 + 0.99 log TW (*E. areolatus*). Similarly the relation between absolute fecundity and total length (TL) was presented as: Log F = 1.993 + 2.236 log TL (*E. chlorostigma*) and Log F = -0.207 + 3.707 log TL (*E. areolatus*). Regression analysis of the fecundity with total length as well as total weight of both the species indicate that absolute fecundity has a significant relation with the total length (r= 0.9298 for *E. chlorostigma* and 0.9201 for *E. areolatus*) and total weight (r= 0.973 for *E. chlorostigma* and 0.945 for *E. areolatus*).

The values of correlation coefficient (r) indicate that fecundity of fish has a stronger relation with weight than the length. This supports the hypothesis set out by Joganzen (1955). According to him the number of eggs produced by the female is mainly dependent on its weight rather than the length. However Bagenal and Erich (1978) have pointed out that the use of weight in such analysis might produce biased results. Ghorab et al. (1986) reported that the absolute fecundity of *E. chlorostigma* increases with increase of both, total length and total weight. Similar results were also recorded in the present study. Direct relationship between the fecundity and total weight and length of the fish was reported by Daoulas and

Kattoulas (1985) and Hussein (1986) and extend considerable support to present results but contrary results were reported by Mikhail (1980).

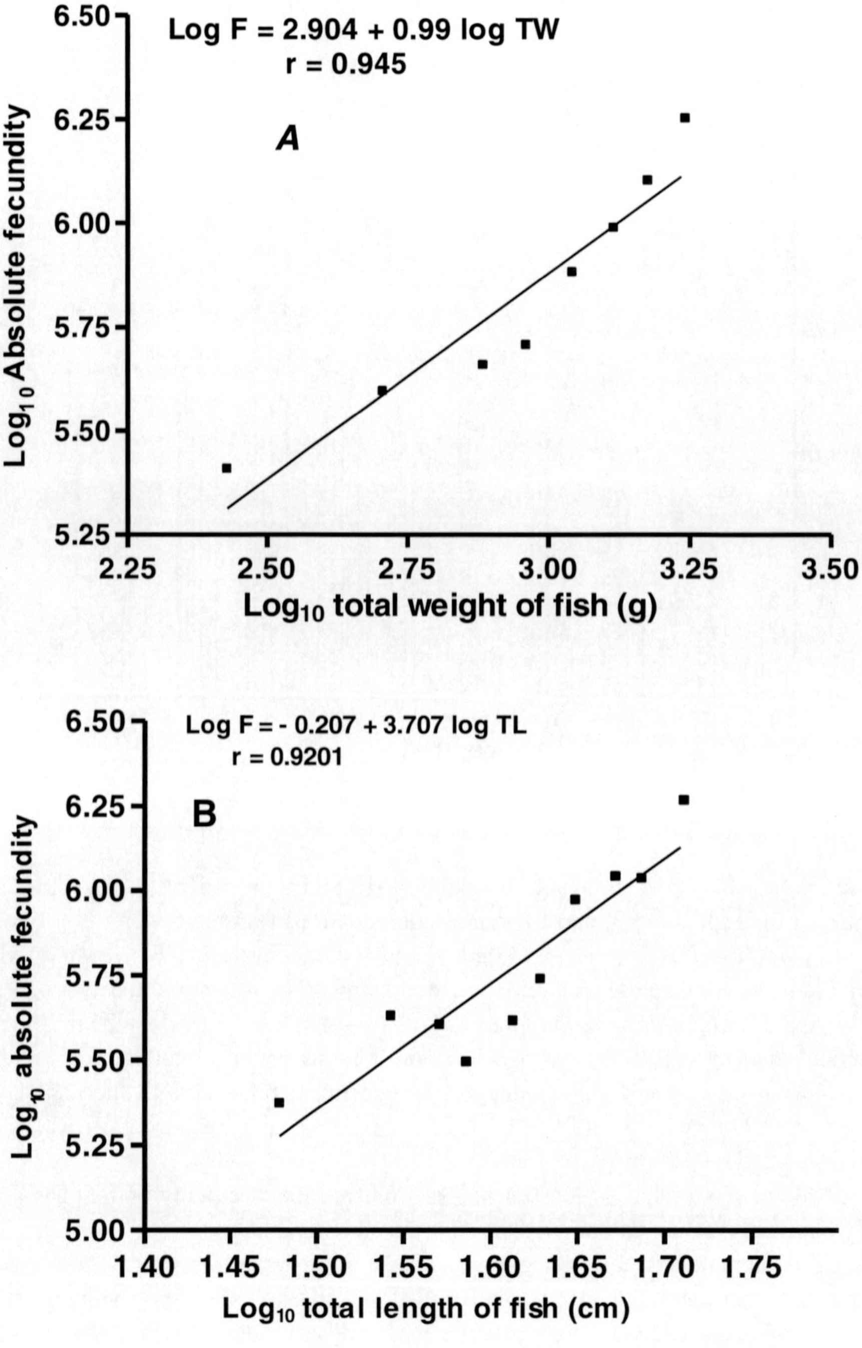

Figure 2. Relationship between absolute fecundity and total weight (A) and total length (B) of Epinephelus areolatus.

The values of gonado-somatic index of both the species registered in the present study indicate that the time of maturation of ova for *E. chlorostigma* extends from June to September and that of *E. areolatus* from March to September.

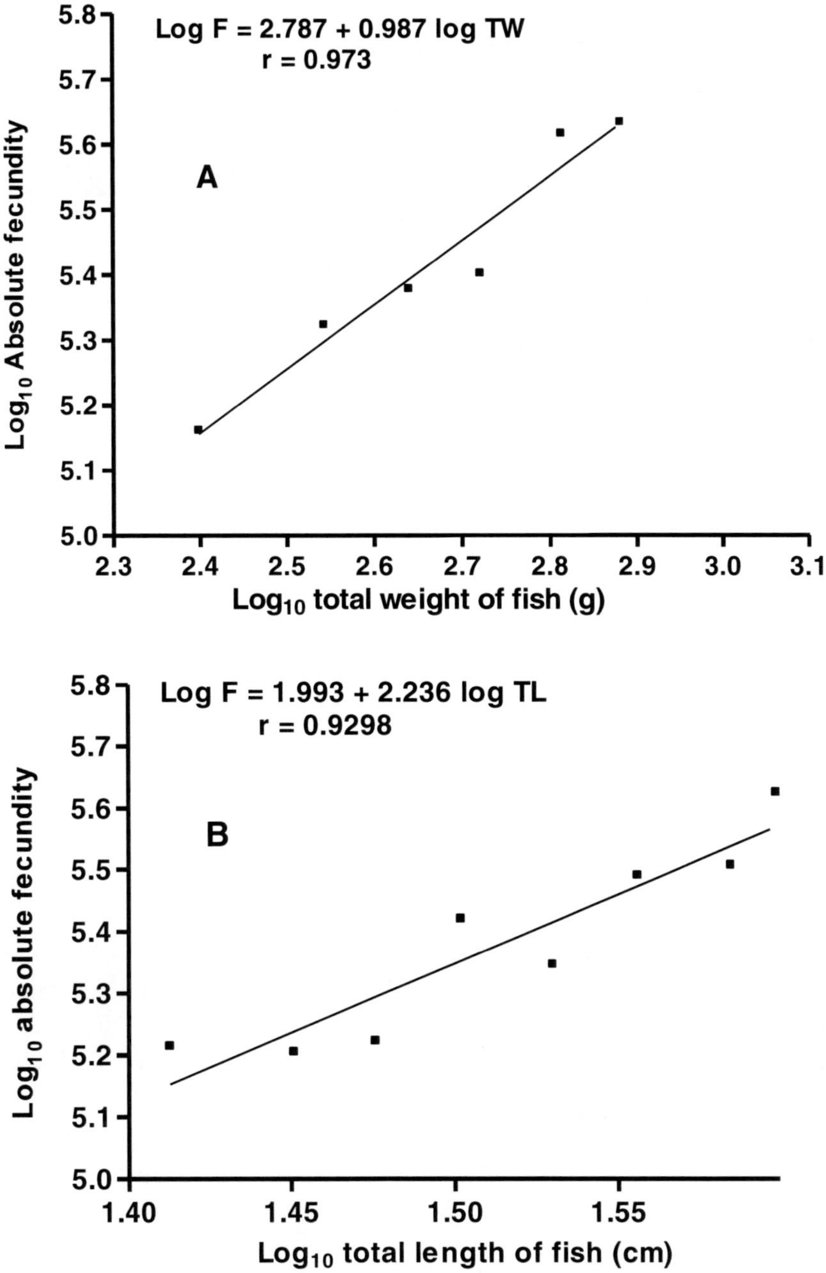

Figure 3. Relationship between absolute fecundity and total weight (A) and total length (B) of *Epinephelus chlorostigma.*

The highest value of this index registered in the month of September (Figure 1) shows the presence of mature eggs where as sharp decline in October is an indication of release of eggs from the ovary. The highest value of this index coincided with peak values of condition factor, thus shows a direct relationship between the two parameters. The sex ratios show a heterogenous distribution of male and female of *E. chlorostigma* and *E. areolatus* throughout the year. Males of *E. areolatus* were more frequent in the months like January, March, September and October while females outnumber male in February, May, June, August, November and December (Table 3).

Table 3. Sex ratios of *Epinephelus chlorostigma* and *Epinephelus areolatus* collected from the Arabian Gulf

Months	*E. chlorostigma*			*E. areolatus*		
	No. of male	No. of female	Sex ratio (M: F)	No. of male	No. of female	Sex ratio (M: F)
January	19	15	1: 0.79	20	13	1: 0.65
February	21	15	1: 0.71	09	23	1: 2.56
March	18	19	1: 1.06	22	16	1: 0.73
April	16	18	1: 1.13	16	17	1: 1.06
May	19	15	1: 0.79	06	33	1: 5.50
June	13	26	1: 2.00	07	28	1: 4.00
July	16	15	1: 0.94	17	18	1: 1.06
August	12	08	1: 0.67	08	13	1: 1.63
September	11	23	1: 2.09	28	09	1: 0.32
October	20	09	1: 0.45	17	10	1: 0.59
November	12	17	1: 1.42	01	05	1: 5.00
December	14	15	1: 1.07	13	23	1: 1.77

Females of *E. chlorostigma* dominated over males in April, June September and November while number of males exceeded to females in the months of January, February, May, August and October (Table 3). In remaining months the sex ratio was same as expected (1:1) in natural habitat. Domination of males over female during spawning season may be probably to ensure the fertilization of eggs. Variations in sex ratio depend mainly on the size of the samples analyzed, spawning migration, exploitation, weather condition at the time of catch, gear used for the catch and sex reversal (Bhatt, 1972, Thompson and Munro, 1978, Hashim and Salamah, 1984, and Ghorab et al., 1986). Disparities in the assumed (1:1) sex ratio were reported in different fish species by different workers. Domination of females in all the months of year except in the peak spawning period was reported by Hashim and Salamah (1984) and Coleman et al. (1996). Males of Epinephelus micodon show domination (4:1) over females throughout year (Tamaru et al., 1996).

REFERENCES

Abu-Hakima, R., Al-Abdul-Elah, K. M., El-Zahr, C., Akatsu, S., Al-Shoushi, M. and Abdullah, M. S. (1982). Spawning season and fecundity of six commercially important food fishes found in Kuwait. *Annual Research Report*, Kuwait Institute of Scientific Research.

Aday, D. D., Kush, C. M., Wahl, D. H. and Philipp, D. P. (2002). The influence of stunted body size on the reproductive ecology of bluegill *Lepomis macrochirus. Ecol. Freshwater* Fish 11:190-195.

Al-Akel, A. S., Al-Misned, F., Al-Kahem Al-Balawi, H. F., Al-Ghanim, K., Ahmad, Z. and Annazri, H. (2010). Reproductive biology of sailfin molly, *Poecilia latipinna* (Lesueur, 1821) in wadi Haneefah stream, Riyadh, Saudi Arabia. *Pakistan J. Zool.* 42: 169-176.

Arlinghaus, R. and Wolter, C. (2003). Amplitude of ecological potential: chub *Leuciscus cephalus* (L.) spawning in an artificial lowland canal. *J. Appl. Ichthyol.* 19: 52-54.

Azaza, M. S., Dhraie, M. N. and Karaiem, M. M. (2008). Effects of water temperature on growth and sex ratio of juvenile Nile tilapia *Oreochromis niloticus* (Linnaeus) reared in geothermal waters in southern Tunis. *J. Therm. Biol.* 33:98-105.

Bagenal, T. B. and Erich, B. (1978). Eggs and early life history. In: Timothy, B. (eds.), *Methods for assessment of fish production in freshwater.* London, Oxford, Blackwell Scientific Publication pp: 165-201.

Bal, D. V. and Rao, K. V. (1984). Marine fisheries. New Delhi, Tata McGraw- Hill, P: 470.

Bhatt, V. S. (1972). Studies on the biology of some freshwater fishes. *J. Bombay Natur. Hist. Soc.* 68: 556-572.

Coleman, F. C., Koening, C. C. and Cellins, L. A. (1996). Reproductive styles of shallow-water grouper in the Eastern Gulf of Mexico and the consequences. *Environ. Biol. Fishes* 47: 129-141.

Daoulas, C. H. and Kattoulas, M. (1985). Reproductive biology of Rutilus Rubilio (Bonaparte, 1837) in lake Trichonis. *Hydrobiologia* 124:49-55.

Dominguez-Petit, R., Korta, M., Saborido-Rey, F., Murua, H., Sainza, M. and Pineiro, C. (2008). Changes in size at maturity of European hake Atlantic populations in relation with stock structure and environmental regimes. *J. Mar. Syst.* 71:260-278.

Ganias, K., Somarakis, S., Machias, A. and Theodorou, A. (2004). Pattern of oocyte development and batch fecundity in the Mediterranean sardine. *Fish. Res.* 67:13-23.

Geertjes, G. J. and Videler, J. J. (2002). A quantitative assessment of the reproductive system of the Mediterranean cave-dwelling triplefin blenny, *Tripterygion melanurus Mar. Ecol.* 23: 327-340.

Ghorab, H. M., Bayoumi, A. R. and Bebars, M. I. (1986). The reproductive biology of grouper, *Epinephelus chlorostigma* (Pisces, Serranidae) from the Red Sea. Bull. Inst. Oceanogr. Fish. ARE. 12: 13-33.

Hashim, M. T. and Salamah, A. J. (1985). The catch and growth rates of *Gerres ruppellii* Klunz, 1884 of the Red Sea. *J. Faculty Mar. Sci.* (King Abdul Aziz University) 4: 213-231.

Heibo, E. and Vollestad, L. A. (2002). Life history variation in perch (*Perca fluviatus*) in five neighbouring Norwegian lakes. *Ecol. Freshwater Fish* 11: 270-280.

Hussein, K. A. (1986). Timing of spawning and fecundity of Mediterranean *Siganus rivulatus* Forsk. Bull. Inst. Oceanogr. *Fish. ARE.* 12: 175-186.

Joganzen, B. V. (1955). Fish fecundity and the factors that influence it. *Vopr. Ikhtiol.* 3: 57-68.

Machado, M. D., Heins, D. C. and Jr Bart, H. R. (2002). Micro-geographical variation in ovum size of the blacktail shiner, *Cyprinella venusta* Girard, in relation to stream flow. *Ecol. Freshwater Fish 11*: 11-19.

Mikhail, M. Y. (1980). *Some biological studies on Epinephelus species in the Egyptian Mediterranean water.* M. Sc. Thesis. Faculty of Science, Alexandria University p- 157.

Moyle, P. B. and Jr Cech, J. J. (1982). Fishes: An introduction to ichthyology. Prentice-Hall International Inc. London, Sydney, Toronto and New Delhi P- 593.

Orlando, E. F., Binczik, G. A., Denslow, N. D. and Jr. Guillette, L. J. (2007). Reproductive seasonality of the female Florida gar, *Lepisosteus platyrhincus. General Comp. Endocrino.* 151:318-324.

Pusey, B. J., Arthington, A. H., Bird, J. R. and Close, P. G. (2001). Reproduction in three species of rainbow fish (Melanotaeniidae) from rainforest streams in northern Queensland, Australia. *Ecol. Freshwater Fish* 10: 75-87.

Qasim, S. Z. (1973). The biology of *Centronotus gunnellus* (L.) teleost. *J. Anim. Ecol.* 26: 389-401.

Tamaru, C. S., Carlstram, T. C., Fitzgeratd, W. J. and Ako, H. (1996). Induced final maturation and spawning of the marbled grouper, *Epinephelus microdon*, captured from spawning aggregation in the Republic of Palau, Micrnesia. *J. World Aquacult. Soc.* 27: 363-372.

Thompson, R. and Munro, J. L. (1978). Aspects of biology and ecology of Caribbean reef fishes, Serranidae, hinds and groupers. *J. Fish Biol.* 12:115-146.

Wooten, R. J. (1973). The effect of size of food ration on egg production in the female three-spined stickleback. *J. Fish Biol.* 5: 89-96.

Zabala, M., Garcia, R. A., Louisy, P. and Sala, E. (1997). Spawning behaviour of the Mediterranean dusky grouper, *Epinephelus marginatus*, in the Medes Islands. *Mar. Reserve. Scientia Marina* 61:65-77.

Zhang, J., Takita, T. and Zhang, C. (2009). Reproductive biology of Ilisha elongata (Teleostei: Pristigasteridae) in Ariake Sound, Japan: Implications for estuarine fish conservation in Asia. *Estur. Coastal Shelf. Sci.* 81: 105-113.

In: Vertebrate and Invertebrate Animal Approaches
Editor: José Rosa Gomes
ISBN: 978-1-63117-102-4

Chapter 25

LABORATORY STUDIES ON THE TERRESTRIAL MARSH SLUG *DEROCERAS LAEVE* (MÜLLER) (AGRIOLIMACIDAE: MOLLUSCA)

M. I. Mohamed and R. F. Ali
Department of Zoology and Agricultural Nematology,
Faculty of Agriculture Cairo University; Giza, Egypt

ABSTARCT

Biological aspects of the terrestrial garden slug *Deroceras laeve* including oviposition, hatching, generation period, life span as well as growth parameters were studied under laboratory conditions. Results revealed that mating is not essential for this species to lay eggs, while self-fertilization is the normal breeding system. The oviposition period averaged 33.6 days, during which the individual laid an average of 34.5 eggs. The incubation period averaged 18.2 days. While the generation period (from egg to egg) lasted 83.3 days in average. On the other hand, the maximum life span durated an average of 135.9 days.

Growth parameters, depending on the increase and decrease of slug body weights, rapidly increased during the first two months of the age, and then gradually decreased.

Keywords: *Deroceras laeve* – self-fertilization - oviposition – generation period – growth parameters

INTRODUCTION

The land slugs have a wide range of feeding habits. Rollo (1983) stated that the terrestrial slug *Arion ater* is general herbivores that eat a wide variety of the leaf litter, green plants and fungus.

Harper (1988) mentioned that the banana slug, *Ariolimax columbianus* feeds on living and decaying vegetation such as roots, fruits, seeds, bulbs, lichens, alage, fungi, animal droppings and carrion. It was also stated that mushrooms such as *Pleurolus ostrealus* and

Boletus edulis are a frequent food source for this slug. Capinera (2001) mentioned that slugs are fed on plant tissue, causing serious defoliation injury to leafy vegetables and its seedlings. Seedlings are often cut off at the soil surface by slugs. They also feed below-ground on roots and tubers of beet, carrot, potato, radish and turnip. Also, slugs usually feed on the surface of potato tubers, tomato and eggplants fruits, but sometimes dig deeply into the flesh of these vegetables. While, Keller and Snell (2002) indicated that most slug species are generalists and are fed on a broad spectrum of plant tissues in nature such as green leaves of living plants, lichens, mushrooms, decaying vegetation and even carrion as well as they are fed on carrots, peas, apples and cabbage that are offered as a sole food source.

In Egypt, the land slugs are becoming a serious pest during the last two decades, where they attack various types of plants. It was found that the marsh slug *Deroceras laeve* (Müller) fed on lettuce leaves (*Lactuca sativa* L.), Egyptian clover (*Trifolium alexandrinum* L.), leaves and pods of broad bean (*Vicia faba* L.), carrot root (*Dacus carota* L.), leaves and fruits of tomato (*Solanum lycopersicum*), fruits of squash (*Cuecurbita pepo*), fruits of strawberry (*Fregria ananassa* Duch.) and fruits of cucumber (*Cucumis sativus* L.) that offer the main sources of food. Thus, the main objective of this investigation is to spot more light upon the biological aspects of the dominant land slug *D. laeve* under laboratory conditions.

MATERIAL AND METHODS

Fifty individuals of the land slug species *Deroceras leave* (Müller) have been hand-collected during February 2008, under different types of ornamental plants such as aluminium plant (*Pilea cadierier* L.), polar palm (*Chamaedorea elegans*) and ficus (*Ficus nitida*) from some nurseries in Cairo Governorate. The individuals were kept in two plastic boxes (27 x 27 x 25 cm) with fresh lettuce leaves (*Lactuca sativa* L.) which offered a sole food source. Boxes were covered with muslin cloth and fixed with rubber band to prevent slugs from escaping, then transferred to the laboratory.

The soil of each box was searched twice daily for batches or clutches (egg-masses) and remoistened as required. The slugs were supplied with fresh lettuce leaves and the remained leaves were removed. The newly deposited batches were left in the same box and time of oviposition was recorded, while slug individuals were only transferred to new prepared box. The batches were observed twice daily till hatching to determine the incubation period and hatchability. The juveniles or offsprings (new hatch slugs) were placed singly in plastic cups (10 x 6) cm filled with moist clay soil to depth of five centimeters with disks of lettuce leaves and each covered with muslin cloth with rubber band to prevent escaping. Fresh lettuce leaves were provided and moisture was added daily until juveniles reached maturity and complete their life span. Juveniles were weighed monthly, using a digital balance, to determine their growth rate. Growth rate was calculated monthly according to the formula:

$$\text{Growth rate} = \frac{W_F - W_I}{W_F} \times 100$$

where W_I and W_F are the initial and final weights of slugs for given periods.

The regression analysis was carried out for weight and growth rate over age according to the model:

$$Y = \alpha + \beta X$$

where Y is the dependent variable (slug weight or growth rate).

α is the intercept

β is the slope of regression of Y over X

X is the age in months.

RESULTS

Incubation Period and Hatchability

Forty-six slugs simultaneously deposited batches, with a range of 3 – 19 eggs for each, were observed twice daily till hatching to determine its incubation period and hatchability. The incubation period ranged from 17 to 21 days with an average of 18.2 ± 1.3 days under laboratory conditions of 23.2 ± 0.7 °C and 50.3 ± 2.8 % R.H. (Table 1).The hatchability ranged from 28 to 100% with an average of 79.2 ± 20.0%.It was noticed that the egg mass may be hatched completely during one day or within few days ranged from two to six days.

Life Cycle and Generation Period

After hatching 70 juveniles were separated each in prepared plastic cups as described before and observed daily till maturity and oviposition.

It was difficult to differentiate between immature and mature slugs by the external shape or size. Thus, the life cycle from hatching till juveniles reach maturity including the pre-oviposition period ranged from 61 to 68 days with an average of 63.5 ± 2.0 days (n=30) at an average of 24.1 ± 0.2°C and 50.0 ± 0.3% R.H. While, the generation period (from egg to egg) lasted an average of 83.3 ± 2.8 days under the same mentioned laboratory conditions (Table 1). It was observed that the mortality ratio was higher during juvenile period (about 43%) than in adult stage (0.0%).

Oviposition

Individuals of *D. laeve* slug are hermaphrodites, each one possessing both male and female sexual organs, thus self-fertilization is dominant occur in this species. Under laboratory conditions of 26.9 ± 3.9°C and 57.8 ± 5.2 % R.H., the oviposition period (from first to last deposited egg) lasted an average of 33.6 ± 7.1 days (Table 1) (n=30) during which the slug individual laid an average of 6.1 ± 2.2 batches each with an average size of 5.6 ± 1.2 eggs. Days elapsed between deposited batch and another averaged 6.6 ± 2.2 days (Table 2). On the other hand, the total number of deposited eggs per individual averaged 34.5± 10.2 eggs during its oviposition period under laboratory conditions (Table 2). The eggs were whitish and translucent, with a dimension averaged 2.0 ± 0.3 X 1.8 ± 0.3 mm per egg.

Table 1. Duration in days of different stages of *D. laeve* under laboratory conditions

Period	Incubation period	Juveniles + pre-oviposition	Generation	Oviposition	Postoviposition	Life span
Average	18.2 ± 1.3	63.5 ± 2.0	83.3 ± 2.8	33.6 ± 7.1	33.3 ± 7.6	135.9 ± 10.5
Range	17 – 21	61 – 68	79 – 89	20 – 46	22 – 45	120 – 164

Table 2. Fecundity of *D. laeve* under laboratory conditions of 26.9 ± 3.9°C and 57.8 ± 5.2 % R.H.

Items	No. of deposited batches / day	No. of eggs / batch	No. of batches / slug	Period between each batch	Total No. of deposited eggs / slug
Average	2.0 ± 0.7	5.6 ± 1.2	6.1 ± 2.2	6.6 ± 2.2	34.5 ± 10.2
Range	1 – 3	4 – 8	3 – 9	4 – 11	21 – 50

Post-Oviposition and Life Span Period

Post-oviposition period was calculated from last laid egg till mortality. It durated an average of 33.3 ± 7.6 days. While the life span period from hatching till death averaged 135.9 ± 10.9 days under laboratory conditions of 26.9 ± 3.9°C and 57.8 ± 5.2 % R.H. (Table 1).

Growth Parameters

The averages of slug weights and growth rates over age are presented in Table (3) and Figure (1and 2). The weight had gradually increased every month and was the maximal by the third month of age, ranging from 0.05 to 0.16 g with an average of 0.091± 0.025 g. The slug weight had declined to 0.061 g in the fifth month.

Table 3. Body weights and growth rates of land slugs *D. laeve* over age

	Body weight (g)		Growth rate (%)	
Age (month)	LSM ± SD	Range	LSM ± SD	Range
Zero time	0.0	0 – 0		
1	0.0053 ± 0.01	0.0 – 0.03	23.33 ±43.02	0.0 - 100
2	0.053 ± 0.02	0.02 – 0.08	88.35 ± 22.0	33.3 – 100
3	0.091 ± 0.03	0.05 – 0.16	40.20 ± 14.06	11.11 – 71.43
4	0.072 ± 0.02	0.05 – 0.1	-26.81 ± 25.5	-100 – 0.0
5	0.061± 0.02	0.04 – 0.09	-20.53 ± 12.0	-60 –11.11
Regression coefficient (ß)	0.01303**		*-20.289***	

**(P ≤ 0.0001).

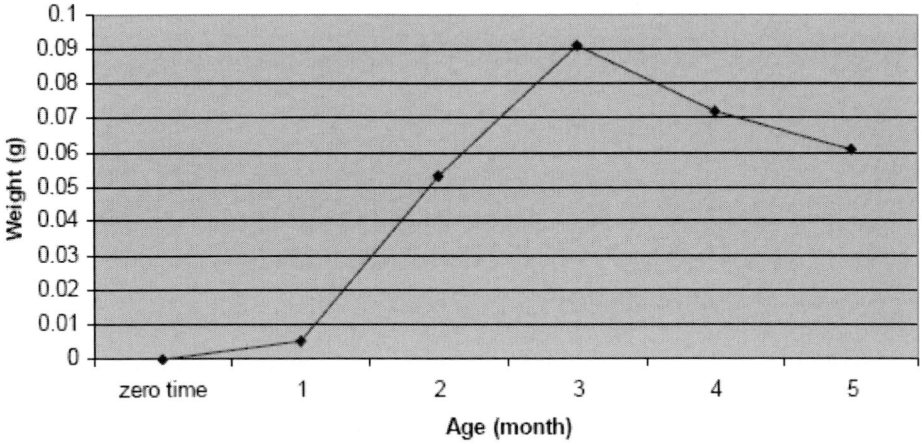

Figure 1. Averages of slug weights over age.

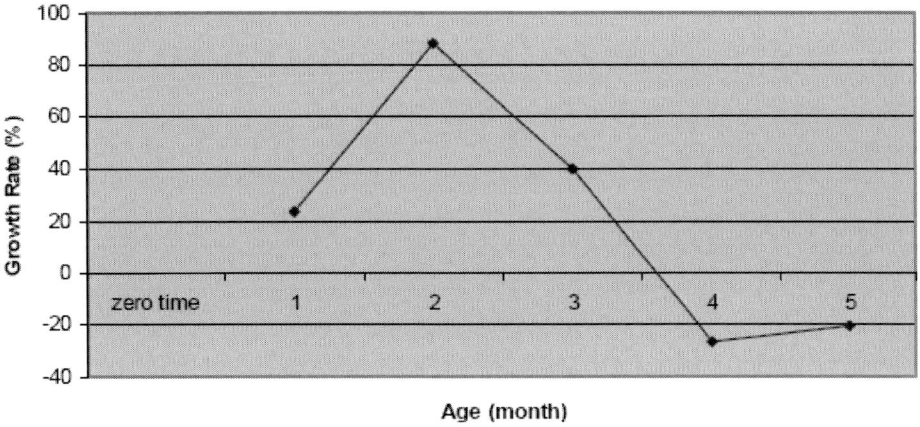

Figure 2. Averages of growth rate of slugs over age.

Figure (3) shows the regression of slug weight over age, where the slug weight at hatching time (zero time) was significantly ($P \leq 0.0001$) estimated to be 0.01750 g, and was significantly ($P \leq 0.0001$) increased by 0.01303 g every month. The results indicate that the slugs had increased in weight during the activity period for up to 3 months of age. The maximum growth rate of the slugs was 88.349% by the end of second month and then retarded with age.

Figure (4) presents a significant ($P \leq 0.0001$) regression slope of growth rate over age, where the growth retardation was by -20.289% every month. It is obvious that the slugs had significantly grown fast during the first month of age and the reduction in growth rate thereafter denotes that the increase in slug weight every month was in decreasing rate. The correlation coefficients (Table 4) indicate that the most important period of the land slug life cycle is between first and second months of age.

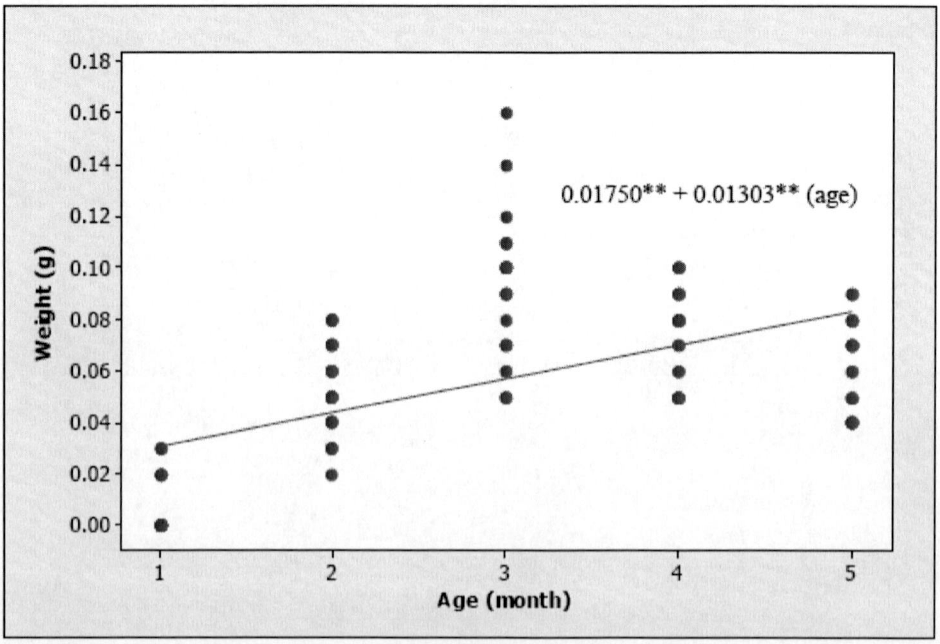

** $P \leq 0.0001$.

Figure 3. Regression of slug weight over age.

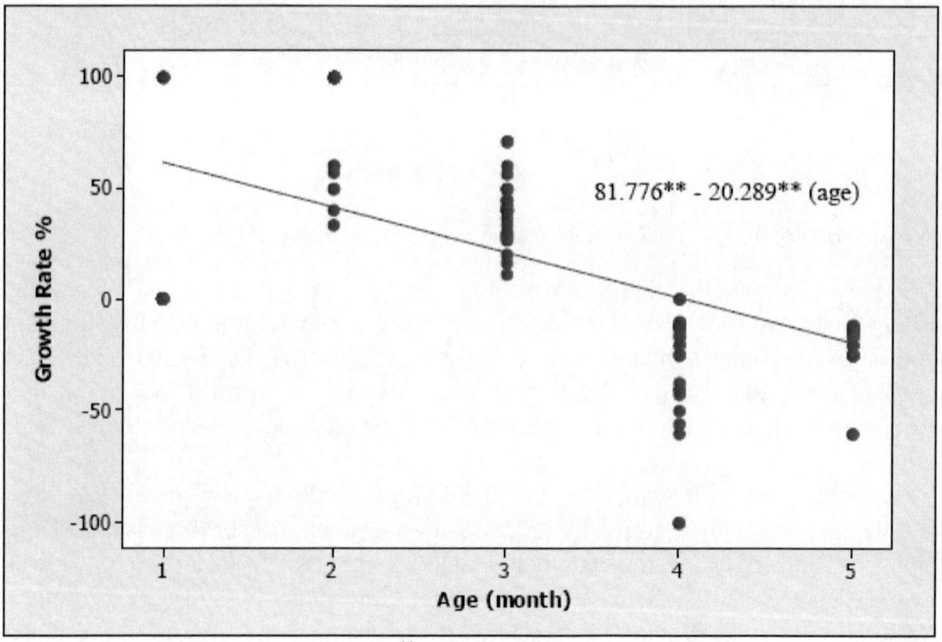

** $P \leq 0.0001$.

Figure 4. Regression of slug growth rate over age.

Table 4. Estimates ates of correlation coefficients among slug weights and growth rates over age

	Weight 1	Weight 2	Weight 3	Weight 4	Weight 5
Weight 1		0.15634 NS	-0.04854 NS	0.00520 NS	0.04642 NS
Weight 2	-0.15634 NS		0.60981**	0.62026**	0.55897**
Weight 3	-0.04854 NS	0.60981**		0.77650**	0.71237**
Weight 4	0.00520 NS	0.62026**	0.77650**		0.95856**
Weight 5	0.04642 NS	0.05589**	0.71237**	0.95856**	
	Growth Rate 1	**Growth Rate 2**	**Growth Rate 3**	**Growth Rate 4**	**Growth Rate 5**
Growth Rate 1		-0.97730**	0.04887 NS	0.17089 NS	0.13483 NS

NS, non significant.
** significant ($P \leq 0.0001$).

The significant positive correlations between slug weight at second month of age and the subsequent weights reveal the significance of the weight accumulated by the end of second month on subsequent slug weights.

Also, the significant positive correlation between growth rates at first and second months of age denotes to the significance of the growth process during the first to second month period. So, it is concluded that the growth rate between first and second month of age was significantly critical to the slug weight at second month and the subsequent weights.

DISCUSSION

McCracker and Selander (1980) reported that many species of terrestrial slugs of several families, including some of the more ubiquitous and familiar forms are monogenic or nearly so, apparently as a consequence of frequent or obligatory self-fertilization. They also reported that many slug species have different breeding systems in other regions. *Arion ater* in Europe may consist of both outcrossing and self-fertilizing strains, as does *A. subfuscus* in North America.

In this work, the laboratory studies showed that copulation in the terrestrial slug *D. laeve* is not essential for oviposition, where self-fertilization was the normal breeding system in all separated individuals.

Rollo (1983) mentioned that the terrestrial slug *Limax maximus* produced about half the number of batches in the monoculture (2.02 batches / slug) compared to slugs caged with *Ariolimax columbianus* (4.0 batches / slug) or *Arion ater* (3.57 batches / slug). While in the laboratory, only one individual of *L. maximus* produced five batches of eggs in one season. On the other hand our study revealed that *Deroceras leave* gave (6.1 batches / slug) under laboratory conditions.

According to Rollo (1983), the mean number of eggs / batch was significantly differed among species, hence in *A. ater* averaged 47.9 eggs / batch, *L. maximus* produced 59.3 eggs / batch and *A. columbianus* laid 17.7 eggs / batch. While in this investigation *D. laeve* recorded an average of 5.6 eggs / batch under laboratory conditions.

The total number of produced eggs per individual slug differed according to species. Capinera (2001) mentioned that individual slugs of *Deroceras reticulatum* produce about 100

– 200 eggs during their life span. While in this work it was found that the total number of deposited eggs per individual slug of *D. laeve* averaged 34.5 eggs during its life span.

The incubation period differs according to species and temperature as well as hatching rate. Sokolve and McCrone (1978) recorded that the incubation period of slug *Limax maximus* lasted one month at 15ºC. While, Faberi et al. (2006) mentioned that the incubation period of the slug *Deroceras laeve* was 35.3 and 16.03 days and the hatching rate 89.2% and 66.4% at 12 and 20ºC, respectively.

In this work it was found that the incubation period of the marsh slug *D.laeve* averaged 18.2 days with hatchability of 79.2% under laboratory conditions of 23.2ºC and 50.3% R.H.

The generation period (from egg to egg) of an slug is differed according to species. Hunter (1968) reported that *D. reticulatum* in England had two generations per year; *A. hortensis* had one generation annually and *Tandonia budapestensis* required two years for a single generation. In this study, it is agreed with Hunter (1968) and Alford (1984) that *A. hortensis* has a single generation each year.

Barker (1991) stated that the maximum life span of *D. reticulatum* durated from 8 to 12 months, while Zotin (2007) mentioned that the life span of the same species under laboratory conditions ranged from 6 to 10 months. In our study, the life span of *D. laeve* ranged from 4.0 to 5.5 months with an average of 4.5 months under laboratory conditions.

Important factors such as temperatutre, humidity and photoperiod influence the growth rate of land slugs. South (1982) showed that the growth rate of *Deroceras reticulatum* increased at 15ºC and reached its maximal value at 18ºC. He observed that there is a negative correlation between temperature and maturation. It has been noticed that the growth rate, for the first phase, was higher when exposed to alternating temperatures than to a constant temperature.

Hommay et al. (2001) mentioned that the slug weight increase in *Limax valentianus* Férussac occurred in three distinct phases: an initial phase of rapid growth (infantile phase), followed by a phase with a smaller increase in weight (juvenile phase) and finally a phase with minimal growth (adult phase). It was also mentioned that maturity ended sometimes with a weight decrease due to senescence of reproductive tract.

Faberi et al. (2006) mentioned that the growth rate was lower at 12ºC than at 20ºC. At the latter temperature, the inflection point was reached when they were 2.5 months old and the mean body weight was 204.7 mg. At the same age at 12ºC, the body mass was 132.9 mg. At the lower temperature the growth increased with time until the animal were 4.5 months old, when they weighed 468.1 mg. However slugs of same age and exposed to 20 ºC weighed 340.4 mg. In addition, slugs reached a greater mean body weight at 12ºC (936.2 mg ± 18.96) than at 20ºC (409.4 mg ± 16.02). In this work, it was observed that growth was rapidly increased in the first two months of age then decreased gradually.

At zero time (hatching time) the individual slug started to increase after one month of age (0.005 gm) and reached to its maximum weight (0.09 gm) during the third month, then gradually decreased during the fourth (0.07 gm) and fifth month (0.06 gm).On the other hand the high average growth rate of body weight was averaged 88.3% in the second month of age, while the lowest rate (-20.5%) was in the fifth month.

The correlation coefficients between slug weight at second month of age and the subsequent weights were significant and reveal the significance of the accumulated weight by the end of the second month.

ACKNOWLEDGMENT

The authors wish to thank Dr. E. A. El-Gendy Professor of Poultry Sciences, Fac. of Agric. Cairo Univ. for the assistance in the estimation of growth parameters and regression analysis. We are also very grateful to Dr. John M. C. Hutchinson and Dr. Heike Reise, Senckenberg Museum of Natural History, Görlitz, Germany for identifying the specimens.

REFERENCES

Alford, D.V. 1984. A colour Atlas of Fruit Pests their recognition, biology and control. *A wolfe science Book,* 320 pp.

Barker, G.M. 1991. Biology of slugs (Agriolimacidae and Arionidae: Mollusca) in New Zealand hill country pastures. *Oecologia* 85: 581 – 595.

Capinera, J.L. 2001. *Handbook of Vegetable Pests*. Academic Press, New York, London. 729 pp.

Faberi, A. J.; Lopez, A. N.; Manetti, P.L.; Clemente, N.L. and Alvarez Castillo, H. A. 2006. Growth and reproduction of the slug *Deroceras laeve*(Müller) (Pulmonata: Stylommatophora) under controlled conditions. Span. J. Agric. Res. 4(4): 345 – 350.

Harper, A. B. 1988. The Banana Slug. Apots, California. Bay Leaves Press. 32 pp.

Hommay, G.; Kienlen, J.C.; Gertz, C. and Hill, A. 2001. Growth and reproduction of slug Limax valentianus Férussac in experimental conditions. *J. Moll. Stud.*, 67: 191 – 207.

Hunter, P. J. 1968. Studies on Slugs of Arable ground II. Life Cycles. *Malacologia* 6: 379 – 389.

Keller, H.W. and Snell, K.L. 2002. Feeding activities of Slugs on Myxomycetes and Macrofungi. *Mycologia*, 94 (5): 757 – 760.

McCracken, G. F. and Selander, R.K. 1980. Self-fertilization and monogenic strains in natural populations of terrestrial slugs. *Proc. Nat. Acad. Sci. USA*. 77 (1): 684 – 688.

Rollo, C.D. 1983. Consequences of competition on the reproduction and mortality of three species of terrestrial slugs. *Res. Popul.* Ecol. 25: 20 – 43.

Sokolove, P.G. and McCrone, E.J. 1978. Reproductive maturation in the slug, *Limax maximus* and the effects of artificial photoperiod. *J.Comp. Physiol.* 125: 317 – 325.

South, A.1982. A comparison of the life cycles of *Deroceras reticulatum* (Müller) and *Arion intermedius* Normand (Pulmonata: Stylommatophora) at different temperatures under laboratory conditions. *J. Moll. Stud.*, 48: 233 – 244.

Zotin, A.A. 2007. Patterns of Individual Growth of Gray Garden Slug *Deroceras reticulatum*. *Biology Bulletin*, 34 (5): 457 – 462.

In: Vertebrate and Invertebrate Animal Approaches
Editor: José Rosa Gomes
ISBN: 978-1-63117-102-4

Chapter 26

LIFE CYCLE AND GROWTH RATES OF THE CONICAL SNAIL *COCHLICELLA ACUTA* (MÜLLER, 1774) (GASTROPODA: COCHLICELLIDAE)

Mahmoud I. Mohamed and Reham F. Ali
Department of Zoology and Agricultural Nematology,
Faculty of Agriculture, Cairo University; Giza, Egypt

ABSTRACT

Mating, pre-oviposition, oviposition, post-oviposition, and generation periods as well as growth rate of the pointed or conical snail *Cochlicella acuta* (Müller, 1774) were studied in Egypt under laboratory conditions during its reproductive and activity season of 2008/ 2009. In spite of this species is hermaphroditic, the laboratory observations indicated that cross mating was essential for oviposition as unmated snails did not lay any egg. Mating occurs when snail individuals reach their sexual maturity as calculated from adulthood till mating. This period was 99.2 ± 23.8 days in average. Pre-oviposition period (from mating till laying first egg) lasted for an average of 17.6 ± 6.1 days. Oviposition period (from laying first egg till last one) averaged 86.3 ± 22.5 days during which snail individual deposited a total of 113.3 ± 58.3 eggs. Post-oviposition period (from last laid egg till mortality) was 140.7 ± 74.3 days. The generation period (from egg to egg) of this species extended to an average of 283.7 ± 29.3 days.

Keywords: *Cochlicella acuta* - Sexual maturity – Oviposition – Generation period – Growth rates

INTRODUCTION

The pointed or conical snail *Cochlicella acuta* (Müller, 1774) is a native of coastal areas of the Mediterranean and Western Europe (Lewis, 1977; Kerney and Cameron, 1979). However, Baker (1986) stated that this species is an introduced agricultural pest in south eastern and south Western Australia. He added that large numbers of *C. acuta* aestivate in

Australia on the ears and stalks of cereal plants and often clog machinery and contaminate grains during harvest.

Although *C. acuta* feed on some agricultural plants in the laboratory (Baker, 1989), significant damage to these plants in the field has not been reported.

In Egypt, *C. acuta* is considered one of the most abundant mollusc pest causing damage to palm trees, citrus orchards and ornamental plants. The biology of this pest has been little studied and restricted to non agricultural habitats. The scope of the study reported herein was to examine several biological aspects of the life history (e.g. mating, life cycle, oviposition as well as growth parameters) of *C. acuta* under laboratory conditions.

MATERIALS AND METHODS

Seventy adult individuals of *C. acuta* were hand picked from nurseries in Cairo, Egypt, during its active period in November 2008 on different ornamental host plants e.g. rose-scented pelargonium (*Pelargonium gravealens*), night blooming cestrum (*Cestrum noctumum*), ficus trees (*Ficus benjamina*) paper flower (*Bougainvillea glabra*) and golden dewdrop (*Duranta plumieri*).

The snails were kept in three plastic boxes (15 x 7 x 11 cm) containing moist soil to eight centimeters deep and were fed on fresh lettuce leaves (*Lactuca sativa*). The boxes were covered with muslin cloth fixed with rubber bands to prevent the individuals from escaping. Boxes were examined daily. Fresh lettuce leaves and moisture (drops of water) were supplied as required, and the soil was searched for egg clutches.

All egg clutches were removed and placed in prepared pots (13x10 cm) with moist soil and then observed twice daily till hatching to determine the incubation period and hatchability. New hatched snails (juveniles) were placed solitary in plastic cups (13x10cm) with moist soil and fresh lettuce leaves.

The cups were covered to avoid escaping of snails, and were examined daily, where fresh lettuce and moisture were added as required throughout the life span.

Pairs of snails of similar weight and shell diameter were put in prepared pots (13 x 10 cm) and observed several times daily till mating to determine the periods of sexual maturity and copulation.

After that, individuals of every pair were separated and each individual was placed singly in the similar pot to determine the pre-oviposition period, oviposition period, number of egg clutches per individual, clutch size and post-oviposition period.

Juveniles were weighed using a digital balance, and shell diameter was measured using a caliper. This was done monthly from hatching time (zero time) till maturity (according to lips growth). The shell volume was calculated following Baur and Raboud (1988):

$$\text{Shell volume} = (\text{width})^2 \times \text{height} / 2$$

The growth rate was calculated monthly according to the formula:

$$\text{Growth rate} = \frac{W_F - W_I}{W_F} \times 100$$

where W_I and W_F are the initial and final weights of snails during a period of month. Also, the rate of change in diameter was calculated using the formula:

$$\text{Rate of change in diameter} = \frac{D_F - D_I}{D_F} \times 100$$

where D_I and D_F are the initial and final shell diameter of snails during a period of one month. The rate of change in shell height was calculated as:

$$\text{Rate of change in height} = \frac{H_F - H_I}{H_F} \times 100$$

where H_I and H_F are the initial and final shell height of snails during a period of month. Also, the following formula was applied to calculate the rate of change in shell volume.

$$\text{Rate of change in volume} = \frac{V_F - V_I}{V_F} \times 100$$

where V_I and V_F are the initial and final shell volume of snails for a period of month. The regression analysis was carried out for each of weight, diameter, height, and volume over age according to the model:

$$Y = \alpha + \beta X$$

where Y is the dependent variable (snail weight, diameter, height or volume)

α is the intercept

β is the slope of regression of Y over X

X is the age in months.

This model was also used to explain the increase and the decrease of weight over one month for the land snails (SAS, 1985).

RESULTS

Mating

Although this species is hermaphroditic, self-fertilization does not occur in solitary individuals; therefore cross mating is essential to lay eggs. When individuals reached their adulthood according to lip signs, then paired. In *C. acuta* mating occurred when its individuals reached sexual maturity during an average period of 99.2 ± 23.8 days (N = 12) after adulthood was attained (Table 1). Mating occurred in day time between 10.30 am to 3.30 pm, during its reproductive season (mid November till the end of February).

During copulation two individuals faced each other and attached together strongly from the fleshy parts with a thick layer of mucus for an average period of 142.5 ± 28.1 minutes (N = 6 pairs) (Figure 1). After, individuals gradually pull out their bodies and separated within an average period of 15.0 ± 4.5 minutes.

Table 1. Duration of different stages of *Cochlicella acuta* under laboratory conditions

	Period in days of								
Items	Incubation	Juveniles	Life cycle	Sexual maturity	Pre-oviposition	oviposition	Post-oviposition	Life span	Generation
Average ± SD	13.7 ±1.3	251.8 ±27.5	265.5 ±27.5	99.2 ±23.8	17.6 ±6.1	86.3 ±22.5	140.7 ±74.3	561.8 ±102.1	283.4 ±29.3
Temp. °C	21.1 ±0.3	24.2 ±5.3	24.0 ±5.1	28.0 ±4.2	20.7 ±1.9	19.7 ±1.5	26.2 ±4.6	25.9 ±4.6	22.4 ±3.8
R.H.%	48.4 ±7.9	60.1 ±6.8	60.0 ±6.9	57.0 ±7.4	61.5 ±7.4	60.4 ±7.3	60.6 ±6.2	60.5 ±6.3	59.1 ±7.3

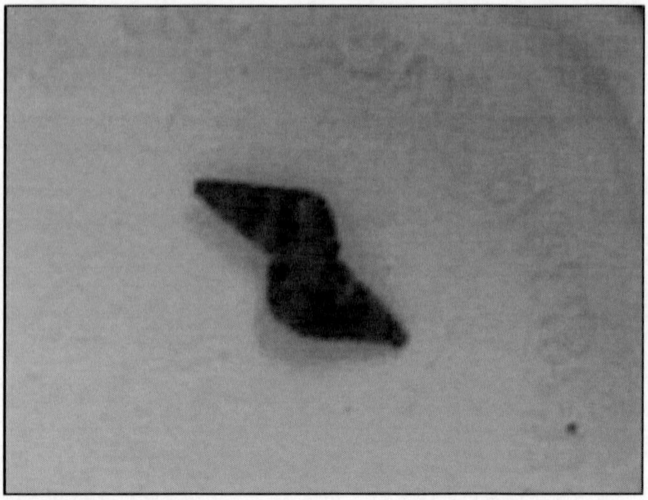

Figure 1. Mating process of *Cochlicella acuta*.

Oviposition

After mating, individuals of *C. acuta* required a pre-oviposition period averaging 17.6 ± 6.1 days before depositing eggs. Oviposition period (from 1st egg to last one) averaged 86.3 ± 22.5 days during which averages of 13.3 ± 4.1 clutches per snail individual were deposited.

The period between two clutches deposit was 6.9 ± 1.7 days. The clutch size ranged from 6.0 to 12.0 with an average of 9.4 ± 2.4 eggs. The total number of eggs deposited by one individual of *C. acuta* during its oviposition period averaged 113.3 ± 58.3 eggs.

Eggs are white, with smooth surface, spherical shape and its diameter ranged from 1.1 to 1.5 mm with an average of 1.3 ± 0.1 mm (N = 21) (Figure 2).

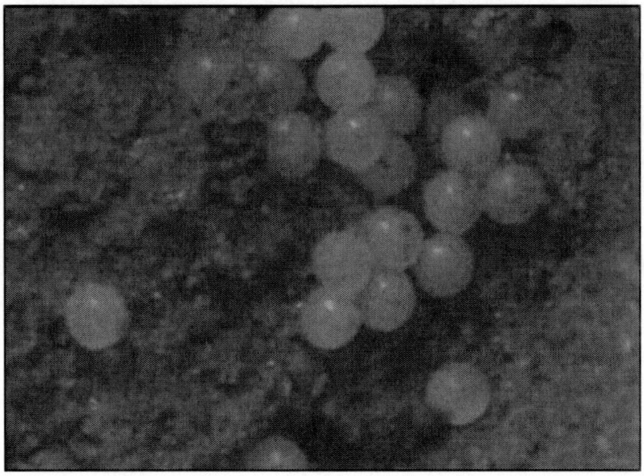

Figure 2. Eggs of *Cochlicella acuta*.

Laboratory observations showed that both mated individual snails or only one deposited eggs. On some cases both mated snails did not lay any eggs and this might be due to failure in copulation.

Post-Oviposition

Post-oviposition period was calculated from last egg deposited till individual death. This period averaged 140.7 ± 74.3 days for *C. acuta*.

Development

Egg incubation period of the conical snail *C. acuta* averaged 13.7 ± 1.3 days under laboratory conditions at temperature of 21.2 ± 0.3°C and 48.4 ± 7.9 % relative humidity with a mean hatchability of 64.8 ± 34.8 % (N = 27). Newly hatching snails (juveniles) are similar to adults but smaller in size and weight. Their average measurements were 1.7 ± 0.3 mm for height and 1.3 ± 0.2 mm for width at zero time (N = 30). They developed slowly in a week activity and reached adulthood after eleven months. At that time their average maximum shell dimensions reached 9.5 ± 1.4 mm for height, 4.3 ± 0.6 mm for width and weighted 0.06 ± 0.012 g (N = 30). The duration of *C. acuta* life cycle from egg to adulthood (incubation period plus juveniles) averaged 265.5 ± 27.5 days. Mating influenced life span of this species (from egg till mortality). For non-mated, it was 528.1 ± 122.8 days, while it was 561.8 ± 102.1 for mated ones (Table 2).

Table 2. Duration of *Cochlicella acuta* life span in its activity season

Average life span period in days ± SD of			
	Mated snails		Unmated
	Laid	Non laid	
Average ± SD	561.8 ± 102.1	272.8 ± 194.88	528.1 ± 122.8
Range	451 - 705	250 - 455	300 - 710
N	7	5	12

Life span of laying snails (life cycle, sexual maturity, pre-oviposition, oviposition and post-oviposition periods) averaged 561.8 ± 102.1 days. On the other hand, this period decreased to half in non-laying snails (life cycle, sexual maturity till death) and averaged 272.8 ± 194.9 days (N = 5).

Generation Period

Generation period (from egg to egg) ranged from 237 to 315 days with an average of 283.4 ± 29.3 days (N = 7).

Snail Growth Performance

Land snail weight increased with age and reached 0.095 g at five months (Table 3 and 4). Weight decreased gradually to 0.06 g at eleven months of age.

This pattern of weight with age was shown by the significant regression coefficient of 0.004 g per month (Figure 3).

The growth rate of snail shell indicated that early weight gain was highly significant, where the snail was grown by a highly significant rate of 69.68 %, then declined significantly by a rate of 9.14 % each month (Figure 5). The change in snail weight due to growth was also expressed in variable changes in width, height and volume. Shell width steadily increased over age and reached 4.32 mm by the ninth month of age and remained constant during the 10th and 11th months of age. This gradual shell width increase was shown in figure (4), however insignificant regression coefficient of 0.24 was found. The pattern of width change rate was high at early age (26.73%) and insignificantly decreased by 3.63% every month (Figure 4).

Shell height was also gradually increased with age to 9.46 mm at nine month of age, with a highly significant regression coefficient of 0.69 mm every month (Figure 4). The change rate in height was highly significant with an intercept of 33.86 mm, indicating that the change occurred in height at early age significantly determined the shell height during life span, although the change rate in height was then regressed insignificantly by 3.65 mm every month (Figure 5). Shell volume was increased from 1.60 mm^3 at hatch to 91.49 mm^3 at nine months of age. It was indicated that volume was highly significantly increased by 9.0 mm^3 every month (Figure 4).

Table 3. Weight, width, height, and volume (Average ± SD) of the land snail *Cochlicella acuta* shell during the life span

Age	Average ± SD			
	Weight	Width	Height	Volume
Hatch time	0	1.3 ± 0.2	1.7 ± 0.2	1.6 ± 0.7
1st month	0.02 ± 0.01	1.88 ± 0.3	2.23 ± 0.5	3.99 ± 1.3
2nd month	0.04 ± 0.02	3.1 ± 0.6	4.8 ± 1.2	23.3 ± 10.8
3rd month	0.08 ± 0.03	3.6 ± 0.6	6.9 ± 1.7	46.2 ± 19.9
4th month	0.09 ± 0.03	3.8 ± 0.7	7.3 ± 1.6	54.9 ± 25.5
5th month	0.09 ± 0.02	3.9 ± 0.7	7.9 ± 1.8	65.3 ± 31.3
6th month	0.09 ± 0.02	4.2 ± 0.7	8.6 ± 1.8	80.2 ± 34.2
7th month	0.08 ± 0.02	4.3 ± 0.7	8.9 ± 1.9	85.0 ± 35.6
8th month	0.07 ± 0.01	4.3 ± 0.7	9.3 ± 1.6	89.4 ± 34.9
9th month	0.07 ± 0.01	4.3 ± 0.6	9.5 ± 1.4	91.5 ± 33.5
10th month	0.06 ± 0.01	4.3 ± 0.6	9.5 ± 1.4	91.5 ± 33.5
11th month	0.06 ± 0.01	4.3 ± 0.6	9.5 ± 1.4	91.5 ± 33.5

Table 4. Monthly change rates (Average ± SD) in shell weight, width, height, and volume of the land snail *Cochlicella acuta* over life span

Age	Average ± SD			
	Weight rate	Width rate	Height rate	Volume rate
1st month	100.00	27.9 ± 12.1	21.7 ± 9.7	58.03 ± 16.28
2nd month	49.8 ± 18.6	35.9 ± 15.2	51.6 ± 12.5	79.19 ± 10.61
3rd month	51.5 ± 18.5	13.7 ± 14.8	29.2 ± 15.9	46.25 ± 21.03
4th month	17.9 ± 25.7	5.1 ± 9.6	5.02 ± 8.2	13.47 ± 19.35
5th month	-3.38 ± 29.1	3.3 ± 13.7	7.00 ± 8.6	11.12 ± 29.32
6th month	-10.03 ± 15.64	5.7 ± 11.1	7.5 ± 9.7	16.71 ± 20.78
7th month	-8.04 ± 16.2	1.5 ± 4.7	3.3 ± 5.7	5.92 ± 10.02
8th month	-11.9 ± 16.8	0.7 ± 2.6	4.5 ± 6.8	5.78 ± 7.66
9th month	-9.2 ± 12.9	0.56 ± 2.11	2.04 ± 5.7	3.11 ± 6.64
10th month	-8.8 ± 12.5	0	0	0
11th month	-5.14 ± 10.2	0	0	0

However the change rate in volume was highly significant throughout the snail shell life span, where it was reduced by 6.88 % every month (Figure 5). Significant positive correlation coefficients were found between snail weight rate and volume rate. However, the correlation coefficients between height or width and weight were in general insignificant.

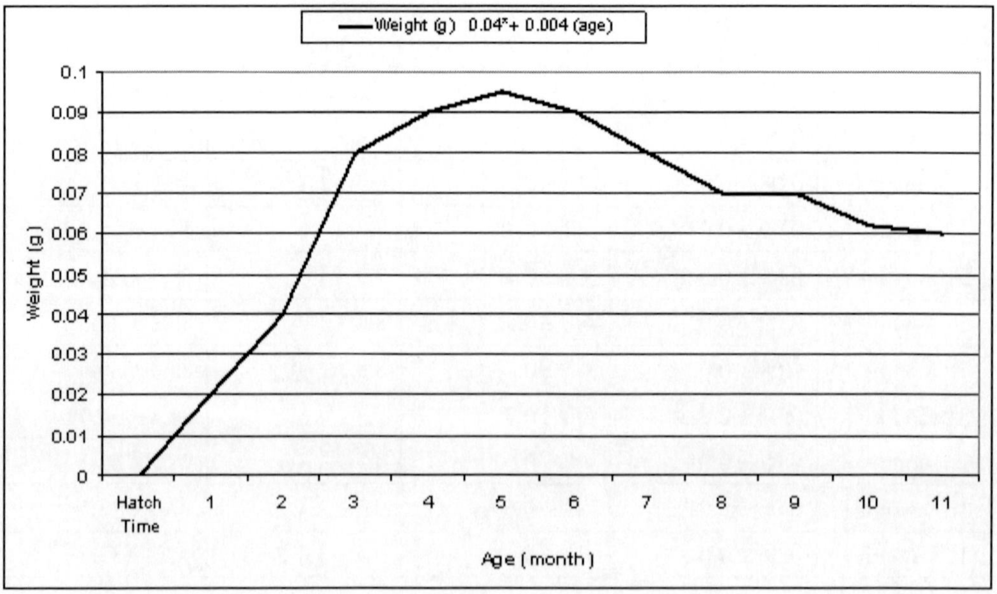

Figure 3. Shell weight of the land snail *Cochlicella acuta*.

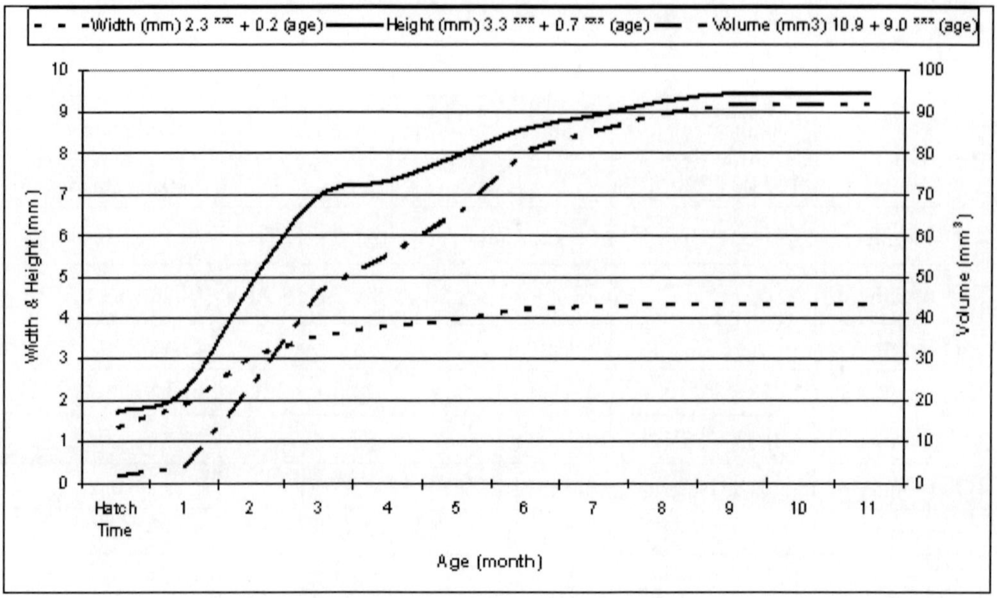

Figure 4. Width, height, and volume of *Cochlicella acuta*.

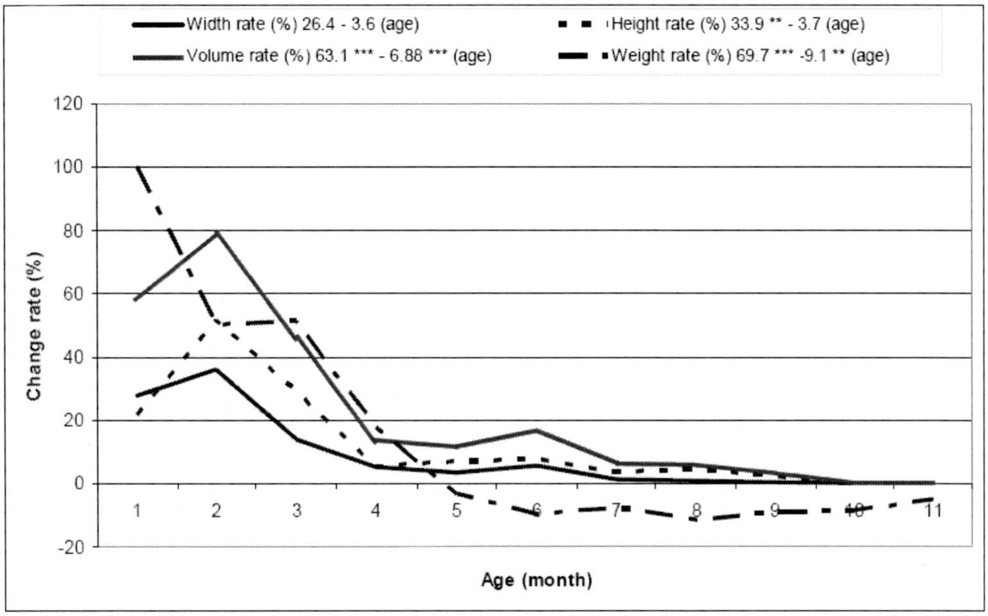

Figure 5. The monthly growth rate and change rates in shell width, height, volume of *Cochlicella acuta*.

DISCUSSION

In spite of the conical snail *C. acuta* is hermaphroditic, self-fertilization does not occur, thus cross mating was essential for oviposition as unmated snails did not lay any egg.

This result agree with Takeda (1983), Bride and Gomot (1991), Bride et al. (1991) and Mohamed and Ali (2009), who stated that egg production in several species of gastropods is stimulated by mating. Also, it was observed that one or both mated snails deposited eggs; it was rarely noticed that both mated snails could not lay eggs and this may be due to copulation failure. These observations are in accordance with Mohamed and Ali (2009) during their studies on the land snails *Monacha cartusiana* (Müller) and *Eobania vermiculata* (Müller). It was noticed that mated individuals of *C. acuta* required an average of 17.6 ± 6.1 days as a pre-oviposition period before depositing eggs.

This agree with Heller (1982) who recorded that *Theba pisana* (Müller) laid its eggs from 9.0 to 14.0 days after copulation and Mohamed and Ali (2009) who mentioned that *M. cartusiana* and *E. vermiculata* laid eggs after 2.7 ± 1.4 and 3.9 ± 1.9 days respectively from copulation.

The results revealed that the growth significance of snail at very early age effect life span thereafter. It was also concluded that the growth of snail shell is significantly explained by weight and volume. However, width and height were both insignificantly contributed to explain the growth pattern of snail life span.

REFERENCES

Baker, G. H. 1986. The biology and control of white snails (Mollusca: Helicidae), introduced pests in Australia CSIRO Australia, *Division of Entomology*, Technical Paper No. 25: 1 – 31.

Baker, G. H. 1989. Damage, population dynamics, movement and control of pest helicid snails in southern Australia. *In "Slugs and Snails in World Agriculture" British Crop Protection Council Monograph* No. 41: 175 –185.

Baur, B. and C. Raboud. 1988. Life History of the Land Snail *Arianta arbustorum* Along an Altitudinal Gradient. *Journal of Animal Ecology*, 57, 71 – 87.

Bride, J. and L. Gomot. 1991. Invitro demonstration of direct stimulation of the *Helix aspersa* albumen gland galactogen synthesis by extracts from the male accessory glands of the genital tract. *C. R. Academic Science Paris Ser.*, III (313): 565 – 572.

Bride, J., L. Gomot and A. S. M. Saleuddin. 1991. Mating and 20-hydroxyectysone cause in increased galactogen synthesis in the albumen gland explants of *Helix aspersa* (Mollusca). *Comparative Biochemistry and Physiology*, 98(B): 369 –373.

Heller, J. 1982. Natural history of *Thepa pisana* in Israel (Pulmonata: Helicidae). *Journal Zoology London*, 196: 475 – 487.

Kerney, M. P. and R. A. D. Cameron. 1979. *A field guide to the land snails of Britain and north – west Europe – Collins*, London.

Lewis, G. 1977. Polymorphism and selection in *Cochlicella acuta*. *Philosophical Transactions of the Royal Society of London*, Series B, 276: 399 – 451.

Mohamed, M. I. and R. F. Ali. 2009. Reproduction and life history in the two land snails *Monacha cartusiana* (Müller) and *Eobania vermiculata* (Müller) (Helicidae: Mollusca) in the laboratory. *Animal Biology Journal*, 1, (2), 99–107.

SAS. 1985, SAS/ STAT user's Guide: statistics. SAS Institute Inc., Cary, NC, US.

Takeda, N. 1983. Endocrine regulation of reproduction in the snail *Euhadre peliomphala*. *Molluscan Neuroen*., J. Lever and H. H. Boer, 106 – 111.

Section III: Studies on Genetic Aspects in Vertebrate and Invertebrate Animals

In: Vertebrate and Invertebrate Animal Approaches
Editor: José Rosa Gomes

ISBN: 978-1-63117-102-4

Chapter 27

KARYOTYPIC DATA DETECT INTERPOPULATIONAL VARIATION IN *PHYSALAEMUS OLFERSII* AND THE FIRST CASE OF SUPERNUMERARY CHROMOSOME IN THE GENUS

Mateus Milani[1], Carla Santana Cassini[2], Shirlei M. Recco-Pimentel[1] and Luciana Bolsoni Lourenço[1*]

[1]Departamento de Anatomia, Biologia Celular, Fisiologia e Biofísica, Instituto de Biologia, Universidade de Campinas – UNICAMP, Campinas, SP, Brazil

[2]Departamento de Biologia, Centro de Ciências Naturais e da Saúde, Universidade Federal do Espírito Santo, Maruípe, Vitória, ES, Brazil

ABSTRACT

Interesting morphological variations previously observed between populations of *P. olfersii* led some authors to believe that this taxon probably represents a species complex. In spite of that, the cytogenetic data available to date for *P. olfersii* were scarce and consisted of the description of the karyotype, with Ag-NOR sites mapped for specimens from one population from the Brazilian state of São Paulo, and a Giemsa-stained karyotype of one male of unknown origin. In the present work, we report an interpopulational NOR variation that differentiates one population from Minas Gerais State from others from the states of São Paulo and Rio de Janeiro. In addition, we describe a supernumerary chromosome found in *P. olfersii*, which is the first supernumerary chromosome reported for Leiuperidae.

Keywords: Cytogenetics, karyotype, *Physalaemus,* NOR, B chromosome

[*] Corresponding author: (bolsoni@unicamp.br/phone:+55 19 3521 6108).

INTRODUCTION

The genus *Physalaemus* (Leiuperidae, Anura), currently composed of 42 species (Frost, 2009), is the most specious genus in the family Leiuperidae. It is characterized by small to large-sized species widely distributed throughout South America (Nascimento et al., 2005) and has been the subject of numerous comparative cytogenetic studies (Silva et al., 1999, 2000; Amaral et al., 2000; Quinderé et al., 2009; Tomatis et al., 2009). The species *Physalaemus olfersii*, grouped with *P. aguirrei*, *P. soaresi,* and *P. maximus* by Nascimento et al. (2005), is widespread in the Atlantic Forest in southern and southeastern Brazil (Frost, 2009; Nascimento et al., 2005) and has been the focus of intriguing taxonomic questions. In describing the species *P. maximus*, Feio et al. (1999) interestingly reported morphological variations between specimens of *P. olfersii* from the Brazilian states of Rio de Janeiro and São Paulo, and specimens from Minas Gerais State, which were referred to as *P.* cf. *olfersii*. More recently, Cruz et al. (2008) analyzed specimens collected from the municipality of Guaratuba, in the state of Paraná, deposited in the MNRJ (Museu Nacional, Rio de Janeiro, RJ, Brazil), which were originally identified as *P. olfersii*. Based on morphological data, they described it as a new species, which was named *P. insperatus*. Additionally, considering the extensive morphological variation between populations of *P. olfersii*, Cruz et al. (2008) and Giaretta et al. (2009) suggested that this taxon probably represents a species complex.

To date, the cytogenetic data available for *P. olfersii* are scarce and consist of the description of the karyotype, with Ag-NOR sites mapped for specimens from one population from the Brazilian state of São Paulo (Silva et al., 2000), and a Giemsa-stained karyotype of one male of unknown origin (De Lucca et al., 1974). Thus, these data are not yet sufficient to contribute to the study of the evolutionary relationships regarding this taxon. Aiming to provide more data for understanding the intraspecific variation in *P. olfersii* and the chromosomal evolution in this genus, herein we report an interpopulational NOR variation that differentiates one population from Minas Gerais State from others from the states of São Paulo and Rio de Janeiro. In addition, we describe a supernumerary chromosome found in *P. olfersii*, which is the first supernumerary chromosome reported for Leiuperidae.

MATERIALS AND METHODS

Chromosome preparations were obtained from intestine and testis cellular suspensions of males of *P. olfersii* from three Brazilian localities: Teresópolis, Rio de Janeiro State (ZUEC 16250 and ZUEC 16251), Núcleo de Santa Virgínia, São Paulo State (ZUEC 16337), and Viçosa, Minas Gerais State (ZUEC 16247-16249). The animals were pre-treated with colchicine (2%) for at least 5 h and deeply anesthetized before removal of the intestines and testes following the method described by Schmid (1978) and Schmid et al. (1979), with some modifications. Immediately after removal, the intestines were opened with surgical scissors and the epithelium was exposed to a 0.9% sodium citrate solution for about 40 min. The intestines were transferred to a Petri dish with (3:1) methanol-acetic acid and the epithelium was scraped vigorously. For a better cell disaggregation, the material was resuspended several times with a Pasteur pipette. The testes samples were immersed immediately in cold distilled water and fixed in (3:1) methanol-acetic acid after about 20 min. In this fixative solution, the testes were cut several times with surgical scissors. The intestine and testicle cellular suspension were centrifuged at 1000 rpm for 5 min, the supernatants were discarded, and the precipitates were resuspended in (3:1) methanol-acetic acid. This operation was repeated twice. All the specimens were sampled under a permit issued by the Instituto Brasileiro do Meio Ambiente e dos Recursos Naturais Renováveis (IBAMA/10678-1) and were deposited in the Museu de Zoologia "Adão José Cardoso" (ZUEC), State University of Campinas (UNICAMP), Campinas, São Paulo State, Brazil

Chromosomes were stained with 10% Giemsa, C-banded (protocol by King, 1980), and impregnated with silver by the Ag-NOR method (according to Howell & Black, 1980). Metaphases submitted to C-banding were destained in 100% ethanol and simultaneously stained with fluorochromes DAPI (500 μg/mL) and mythramycin (MM). The metaphases were analyzed and photographed under an Olympus microscope. The chromosomes were measured with the software Micromeasure 3.3 (Dept. of Biology, Colorado State University) and classified according to Green & Sessions (1991).

RESULTS

Physalaemus olfersii from Teresópolis (Rio de Janeiro State) and Santa Virgínia (São Paulo State).

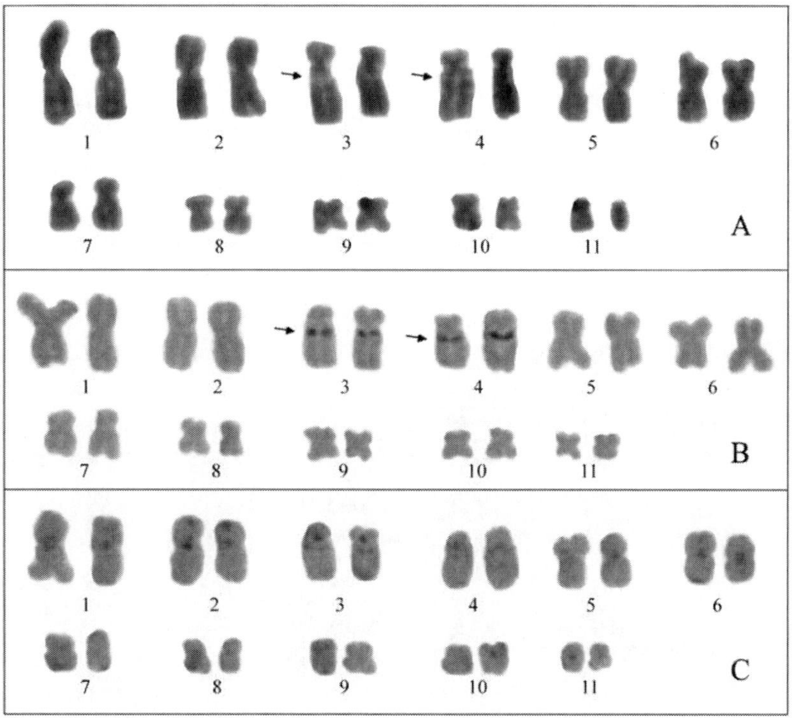

Figure 1. Karyotype found in *P. olfersii* from Teresópolis (Rio de Janeiro State) and Santa Virginia (São Paulo State) submitted to Giemsa staining (A), AgNOR impregnation (B), and C-banding (C).

The karyotype of these individuals had 2n=22, with two pairs of large metacentric chromosomes (pairs 1 and 2), five pairs of medium-sized metacentric (pairs 5 and 6) or submetacentric chromosomes (pairs 3, 4, and 7), and four small pairs of metacentric/submetacentric chromosomes (pairs 8-11) (Figure 1A). The NORs were located in the long arm of chromosomes 3 and 4, near the centromere (Figure 1B), and coincided with secondary constrictions observed in Giemsa-stained metaphases. C-banding showed heterochromatic bands in the centromeric region of all chromosomes and a pericentromeric band coincident with the NOR in chromosome 3. In some metaphases, terminal heterochromatic bands were also observed in all chromosomes, as well as faint pericentromeric C-bands in chromosomes 5, 6, 8, and 10 (Figure 1C). Simultaneous staining of C-banded metaphases with DAPI and mythramycin revealed that all the heterochromatin regions that were detected by the C-banding method were rich in AT; therefore they stained with DAPI (Figure 1D), but not with mythramycin (data not shown).

Physalaemus olfersii from Viçosa (Minas Gerais State)

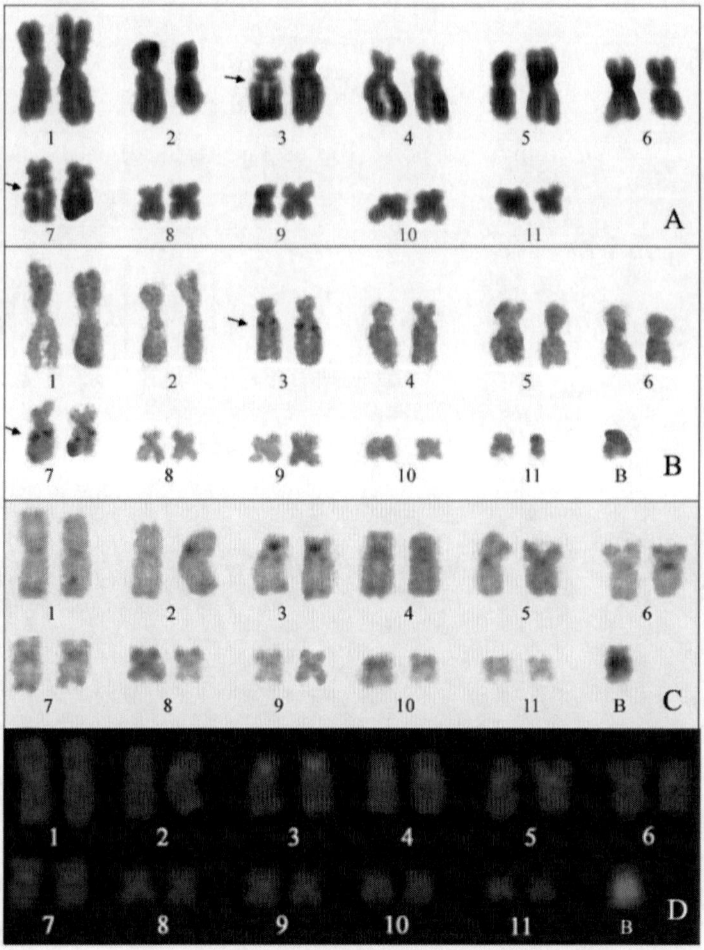

Figure 2. Karyotypes of *P. olfersii* from Viçosa (Minas Gerais State) submitted to Giemsa staining (**A**), AgNOR impregnation (**B**), C-banding (**C**) and DAPI staining after C-banding (**D**). The karyotype with 2n=22 shown in **A** is from the specimen ZUEC 16247. In **B-D**, karyotypes with 2n=23 found in the specimens ZUEC 16248 and ZUEC 16249. Note the supranumerary element (B chromosome) of these complements.

The karyotype of the specimen ZUEC 16247 had 2n=22 (Figure 2A), while two other specimens (ZUEC 16248 and ZUEC 16249) from Viçosa had a diploid complement with 23 chromosomes, with a small acrocentric supernumerary chromosome (Figure 2B-D). This extra chromosome had a heterochromatic DAPI-positive long arm (Figure 2D) and was mitotically stable, being observed in all 20 of the intestinal metaphases of ZUEC 16248 and in 37 metaphases of ZUEC 16249. In meiotic metaphases, this supernumerary chromosome was not associated with any other chromosome (Figure 3).

The A complement of this karyotype and the 2n=22 karyotype of specimen ZUEC 16247 had two pairs of large metacentric chromosomes (pairs 1 and 2), five pairs of medium-sized metacentric (pairs 5 and 6) or submetacentric chromosomes (pairs 3, 4, and 7), and four small pairs of metacentric/submetacentric chromosomes (pairs 8-11) (Figure 2A). In all these karyotypes, the NORs were located near the centromere in the long arm of chromosome pairs 3 and 7 (Figure 2B), and coincided with secondary constrictions in Giemsa-stained metaphases. In C-banded metaphases, all the centromeric regions were strongly stained, as was a pericentromeric region in the long arm of chromosome 3, which coincided with a NOR. In some metaphases, faint pericentromeric C-bands in chromosomes 5, 6, 8, and 10, and terminal bands in all the chromosomes were also observed (Figure 2B). DAPI fluorochrome revealed the same heterochromatic regions stained by Giemsa in C-banded metaphases (Figure 2D).

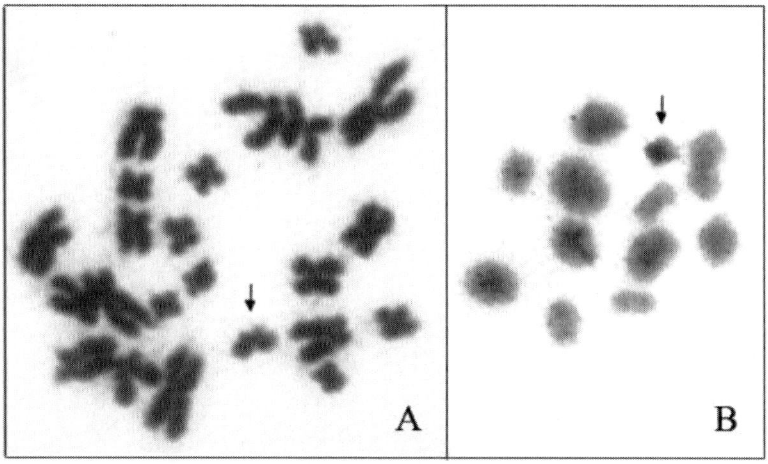

Figure 3. Giemsa-stained mitotic metaphase (**A**) and C-banded meiotic metaphase I from the specimen ZUEC 16249 (**B**). The arrows indicate the supranumerary chromosome.

Discussion

The chromosomal morphology and the karyotype NOR sites found in specimens from the municipalities of Teresópolis (Rio de Janeiro State) and Santa Virgínia (São Paulo State) were the same described previously for specimens from the municipality of Rio Branco (São Paulo State) by Silva et al. (2000). On the other hand, the karyotype of the specimens from the municipality of Viçosa (Minas Gerais State) differed from all of the other *P. olfersii* karyotypes by presenting the NORs in chromosomes 3 and 7, and not in chromosomes 3 and

4. Although chromosomes 3 and 7 of all these karyotypes were very similar in morphology, the NOR-bearing chromosome 7 of the Viçosa karyotype was conspicuously smaller than the NOR-bearing chromosome 4 of the other *P. olfersii* karyotypes. This seems, therefore, to be a real difference between the Viçosa and the other *P. olfersii* karyotypes, but the absence of other cytogenetic markers in chromosomes 3 and 7 prevents a proper interpretation of the rearrangement responsible for this.

This cytogenetic character is not the only trait which distinguishes Viçosa specimens from the *P. olfersii* specimens from Teresópolis and other localities. In the description of *P. maximus*, Feio et al. (1999) referred to specimens from the municipalities of Viçosa and Arapongas (Minas Gerais State), as *Physalaemus* cf. *olfersii* instead of *P. olfersii*, and reported the presence of an inner tarsal tubercle in this specimen, which is absent in *P. olfersii*. A detailed study involving these populations and their taxonomy has been done by one of the authors of this study (C.S.C.) and will be published soon.

Moreover, other interesting finding reported here is the occurrence of the supernumerary chromosome in the Viçosa population. This chromosome, the smallest in the complement, is highly heterochromatic, a characteristic commonly observed in supernumerary chromosomes (for review, see Camacho et al., 2000), and mitotically stable. Apparently the B chromosome that we have found is not related to sex determination, since a male without the B chromosome was found together with two males that had it. This is the first case of the supernumerary chromosome reported in the genus *Physalaemus*. Its frequency in the Viçosa population may be very high, since among the three specimens analyzed, two had one extra chromosome. Further investigations on this supernumerary chromosome, including a better characterization of its heterochromatin, may reveal interesting data for an evolutionary study of these anurans' karyotypes.

Considering chromosomal morphology, the karyotypes of *P. olfersii* are very similar to those found in two other species of *P. olfersii* group, *P. aguirrei* and *P. soaresi* (De Lucca et al., 1974), and in species of *P. albifrons* (Denaro, 1972, Amaral et al., 2000, Silva et al., 2000), *P. cuvieri* (Denaro, 1972, De Lucca et al., 1974, Silva et al., 1999, Quinderé et al., 2009) and *P. gracilis* groups (De Lucca et al., 1974), and also in *P. henselii* and its related species, *P. riograndensis* (Tomatis et al., 2009). On the other hand, the karyotype of *P. olfersii* clearly differs from that of the other species currently assigned to the *P. henselii* group, *P. fernandezae* (Tomatis et al., 2009), and from species of the *P. signifer* group (Silva et al., 2000), in which an uniarmed chromosome 11 is present.

In the majority of the *Physalaemus* species that have been cytogenetically studied to date, the NOR-bearing chromosomes are among the smaller chromosome pairs of the diploid complement. In the karyotypes found in the *P. albifrons*, *P. biligonigerus,* and *P. marmoratus* species (*P. albifrons* group), in *P. riograndensis* (*P. henselii* group) and *P. crombiei* (*P. signifer* group), and in karyotypes found in several populations of the *P. cuvieri* species (*P. cuvieri* group), biarmed chromosomes, classified as chromosomes 8, 9, or 11, had a distal NOR site (Amaral et al., 2000; Silva et al., 1999, 2000; Tomatis et al., 2009; Quinderé et al., 2009). On the other hand, in *P. spiniger* (*P. spiniger* group) and *P. fernandezae* (*P. henselii* group) karyotypes, the NOR is located in the uniarmed chromosome 11 (Silva et al, 2000; Tomatis et al., 2009). A similar NOR-bearing chromosome 11 is also present in *Eupemphix nattereri*, a species previously considered of the *Physalaemus* genus (see Lynch, 1970). Despite the similarities pointed out above, a proper polarization of the characters related to the NOR occurrence is not yet possible.

In the karyotypes of *Edahorhina perezi* (Lourenço et al., 2000) and of several species of *Pleurodema* (Barrio & Rinaldi de Chierri, 1970; Lourenço et al., 2006), two genera considered to be phylogenetically closely related to *Physalaemus* (Lynch, 1970), the NOR is also located in the long arm of the biarmed chromosome 8, as observed in some species of *Physalaemus*. Therefore, it is possible to suppose that this may be a plesiomorphic condition related to the NOR location, while the NORs in large (chromosomes 3 and 4) and medium-sized (chromosome 7) chromosomes of *P. olfersii* are apomorphic conditions. A similar inference was also considered by Tomatis et al. (2009) for the NOR-bearing medium-sized biarmed chromosome of *P. henselii*. In the same way, the NOR-bearing uniarmed chromosome 11 observed in some species of *Physalaemus* may represent one or more apomorphic conditions, but for a proper interpretation of these characters, a phylogenetic analysis of the *Physalaemus* genus is absolutely necessary.

In conclusion, the chromosomal data presented herein amplify our cytogenetic knowledge of the genus *Physalaemus* and may help further the investigation of the karyotypic divergence observed in this group.

ACKNOWLEDGMENTS

This work was supported by FAPESP (Fundação de Amparo à Pesquisa do Estado de São Paulo), Capes (Coordenação de Aperfeiçoamento de Pessoal de Nível Superior) and CNPq (Casadinho-620163/2008-9).

REFERENCES

Amaral, MJLV; Cardoso, AJ; Recco-Pimentel, SM. Cytogenetic analysis of three *Physalaemus* species (Amphibia, Anura). *Caryologia*, 2002 53, 283-288.

Barrio, A; Rinaldi De Chieri, P. Estudios citogenéticos sobre el género *Pleurodema* y sus consecuencias evolutivas (Amphibia, Anura, Leptodactylidae). *Physis*, 1970 30, 309–319.

Camacho, JPM; Sharbel, TF; Beukeboom, LW. B-chromosome evolution. *Phil. Trans. R. Soc. Lond.* B, 2000 355, 163-178.

Cruz, CAG; Cassini, CS; Caramaschi, U. A new species of the genus *Physalaemus* Fitzinger, 1826 (Anura, Liuperidae) from southern Brazil. *South Am. J. Herpetol.*, 2008 3, 239-243.

De Lucca, EJ; Jim J; Foresti F. Chromosomal studies in twelve species of Leptodactylidae and one Brachycephalidae. *Caryologia*, 1974 27, 183-191.

Denaro, L. Karyotypes of Leptodactylidae Anurans. *J. Herpetol.* 1972 6, 71-74.

Feio, RN; Pombal-Jr, JP; Caramaschi, U. New *Physalaemus* (Anura: Leptodactylidae) from the Atlantic Forest of Minas Gerais, Brazil. *Copeia* 1999 1999, 141-145.

Frost, DR. Amphibians of the world: an on line reference. V5.3. February 12, 2009. Avaiable from: http://research.amnh.org/herpetology/amphibia/index.html.

Green, DM; Sessions, SK. Nomenclature for chromosomes. In: Green DM and Sessions SK, editors. *Amphibian cytogenetics and evolution*. San Diego: Academic Press; 1991; 431-432.

Howell, WM; Black, DA. Controlled silver staining of nucleolus organizer regions with a protective colloidal developer: a 1-step method. *Experientia*, 1980 36, 1014-1015.

King, M. C-banding studies on Australian hylid frogs: secondary constriction structure and the concept of euchromatin transformation. *Chromosoma*, 1980 80, 191-217.

Lourenço, LB; Recco-Pimentel, SM; Cardoso, AJ. Cytogenetics of *Edalorhina perezi* (Anura, Leptodactylidae). *Cytologia*, 2000 65, 359-363.

Lourenço, LB.; Nascimento, JAA; Andrade, GV; Rossa-Feres, DC; Recco-Pimentel, SM. Chromossomal analysis of the leptodactylids *Pleuroderma diplolistris* and *Physalaemus nattereri* (Amphibia, Anura). *Amphibia-Reptilia*, 2006 27, 481-489.

Lynch, JD. Systematic status of the american leptodactylid frog genera *Engystomops, Eupemphix*, and *Physalaemus*. *Copeia,* 1970 1970, 488-496.

Nascimento, LB; Caramaschi, U.; Cruz, CAG. Taxonomic review of the species groups of the genus *Physalaemus* Fitzinger, 1826 with revalidation of the genera *Engystomops* Jiménez-de-La-Espada, 1872 and *Eupemphix* Steindachner, 1863 (Amphibia, Anura, Leptodactylidae). *Arq. Mus. Nac., Rio de Janeiro*, 2005 63, 297-320.

Schmid, M. Chromosome banding in amphibia.I Constitutive heterochromatin and nucleolus organizers regions in *Bufo* and *Hyla. Chromosoma*, 1978 66, 361-388.

Schmid, M; Olert, J; Klett, C. Chromosome banding in Amphibia. III. Sex chromosomes in *Triturus*. *Chromosoma*, 1979 71, 29-55.

Silva, APZ; Haddad, CFB; Kasahara, S. Nucleolus organizer regions in *Physalaemus cuvieri* (Anura, Leptodactylidae), with evidence of a unique case of Ag-NOR variability. *Hereditas,* 1999 131, 135-141.

Silva, APZ; Junior, FAB; Haddad, CFB; Kasahara, S. Karyotypes and Nucleolus organizer regions in four species of the genus *Physalaemus* (Anura, Leptodactylidae). *Ilheríngia, Sér. Zool.*, 2000 88, 159-164.

Tomatis, C; Baldo, D; Kolenc, C; Borteiro, C. Chromossomal variations in the species of the *Physalaemus hanselii* group (Anura: Leiuperidae), *J. Herpetol.*, 2009 43, 555-560.

In: Vertebrate and Invertebrate Animal Approaches
Editor: José Rosa Gomes
ISBN: 978-1-63117-102-4

Chapter 28

GENETIC DIVERSITY AND DIFFERENT ECOLOGICAL CONDITIONS IN *SALAMANDRA INFRAIMMACULATA* LARVAE FROM VARIOUS BREEDING SITES

***Tali Goldberg*[1, 2]*, Eviatar Nevo*[2] *and Gad Degani*[1*]**
[1]School of Science and Technology, Tel Hai Academic College, Upper Galilee, Israel
[2]Institute of Evolution, Faculty of Sciences and Science Education, University of Haifa, Haifa, Israel

ABSTRACT

The present study examined the adaptive differentiation of ten populations of *Salamandra infraimmaculata* larvae in various breeding sites in northern Israel. Their ecological parameters were compared to the genetic variation in their DNA, using the random amplification of polymorphic DNA polymerase chain reaction (RAPD PCR) method. *S. infraimmaculata* larvae remained for 6-10 months in stable water bodies (streams and springs) and rock pool holes, where water temperatures were relatively constant. In temporary water bodies, e.g., winter ponds, the larvae grew faster, completing metamorphosis within 2-5 months. Temperature and oxygen quality parameters of stable water bodies differed from those of temporary water bodies, even when the ponds were situated in the same geographic area. DNA variation among salamander larvae was assessed by a number of different primers, but the best results were obtained with the primers, OPA9 (5' GGGTAACGCC 3') and OPA11 (5' CAATCGCCGT 3'). Populations from a variety of stable water bodies were found to be similar, although they were located at sites with different ecological conditions, i.e., Navoraya spring and Tel-Dan stream (primer OPA9), and Balad spring and Humema spring (primer OPA11). Moreover, a high similarity was found among the populations of three temporary water bodies: Fara pond, Manof pond and Sasa pond. On the other hand, although some breeding sites were located relatively close to each other, Dovev pond and Humema spring (primers OPA9, OPA11), they had different ecological conditions and a

* Corresponding author: MIGAL-Galilee Technology Center, P.O. Box 831, Kiryat Shmona 11016, Israel. Tel: +972-4-6953544. Fax: +972-4-6944980. e-mail: gad@migal.org.il.

relatively low genetic similarity. The average rate of increase in water temperature in temporary water bodies was higher than in stable water bodies.

In conclusion, the results of this study, along with the results of our former studies, support the hypothesis that the genetics of the populations from stable breeding sites (relatively constant water temperatures) differed from that of the populations from other less predictable breeding sites.

Keywords: Genetics; mitochondria; xeric habitat; Salamandra infraimmaculata; adaptation; breeding site

1. Introduction

Under appropriate conditions, adaptive and genetic divergence of sympatric populations can occur relatively quickly, on the order of hundreds to thousands of generations. Degani et al., 1999 and Steinfartz et al., 2000. support the hypothesis that habitat selection occurs within the genus of *Salamandra*, even in a relatively small area where the ecological conditions vary.

Salamanders of *Salamandra* are present throughout Europe, North Africa and East Asia, surviving in various habitats and climates (reviewed by Alcobendas 1994). Based on the morphological data, the species *S. salamandra* is subdivided into 16 subspecies found throughout Europe, the Middle East and North Africa (Escoriza. et al., 2006; Veith 1994) tudies have examined genetic variation relative to ecological conditions, according to this systematic variation, and concluded that there is a correlation between molecular variations and geographic location. In the present study, we investigated the genetic variation at molecular level in ten different populations of *S. infraimmaculata* found at various breeding sites in a relatively small area in northern Israel.

The subspecies, which may be differentiated on the basis of plasma protein electrophoresis and allozyme data, dramatically vary one from the another. Mizuno and MacGregor (1974) and Mizuno et al. (1976) used chromosomes and DNA sequences to study the evolution of the salamanders belonging to the genus *Plethodonital*. Based on the sequence analysis of the mitochondrial D-loop region and geological dates, Steinfartz et al. (2000) suggested that five major monophyletic groups exist in Europe (*S. salamandra, S. infraimmaculata, S. corsica, S. atra* and *S. lanzai*) and one, *S. algira,* occurs in Africa. The salamanders in Israel belong to *S. infraimmaculata*. (Goldberg et al., 2009a) studied the molecular DNA variation of the mitochondrial D-loop region and cytochrome *b* among *S. infraimmaculata* larvae populations, representing eight breeding sites in Israel, and showed that although the molecular polymorphisms in mitochondrial DNA are small, they reflect a sharp ecological separation between seasonal breeding sites and permanent water sources. These are presumably adaptive changes caused by natural selection. In the present study, the same hypothesis of habitat selection was tested on nuclear DNA by a different method.

S. infraimmaculata is distributed as isolated populations in breeding sites in northern Israel. There are physiological, morphological and biological differences among its isolated populations, which are affected by habitat conditions. The various types of breeding sites are situated in three different regions of northern Israel: the relatively drier Mount Carmel, central Galilee and foothills of Mount Hermon, where predictable (perennial springs and streams) and unpredictable (seasonal water bodies) breeding sites occur. There are many types of breeding sites in xeric areas, including streams, springs, rock pools, rain pools and large ponds, where water is available during different periods of time and under various conditions. No variations in the dorsal pattern, serum protein or ten enzyme systems with 14 loci have been found among the populations of these areas. The study of the hypothesis, proposing that habitat selection in a relatively small area has an effect on genetic variation, is appropriate to be conducted in Israel, since there are various habitats that have very extreme conditions for *S. infraimmaculata*. Salamanders are known to breed in various habitats. The molecular DNA variation in salamanders from two habitats has been studied, using the random amplification of polymorphic DNA polymerase chain reaction (RAPD PCR) method. Genetic variation was discovered between the populations of the two habitats, but not between the two populations in the semi-arid habitat. In order to examine the hypothesis that habitat selection has an effect on the genetic level, genetic variation was examined by RAPD PCR on DNA obtained from the samples of ten populations, collected from streams, springs, rock pool holes and rain ponds in northern Israel. The habitats contained both larvae and mature individuals, which had been exposed to various conditions.

2. Material and Methods

2.1. Samples

Forty nine tissue samples of *S. infraimmaculata* larvae from different populations were obtained from ten locations on Mount Carmel, in the Galilee and on the foothills of Mount Hermon, all located in northern Israel. The habitats include unpredictable and predictable breeding sites within a relatively small area (Figure 1, Tables 1 and 2).

Water is available from winter to next summer (October to July) in the unpredictable water bodies, including winter ponds and rock pool holes (Pi1, Pi2, Po1, Po2, Po3 and Po4). The predictable water bodies are available all year round, including springs and one stream (Sp1, Sp2, Sp3 and St1).

Table 1. Ten breeding sites of S. infraimmaculata in northern Israel, examined in this study: Streams and springs are permanent, while ponds and pits are temporary

Label	Name	Type	Longitude	Latitude	Height (m.a.s.l)
Sp1	Balad	Spring	32°43'13" N	35°04'17" E	446
Sp2	Humema	Spring	33°00'28" N	35°23'43" E	900
Sp3	Navoraya	Spring	32°59'47" N	35°30'39" E	663
St1	Tel-Dan	Stream	33°14'53" N	35°39'10" E	190
Pi1	Maalot	Rock pool hole	33°00'06" N	35°16'26" E	596
Pi2	Nimrod	Rock pool hole	33°15'05" N	35°42'48" E	760
Po1	Manof	Pond	32°50'58" N	35°13'52" E	340
Po2	Dovev	Pond	33°03'05" N	35°24'54" E	765
Po3	Sasa	Pond	33°01'58" N	35°23'30" E	810
Po4	Fara	Pond	33°03'58" N	35°27'39" E	676

Figure 1. Locations where salamanders were sampled- Sp1: Balad spring, Sp2: Humema spring, Sp3: Navoraya spring, St1: Tel-Dan stream, Pi1: Maalot rock pool hole, Pi2: Nimrod rock pool hole, Po1: Manof pond, Po2: Dovev pond, Po3: Sasa pond and Po4: Fara pond.

Table 1. Ten breeding sites of *S. infraimmaculata* in northern Israel, examined in this study: Streams and springs are permanent, while ponds and pits are temporary

Label	Name	Type	Longitude	Latitude	Height (m.a.s.l)
Sp1	Balad	Spring	32°43'13" N	35°04'17" E	446
Sp2	Humema	Spring	33°00'28" N	35°23'43" E	900
Sp3	Navoraya	Spring	32°59'47" N	35°30'39" E	663
St1	Tel-Dan	Stream	33°14'53" N	35°39'10" E	190
Pi1	Maalot	Rock pool hole	33°00'06" N	35°16'26" E	596
Pi2	Nimrod	Rock pool hole	33°15'05" N	35°42'48" E	760
Po1	Manof	Pond	32°50'58" N	35°13'52" E	340
Po2	Dovev	Pond	33°03'05" N	35°24'54" E	765
Po3	Sasa	Pond	33°01'58" N	35°23'30" E	810
Po4	Fara	Pond	33°03'58" N	35°27'39" E	676

Table 2. Distances (km) between the ten breeding sites of *S. infraimmaculata*

	Po1	Pi1	Sp2	Po2	Po4	Sp3	St1	Po3	Pi2
Sp1	21.5	36	44.5	48.5	51	50.5	79	46.5	84
Po1		16.5	23	28	31.5	31	59	27	64
Pi1			12	14	18.5	21.5	44.5	14	49.5
Sp2				5	8	10.5	35.5	6	40.5
Po2					5.5	11	31.5	1	36.5
Po4						9	27	4.5	32
Sp3							30	12.5	35
St1								36	5.5
Po3									37

The water temperatures of the springs, stream and rock pool holes remain relatively constant. Once every two weeks, during four seasons throughout the years 2005-2008, the water temperature, dissolved oxygen, conductivity, pH and ammonium of each wetland were determined, using the methods described by Goldberg et al. (2009b). In addition, larvae of *S. infraimmaculata* were collected by net from each breeding site and identified.

These measurements revealed that the variations among habitats, to which the larvae are exposed, might have an effect on their molecular variations.

2.2. DNA Extraction, Amplification and Sequencing

Genomic DNA was extracted from the clipped or whole tails of larvae with the QIAamp DNA Mini Kit (Qiagen, Germany). The kit includes proteinase K lysis of tissue and specific DNA binding to the QIAamp silica-gel membrane through which contaminants pass, as described previously by Pearlson and Degani (2007). In order to find the primers that would suit the study on DNA variation among salamander larvae, different primers were tested (Table 3). Among the primers that gave the best results were OPA9 (5' GGGTAACGCC 3') and OPA11 (5' CAATCGCCGT 3'), and therefore they were chosen for this study.

PCR amplification was performed in a 50 μl solution containing 10 mM Tris-HCl, 50 mM KCl, 2.5 mM $MgCl_2$, 0.5 mM of each dNTP, 2 μM of primer, 10-500 ng gemonic DNA and 2.5 units of Taq DNA polymerase (Promega, USA). The reaction, 3 min denaturation at 94 °C, followed by 32 cycles of 1 min at 94 °C, 1 min at 35 °C and 1 min at 72 °C, with an additional 5 min of elongation at 72 °C following the last cycle, was performed on a PTC-150 MiniCycler (MJ Research, USA). After amplification, the PCR products were separated by electrophoresis on a 1.5 % agarose gel, stained with ethidium bromide. For each primer, samples were run on one gel in order to eliminate variations among gels.

Table 3. Primers used in this study

Primer	Sequence 5'—3'
OPA-06	GGTCCCTGAC
OPA-09	GGGTAACGCC
OPA-11	CAATCGCCGT
OPA-16	AGCCAGCGAA
OPA-17	GACCGCTTGT
OPA-19	CAAACGTCGG
OPA-20	GTTGCGATCC

2.3. Analysis

The RAPD patterns were scored for absence/presence of bands and coded as 1/2, respectively. The level of genetic distance was estimated by Nei and Li's (1979) distance matrix, using Tools for Population Genetic Analyses (TFPGA) software. The evolutionary history was inferred using the Neighbor-Joining method. Phylogenetic analyses were conducted by MEGA4 software (Tamura 2007).

Analysis of molecular variance (AMOVA, Excoffier 1992) is a method used to estimate population differentiation directly from molecular data. It tests hypotheses with respect to such differentiation. Types of molecular data, such as those from molecular markers, direct RAPD data or phylogenetic trees, based on such data, may be analyzed using this method.

3. RESULTS

The time periods during which the salamander larvae were found in the various habitats are presented in Figure 2. In the stable water bodies, *S. infraimmaculata* larvae grew for 6-10 months, while in the temporary ones, the larvae grew faster and completed metamorphosis within 2-5 months. Differences in temperature parameters were detected between stable and temporary water bodies, although the ponds were situated in the same geographical area and not far away from each other (Tables 1 and 2). Among the abiotic factors of the water bodies, temperature seemed to be the most important parameter. *S. infraimmaculata* larvae were found in water bodies at temperatures ranging about 2–19 °C (Figure 2, Table 4). Temperatures rose from winter to summer in all water bodies, but the average increasing rate of the water temperature in temporary water bodies (ponds) was higher than that in stable water bodies (streams and springs).

Table 4. The average (± SD) abiotic parameters during larval growth period in the various water bodies

Label	Site	Water temp. (°C)	Oxygen (mg/L)	EC (uS/cm)	pH	Ammonium (mg/L)
Sp1	Balad spring	16.4±2.6	10.7±5.0	1286±129	8.1±0.3	0.126±0.11
Sp2	Humema spring	13.5±2.8	8.2±2.8	852±76	8.5±0.4	0.094±0.09
Sp3	Navoraya spring	9.7±3.4	4.9±3.1	688±148	7.6±0.4	0.392±0.50
St1	Tel-Dan stream	16.0±1.3	10.3±4.4	335±21	8.3±0.3	0.059±0.03
Pi1	Maalot rock pool hole	12.8±2.8	2.2±2.4	640±102	7.8±0.4	0.373±1.00
Pi2	Nimrod rock pool hole	12.6±4.5	8.3±4.7	539±97	8.6±0.6	0.223±0.20
Po1	Manof pond	12.2±3.3	11.3±3.9	254±95	9.0±0.9	0.168±0.17
Po2	Dovev pond	10.4±3.9	10.0±3.0	334±48	8.5±0.6	0.210±0.50
Po3	Sasa pond	12.5±4.3	12.1±2.6	420±43	9.4±0.7	0.336±0.35
Po4	Fara pond	10.1±5.8	11.5±4.2	861±155	8.3±0.7	0.919±0.80

For example, the water temperature of Balad spring (Sp1) ascended at an average rate of 0.5 °C per month during the eight months of the breeding period, while that of Dovev pond (Po2) was 3 °C per month during the short three-month breeding period (Figure 2). The differences in temperatures among various breeding sites were significant (ANOVA, $P<0.05$).

The oxygen concentration reflected the water temperature, sun-light exposure, and flora and organism densities in a water body. In this study, stable water bodies and pits usually had oxygen concentrations under 10 mg/L (Table 4). The extreme values of electric conductivity (EC), pH and ammonium concentrations (Table 4) might hint to water contamination, which could produce mutations.

A phylogenetic analysis, which was conducted by MEGA4 in order to obtain the similarity values (F) between the RAPD-PCR products (Figure 3) of one breeding site and the others, is presented in Figure 4.

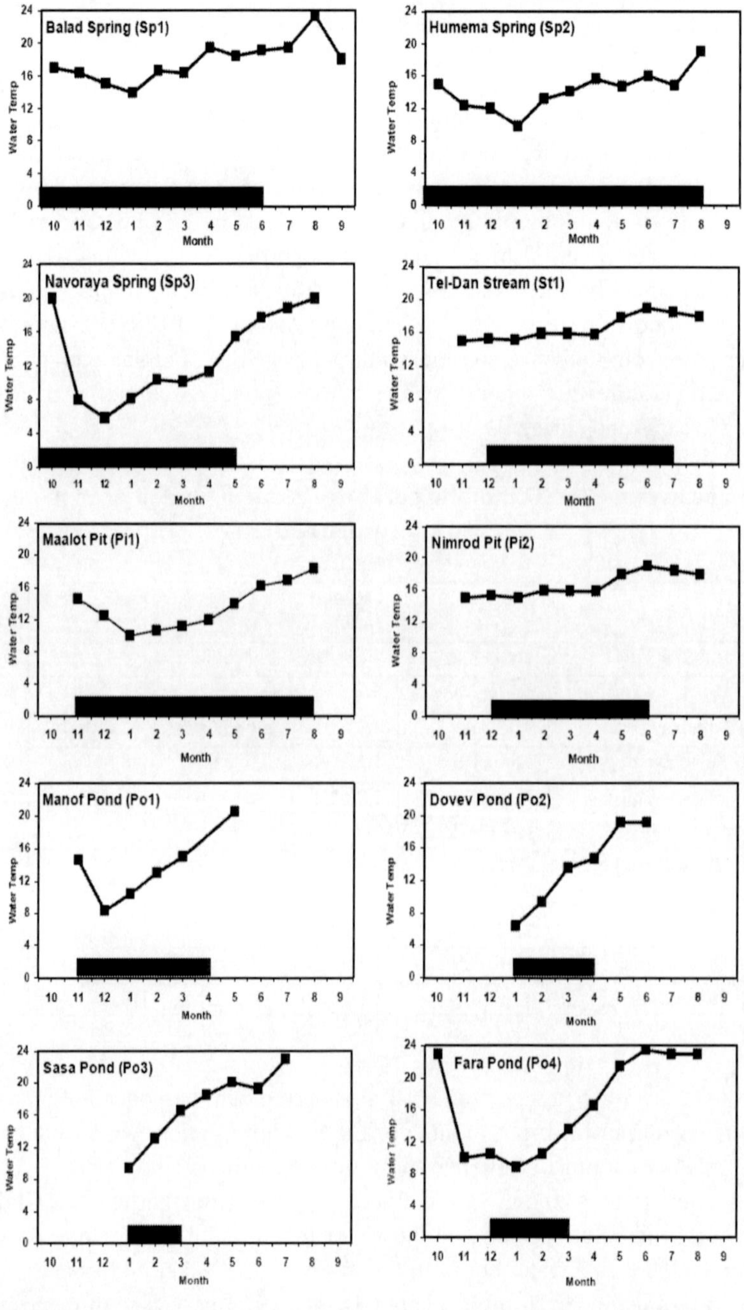

Figure 2. Temperature changes and presence of *S. infraimmaculata* larvae at different breeding sites throughout the year.

The analysis showed a high similarity among the stable water bodies, Sp3 and St1, using primer OPA9 (Figure 4A) and Sp1, Sp2 and Po2, using primer OPA11 (Figure 4B). Moreover, when using these two primers, a high similarity was found among the populations of the three temporary water bodies, Po1, Po3 and Po4 (Figure 4).

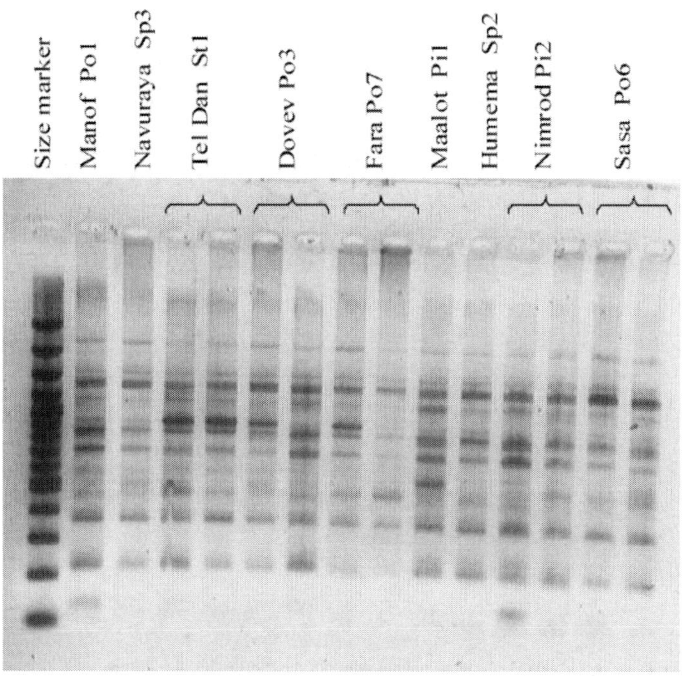

Figure 3. An example of a RAPD-PCR band pattern obtained using the OPA11 primer. Only the larvae found to be different are represented.

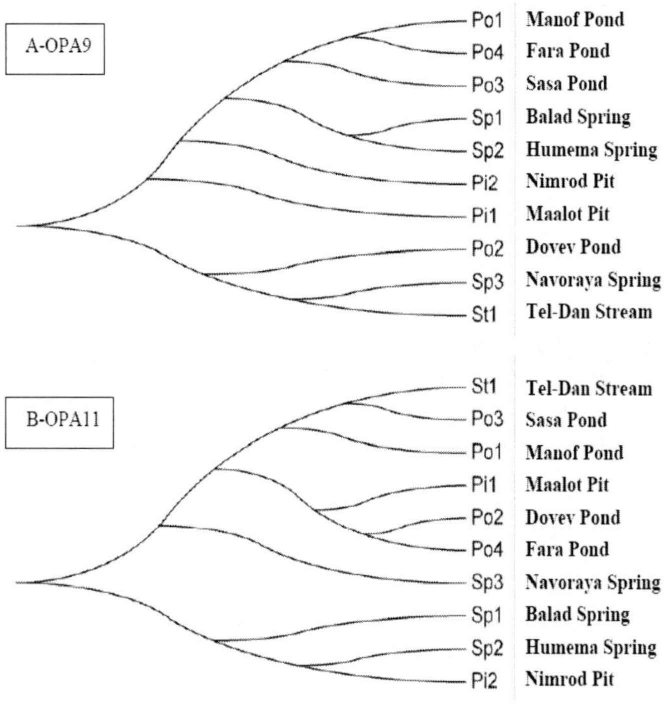

Figure 4. Phylogenetic trees of *S. infraimmaculata* populations from ten breeding sites in northern Israel. The optimal trees, with the sum of branch lengths = 0.28609062 (A) and lengths = 1.30471250 (B), are shown.

S. infraimmaculata larvae could be observed for 6-10 months in stable water bodies and pits, where water temperatures remained relatively constant (and not according to the altitudes of these water bodies). In other words, the analysis supported the hypothesis that the populations of the most permanent breeding sites varied genetically from those of less predictable breeding sites.

The AMOVA test, which was employed to examine the variation among and within the populations, demonstrated that the difference was significant (64% among populations and 36% within the populations, P<0.01). A relatively high similarity was also found among springs (Sp1 and Sp2), and rock pool holes (Pi1 and Pi2) (Figure 4). The rock pool holes, per definition, are temporary water bodies, but according to their ecological conditions and their branches in the phylogenetic trees, they were characterized as stable breeding sites.

4. Discussion

The hypothesis, that habitat selection occurs among the isolated populations of *S. infraimmaculata* larvae, growing in various breeding sites under a variety of ecological conditions (e.g., temperature), is supported, at least in part, by the present study, which examined genetic variation among *S. infraimmaculata* larvae by means of RAPD PCR. The evolutionary genetic variation among *S. infraimmaculata* larvae from different populations in Israel, as demonstrated in this investigation, is lower than that found between *S. salamandra* populations from Europe and Israel. The genetic variation in *S. salamandra* has been studied extensively (See the review by Steinfartz et al., 2000). However, these earlier studies have focused on geographical and taxonomical aspects, examining the differences among subspecies, rather than among different populations of a single species, as done here. The parameters by which the differences among *S. infraimmaculata* populations in Israel may be determined, include morphology, plasma proteins, isozymes, DNA by RAPD PCR and mitochondrial DNA analysis. In the present investigation, not only the genetic variation among the populations of various breeding sites was assessed by a sensitive method, RAPD PCR, but also the ecological parameters of these habitat populations were evaluated, to show the variety of different habitats existing in a relatively small area.

Mitochondrial genes demonstrated a very low variation among populations, but a high similarity among the larvae in springs and streams. This finding is in agreement with the present study, which used a different method.

Plasma protein patterns and enzyme system analyses are limited to protein variation, so it is difficult to use them to determine genetic variations within populations or between populations of the same subspecies. Moreover, it is less sensitive than the method used here. However, it seems that at least some of the physiological (reviewed by Degani, 1994) and biological (reviewed by Degani, 1996) variability results from the adaptation to different habitats. The present study shows the differences among *S. infraimmaculata* belonging to the same population. This result is in contrast to that of Degani (1986) and Veith et al. (1992), who found that individual populations were completely monomorphic. However, it does coincide with Degani et al. (1999). The differences depend on the method employed, which in this case was based on genetic markers. The DNA variation, revealed in this study, indicates that there are differences among salamanders within a population. The geographical distances, even short ones, among the breeding sites seem to have influenced genetic variation as well as ecological conditions in the habitats (Degani, 1996; Goldberg et al., 2007, 2009b), who used mitochondrial DNA analysis on *S. infraimmaculata* larvae from the habitats in northern Israel, found genetic variation among eight populations based on two sequences, the control region of the mitochondrial DNA and the gene, cytochrome *b*. The variation among the salamanders in Israel, based on the cytochrome *b* sequence, was even greater than that among those in Europe. The highest variation was discovered between the salamanders from winter ponds and perennial water sources. These results are supported by the investigations that have been carried out recently. Buckley et al. (2007) and Steinfartz et al. (2007) studied the genetic variation of the salamanders in Europe. Many biological differences among the salamanders from different habitats in northern Israel, which is the southern limit of their distribution, have been described. These biological differences include body size (Degani, 1986), number of larvae per parturition (Degani and Warburg, 1995), survival under dry conditions (Degani, 1981a) and reproductive cycle (Degani et al., 1997; Sharon et al., 1996). The ten habitats investigated in this study are geographically close to each other, with minor differences in rainfall, photoperiod and other meteorological conditions. However, the semi-arid conditions in four of them, where the water source dries up during the summer, differ greatly from the moist conditions in the other sites, where water is available almost in a year. Evidently, this is the difference, which gives rise to the biological, morphological and physiological variations in these parameters, as mentioned above, as well as to the genetic variations found here for the first time.

In summary, the results obtained from this study, using RAPD PCR and data from previous studies that used other methods (Goldberg et al., 2009a), support the hypothesis that in a relatively small area habitat selection occurs according to the influence of the ecological parameter, e.g. temperature, in the habitats. A number of different methods have been used to examine genetic variation, but RAPD PCR is the only one that was able to detect the genetic variation, resulting from habitat selection in the populations of the genus *Salamandra*.

REFERENCES

Alcobendas, M. M., Dopazo, H. and Alberch, P. Genetic structure and differentiation in Salamandra salamandra populations from the northern Iberian peninsula. *Mertensiella*. 1994.:7-23.

Buckley, D., Alcobendas, M., Garcia-Paris, M. and Wake, M. H. Heterochrony, cannibalism, and the evolution of viviparity in Salamandra salamandra. *Evolution and Developmen*. 2007*t*. 9:105-115.

Degani, G.. The adaptation of Salamandra salamandra (L.) from different habitats to terrestrial life. *British Journal of Herpetology* 1981a. 6:169-172.

Degani, G. Plasma protein and morphology of Salamandra salamandra in Israel. *Amphibia-Reptilia*. 1986. 7:105-114.

Degani, G. Ecophysiology of Salamandra salamandra at the southern limit of its distribution. *Mertensiella* 1994. 4:111-124.

Degani, G. *The salamander at the southern limit of its distribution*. Israel: Laser Pages Publishing. 1996. 154 p.

Degani, G., Jackson, K., Dosoretz, C. and Plotzky, Y. Molecular DNA variation in Salamandra s. infraimmaculata from different habitats. *Israel Journal of Zoology* 1999. 44:239-246.

Degani, G., Sharon, R. and Warburg, M. Ovarian steroid levels in Salamandra salamandra infraimmaculata during the reproductive cycle. *General and Comparative Endocrinology* 1997. 106:356-360.

Degani, G. and Warburg, M. Variation in brood size and birth rates of Salamandra salamandra (L.) (Amphibia, Urodela) from different habitats in northern Israel. *Amphibia-Reptilia* 1995. 16:341-349.

Escoriza, D., Comas, M.M., Donaire, D. and Carranza, SRediscovery of Salamandra algira Bedriaga, 1833 from the Beni Snassen massif (Morocco) and phylogenetic relationships of North African salamandra. *Amphibia-Reptilia.* 2006. 27:448-455.

Excoffier, L., Smouse, P.E. and Quattro, J.M. Analysis of molecular variance inferred from metric distances among DNA haplotypes: application to human mitochondrial DNA restriction data. *Genetics* 1992. 131:479-491.

Goldberg, T., Pearlson, O., and Nevo, E. and Degani, G.. Mitochondrial DNA analysis of Salamandra infraimmaculata larvae from habitats in northern Israel. *Progrese şi Perspective in Medicina Veterinară" - Lucrări ştiinţifice* 2007. 50:23-31.

Goldberg, T., Pearlson, O., Nevo, E. and Degani, G. Sequence analysis of mitochondrial DNA in Salamandra infraimmaculata larvae from populations in northern Israel. *South American Journal of Herpetology* 2009a. 4:268-274.

Goldberg, T., Nevo, E.and Degani, G. Breeding site selection according to suitability for amphibian larval growth under various ecological conditions in the semi-arid zone of northern Israel. *Ecologia Mediterranea.* 2009b. 35:65-74.

Mizuno, S., Andrews, C. and Macgregor, H.C. Interspecific "common" repetitive DNA sequences in salamanders of the genus Plethodon. *Chromosoma* 1976. 58 (1):1-31.

Mizuno, S. and Macgregor, H.C. Chromosomes, DNA sequences, and evolution in salamanders of the genus Plethodon. *Chromosoma* 1974. 48 (3):239-296.

Nei, M. and Li, W. H. Mathematical model for studying genetic variation in terms of restriction endonucleases. *Proceedings of the National Academy of Sciences USA* 1979.76:5269-5273.

Pearlson, O. and Degani, G. Molecular DNA variations among Triturus vittatus vittatus (Urodela) from different breeding sites at the southern limit of its distribution. *Acta Herpetologica* 2007. 2:69-77.

Sharon, R., Degani, G. and Warburg, M. *Environmental effects on reproduction in Salamandra salamandra infraimmaculata in north Israel*; 1996. 527-529p.

Steinfartz, S., Veith, M. and Tautz, D. Mitochondrial sequence analysis of Salamandra taxa suggests old splits of major lineages and postglacial recolonizations of central Europe from distinct source populations of Salamandra salamandra. *Molecular Ecology* 2000. 9:397-410.

Steinfartz, S., Weitere, M. and Tautz, D. Tracing the first step to speciation: ecological and genetic differentiation of a salamander population in a small forest. *Molecular Ecology.* 2007.16: 4550-4561.

Tamura, K., Dudley, J., Nei, M. and Kumar, S. MEGA4: Molecular Evolutionary Genetics Analysis (MEGA) software version 4.0. *Molecular Biology and Evolution* 2007.10.1093/molbev/msm092.

Veith, M. Morphological, molecular and life history variations in Salamandra salamandra (L.). *Mertensiella* 1994. 4:355-397.

Veith, M., Degani, G, and Seitz, A. Discordance of genetical and morphological variation of Salamandra salamandra (L) in Israel. *Zool Anz* 1992. 229:63-72.

SECTION IV: REVIEWS RELATED TO VERTEBRATE AND INVERTEBRATE ANIMALS AND HUMANS

In: Vertebrate and Invertebrate Animal Approaches
Editor: José Rosa Gomes
ISBN: 978-1-63117-102-4

Chapter 29

HISTORICAL EPIDEMIOLOGICAL ASPECTS OF AMERICAN VISCERAL LEISHMANIASIS IN BRAZIL

Kárita Cláudia Freitas-Lidani[1]*, Jhulie Mirandola[2], Jimena Ferreira[2], Raphaela Mordaski[2] and Edna Aoba Yassui Ishikawa[3]

[1]Hospital das Clínicas - Lab. de Imunopatologia Molecular, Universidade Federal do Paraná, Curitiba-PR, Brasil.
[2]Faculdades Integradas do Brasil, Unibrasil, Curitiba- PR, Brasil
[3]Núcleo de Medicina Tropical da Universidade Federal do Pará, Belém-PA, Brasil

ABSTRACT

The American Visceral Leishmaniasis (VL) is a serious chronic disease, potentially fatal to humans; its lethality can reach 10% if no proper treatment is used. The growing urbanization of the disease occurred in the past 20 years and puts the discussion on control strategies because Brazil is considered a major outbreak of the disease, which originated approximately 90% of the cases in the Americas. In recent years, the lethality of VL is increasing gradually, pointing to infectious complications and bleeding as the major factors of risk for death. In this sense, this works aims a revision of the Brazilian historic of the disease, infection ways, symptoms and fatality rates since it was discovery in Brazil.

Descriptors: American visceral Leishmaniasis, VL and Leishmaniasis

* Mail address of corresponding author: kari.lidani@gmail.com; Telephone number: 55 (41) 3360-1800 r.6537 – 55 (41) 88935097 ; Fax number: 55 (41) 3360-1800 r.6537 Postal address: Rua General Carneiro, 181. CEP: 80060-900 Curitiba, PR.

INTRODUCTION

American visceral leishmaniasis (VL) is a generalized infection of mononuclear phagocytic system (MPS) – Mobile Defense System – caused by protozoa of the gender *Leishmania* subgener Leishmania, but only *Leishmania (Leishmania) infantum chagasi* causes visceral leishmaniasis, and the only present in the Americas [1, 2]. The worldwide incidence of leishmaniasis is estimated in 2 million new cases per year, these numbers are probably underestimated, since it is assumed that for every one case diagnosed in the Americas, five more were not identified [3].

According to Desjeux (2001) [4], migration, human pressure, urbanization of the cycles of circulation of *Leishmania*, immunosuppression and malnutrition of the host are considered as risk factors that would explain the expansion, the spread and increase of incidence of leishmaniasis cases in many regions of the world. Roger, 1908 cited in Rey, 2001 [5] demonstrated that the earliest references of VL are from 1882, when it was described a special form of tropical splenomegaly, which occurred endemically n parts of India, associated then with hookworm or a special type of malaria.

VL has a wide distribution in the Old and New World, located among one of the seven endemics priority attention of World Health Organization. VL affects 500000 people each year and approximately 90% of all cases reported worldwide come from: Bangladesh, Brazil, India, Nepal and Sudan. The disease affects primarily the poorest countries, associated with malnutrition, sanitation and disorderly occupation of the environment. In the new world this disease is known as kalazar or neo-tropical American visceral leishmaniasis [3].

In Brazil, the first known case was diagnosed in 1913 by Migone [6, 7], it was a clinical case of an Italian immigrant who, after living 13 years in Santos, travelled to Mato Grosso where he got ill [6]. In 1936, at a commission headed by researcher Evandro Chagas, who investigated the occurrence of several cases of VL, Deane and Mangabeira [7] incriminated *Lutzomyia longipalpis* mosquito as the transmissor, and in 1937, Chagas found the first case of canine VL in Ceará, Castro and Ferreira [1, 6].

Until 1953, there was only 43 confirmed cases of VL in vivo had been reported against 332 cases found by visceroctomy by Madureira Pará [7], distributed in the states of Ceará. Bahia, Sergipe, Pará, Piauí, Rio Grande do Norte, Pernambuco, Mato Grosso, Maranhão, Paraíba and Minas Gerais [1].

In 1953, began to emerge many cases of the disease, especially in Ceará. In 1954, Deane and Deane [8] incriminated the Fox *Lycalopex vetulus* as the wild reservoir in Brazil [1]. Other reservoirs, in addition to *L. vetulus*, are also pointed as the man, the fox *Cerdocyon thous* (wild reservoir), and the dog *Canis familiaris* (domestic reservoir) [9].

L. longipalpis is the only vector species with epidemiological importance involves in the disease transmission [2]. The transmission mechanism to humans is through the bite of the sandfly, which inoculates the agent on the surface of the dermis where it multiplies in the cytoplasm of local FMS cells.

It spreads by hematonenous route, reaching greater intensity the viscera with a bigger number of FMS cells: liver, spleen, bone marrow and lymph nodes. Other systems are also infected such as the digestive tract, lungs, skin, blood vessels and kidneys, with nephropathy related to immune mechanisms, once this condition has been included in the group of autoimmune diseases [1, 10].

The diversity of vectors and reservoirs is large due to the extent of the endemic region and the fact that the disease often occurs in focus separated by geoclimatic barriers [2]. The sandflies easily colonize the canyons, mountains, foothill of mountains, in addition to the interior of river valleys, having to high hability to adapt temperatures and low degree of humidity, which is the period of greatest transmissibility in the rain season. VL especially in a rural endemic disease, but also affecting the urban periphery, affects about 3000 people every year in Brazil [1, 9, 10]. Accordingly, the present study presents a review on the Brazil's history of VL since its discovery until the means of infection, symptoms and death rates.

EPIDEMIOLOGY

VL is a parasitic disease found in rural areas of almost all tropical and subtropical regions at the global [11, 12, 13, 14]. The epidemiology of visceral leishmaniasis in the world varies largely according to the *Leishmania* species causing the infection and the vector species involved in transmission. VL can be divided in didactic terms in two large clinical epidemiological groups: VL caused by *L. (L.) infantum* and *L. (L.) chagasi (considered as the same species)* and VL caused by *L. (L.) donovani* [2].

The infection caused by *L. (L.) chagasi* is restricted to the Americas. Despite this extensive endemic area, *Lutzomyia longipalpis* is the only vector involved in the disease transmission, and the dog the main reservoir [1, 2, 13, 15, 16]. Deane et al. [7] demonstrated that 24.8% of sandflies become infected by biting an infected dog whereas to 14.8% of them became infected by biting a human. Therefore, canine VL is more prevalent than human and above the endemic [1]. The first study reporting canine infection in the Americas was published in Brazil in 1930 by Evandro Chagas [17] who found 4.1% of infected dogs in the towns of Moju and Abaetetuba in the Pará state. On the other hand the first reservoir to be described was the fox *Lycalopex vetulus*, a common canid frequent in rural areas of Brazil. This is an important reservoir due to its intense parasitism, higher than the dog, as to be responsible to the spread of *E. hystolytica* infection to distant sited due to migratory habits [1]. Other animals were also reported, including the fox *Cerdocyon thous* and some marsupials, *Didelphis marsupialis* and *Didelphis albiventris* [2, 9].

VL has a wide distribution in our continent, but Brazil represents the mains focus, with 90% of the reported cases in the Americas (6). One worrying aspect of VL epidemiology in Brazil is the urbanization of the disease that has expanded its areas of transmission to the suburb of large urban centers as shown in outbreaks in Teresina, São Luiz, Fortaleza, Camaçari-BA, Montes Claros, Minas Gerais, Belo Horizonte and Rio de Janeiro [2, 11, 12]. The main focuses of this parasitosis in Brazil are located in semi-arid and arid regions, particularly in Northeast and part of Southeast, and the transmission is observed in the foothills of mountains and ditches, rocky terrain with caves and creeping vegetation (shrubs). In Minas Gerais the highest prevalence is observed in the Drought Polygon (Montes Claros, Januária) and Vale do Rio Doce – São Francisco and Jequitinhonha [11]. In these areas, the local ecological factors, such as hot, humid climate and abundant vegetation in shrub are favorable factors for the vector [6].

In Pernambuco, were reported 336 LVL cases, during the period of 1934 to 1984, 51.78% were found in an area from hinterland region (Araripina, Salgueiro) and 31.25% were concentrated in the metropolitan area of Recife and a small region of Decicuous Forest. In the

90s, about 90% of the reported cases of VL occurred in the Northeast. However, in the period of 2000-2002, the Northeast has showed a reduction of 77% of the reported cases in the country [1, 6, 11, 18].

According to the Brazilian Ministry of Health in 19 years of notification (1984-2002), 48455 cases of VL have been reported, with approximately 66% of them occurring in the states of Bahia, Ceará, Maranhão and Piauí [18]. Over the past ten years about 3000 cases per year have been registered in the country, with an annual average of 3156 cases (Figure 1) [20]. In the period of 2000 to 2008, there was an average of around 200 deaths annually, with a mortality rate of 6.5% (Figure 2) [21].

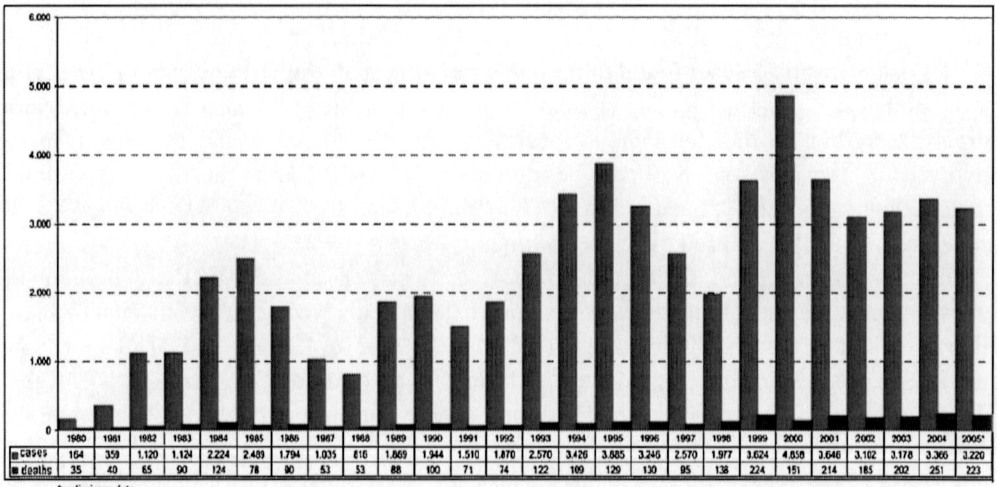

	1980	1981	1982	1983	1984	1985	1986	1987	1988	1989	1990	1991	1992	1993	1994	1995	1996	1997	1998	1999	2000	2001	2002	2003	2004	2005*
cases	164	359	1.120	1.124	2.224	2.489	1.794	1.035	816	1.869	1.944	1.510	1.870	2.570	3.426	3.885	3.246	2.570	1.977	3.624	4.858	3.646	3.102	3.178	3.366	3.220
deaths	35	40	65	90	124	78	80	53	53	88	100	71	74	122	109	129	130	95	138	224	151	214	185	202	251	223

*preliminary data

Figure 1. Distribution of confirmed cases of Visceral Leishmaniasis from 1980 to 2005, Brazil [20].

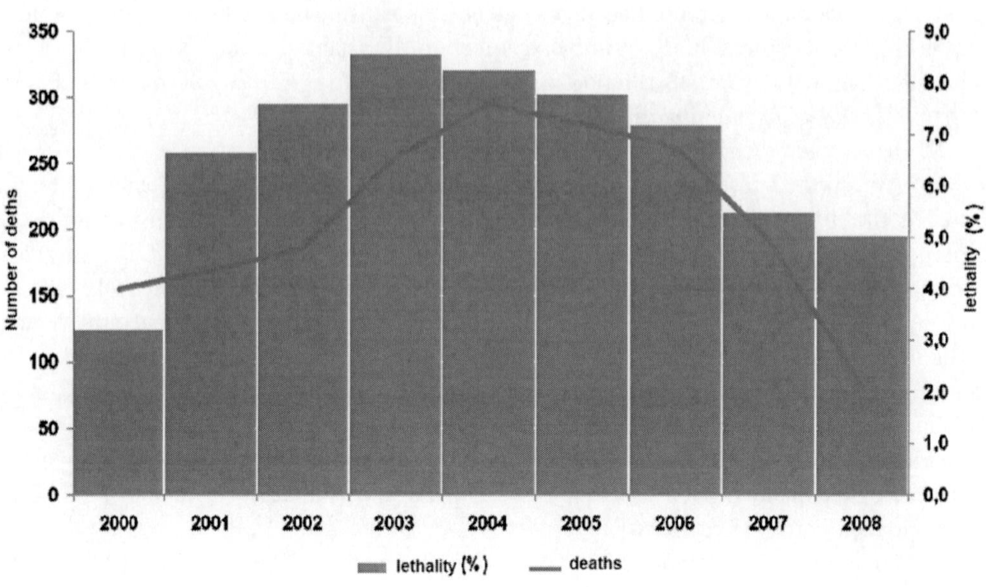

Figure 2. Number of deaths and lethality by Visceral Leishmaniasis in Brazil, during 2000-2008 [21].

The spread of endemic areas and the emergence of new focus are caused by changes in the environment, which are caused by intense migratory process due social or economic pressures, the poverty, the rising of urbanization process, the emptying of countryside and periodic droughts. This leads to a reduction in the ecological space of the disease, facilitating the occurrence of epidemics [18].

Reservoir

In urban areas the dog (*Canis familiaris*) is the main source of VL infection, being the domestic reservoir, responsible for maintaining the disease cycle in the home environment [18]. Once infected, the dog presents intense cutaneous parasitism by *L. (L.) chagasi*, which makes it an excellent source of infection for the sandfly vector. The enzootic canine has preceded the occurrence of human cases, and infection in dogs has been more prevalent then in man. However, like the man, the dog is just an accidental host of the parasite. In sylvatic environment, reservoir is foxes (*Dusicyon vetulus*, *Cerdoyon thous* and *Lycalopex vetulus*) and marsupials (*Didelpis albiventris*) [9, 18]. The fox is the primitive wild reservoir of VL, in which the infection by *L. (L.) chagasi* causes no harm to its health of this animal [6, 19]. In Brazil, the foxes were found infected in the Northeast, Southeast and Amazon [6]. Marsupials didelphids were found infected in Brazil and Colombia [1, 11, 15, 16].

Vectors

The vectors of VL are insects known as sand flies, popularly known as straw mosquito, tatuquira, birigui, among others. In Brazil, so far two species have been related to the disease transmission, *Lutzomyia longipalpis* and *Lutzomyia cruzi*. The first reported is considered the main transmitter of *L. (L.) chagasi* in Brazil and the latter has been reported as a vector in the State of Mato Grosso do Sul [6, 11, 18].

In Brazil, the geographic distribution of *L. longipalpis* is broad and seems to be expanding. This species are found in four geographic regions: Northeast, North, Southeast and Midwest. In the North and Northeast, *L. longipalpis* was originally found in the forest participating in the primary cycle of disease transmission. Gradually the insect has adapted to the rural environment. At the end of the 80s, there was an adaptation of this vector in urban areas, on the suburbs of big cities, mainly in the Southeast, can be found in animals sheds, henhouse, kennel, barns and other environments and in home [18].

These insects are small, measuring 1-3 mm in length. They have their body covered by hair and are light colored (ten or straw color). They are easily identified by their behavior, when flying in little jumps with wings half open. These insects, in adulthood, are adapted to different environments, but during larval it develops in moist environments, rich in organic matter and low light incidence. Both genders need carbohydrates as energy source and the females feed on blood for egg development [18].

The life cycle of *L. longipalpis* includes four phases: egg, larva (with four stages), pupa and adult. After copulation the female lay their eggs on moist soil with high organic material content, to ensure the feeding of the larvae. The eggs usually hatch 7 to 10 days after laying. Larvae develop between 20 and 30 days, according to environment conditions and adverse conditions the fourth stage can enter in diapauses. After this period the fourth stage larvae turn into pupae, in this period they are more resistant to changes in humidity than the eggs and larvae. The pupae do not feed and have air breathing. The pupal period under favorable conditions lasts on average one to two weeks [18].

The egg development to adult insect takes a period of about 30 to 40 days according to the temperature. Females feed on blood mandatory, present eclectic habits and can feed on blood of several species of vertebrates, including humans. The longevity of females is estimated in around 20 days [18].

The sandflies are active during twilight and night. During the day, these insects are at rest in humid places, protected from wind and predators. The infection of vector occurs when the females feed on blood of infected mammals, ingesting macrophages infected by amastigotes forms of *Leishmania.* In the anterior digestive tract occurs the first disruption of macrophages releasing these amastigotes. They multiply by binary division and differentiate rapidly into flagellated forms named promastigotes, which also reproduce through successive processes of binary division. The promastigotes develop into paramastigotes which colonize the pharynx and esophagus of the vector, which remain adhered to the epithelium by the scourge, when they differentiate into promastigote metacyclic infective form. The cycle of the parasite in the insect is completed around 72 hours [18].

The infective females release the metacyclic promastigote forms in her saliva during the next feed on a vertebrate. In the epidermis of the host, these forms are phagocytosed by immune cells in the vacuole of macrophages. The parasite differentiate into amastigotes and multiply intensively until breaking off the cell, releasing these forms to be phagocytosed again by others macrophages in a new continuous process, occurring then the spread by the blood to other tissues, such as lymph nodes, liver, spleen and bone marrow [18].

TRANSMISSION MODE

In Brazil, the mode of transmission is through the bite of the sandfly vector, *L. longipalpis* or *L. cruzi* infected by *Leishmania (L.) chagasi* [12, 19, 22].

There is no evidence of a man to man transmission. Therefore, the majority of human infections would be acquired by the bites of sandflies that have previously bitten infected dogs. It was demonstrated experimentally that the sandfly can transmit the disease 7 days after feeding on infected blood [6, 18, 22].

And that transmission occurs while there is parasitism in the skin or peripheral blood of the host. There are described cases of congenital transmission (six cases), sexual intercourse, blood transfusion or laboratory accident [11, 12, 18, 19, 22].

It is not evident whether there is a certain time of year especially, in which the transmission rate is higher among dogs and people. The main reason is that the exact date of infection cannot be determined. However, there is evidence that the period of highest transmission of VL occurs right after the rainy season, when an increased in insect population density happens [6, 18].

INFECTION

When the infected vector bites a person, the peptides responsible for the inflammatory action attract phagocytic cells to the site of the bite, and the promastigotes penetrate these cells, transforming into amastigotes, escaping the first nonspecific defense mechanism such as lysis by the complement system. From this point on most people can contain the infection (unapparent infection) and establish a lasting immunity. Others develop an oligosymptomatic form of infection that can also resolve spontaneously or develop into the classic disease. An unknown percentage of individual, probably most, develops immunity to reinfection, but without eradicating the parasite from the body, which may again multiply in situations of immunodeficiency [9, 18].

In the individuals who progresses of the disease, the infected macrophages burst releasing the amastigotes forms which leads to invasion of new macrophages. Infected macrophages produce a chemotaxins that will attract new macrophages that become infected creating a vicious cycle that will culminate in the exponential multiplication of the parasite [2].

Unapparent or asymptomatic infections are those in which there is no evidence of clinical manifestations. Diagnosis is done by serological or Montenegro tests or antibody titers are generally low and may remain positive for a longer period. It is important to point out that patients with clinical cure or with cutaneous leishmaniasis (cutaneous and mucosal forms) may present reactivity in serological and Montenegro tests. Therefore, the asymptomatic forms are observed in patients from endemic areas, where there is epidemiological and immunological evidence of infection [18].

SYMPTOMS

The main symptoms and signs of VL include prolonged irregular fever, hepatosplenomegaly, pancytopenia, weight loss, malnutrition, immunosuppression, cachexia, stomatitis, skin peeling, earthy pale, pellagra dermatitis and angular cheilosis [1, 6, 9, 12, 16, 22]. There is a decreased function of mononuclear phagocytic system (MPS) organs, on the site of parasite multiplication that suffers hyperplasia and hypertrophy in response to the infection [2]. It is possible to observe organic structural or tissue changes such as splenomegaly, hepatomegaly, adenopathy, anemia, leukopenia and increase of serum gammablobulin [11, 22].

The period of infection is difficult to be determined because it may vary from three to six months [1]. The skin may show a small ulceration, which may go unnoticed or a small granuloma at the site of insect bite and is constituted by histiocytes filled with amastigotes cells surrounded by epithelial cells followed by giant cells [11].

After 120 to 180 days, there is dissemination of parasites, primarily in the lymph nodes and then through blood to the organs [11]. The lymph nodes are presented systematically and moderate with infarction, predominantly presenting inguinal and axillary micropoliadenias. It is possible to see myeloid transformation of medullar cords and numerous histiocytes rich in amastigote cells [1, 11]. In the bone marrow occurs a progressive substitution of hemopoietic tissue and the fat mass for cells of MPS infected with amastigotes. Initially occurs a hyperplasia of endothelial reticule, until developing into a bone marrow failure, it also occurs a loss of hemopoietic function in precursor cell of plasmocyte due to the production of bloodline cells [2, 11]. The lungs are enlarged and congested. Interaveolar septa are thickened due to inflammatory infiltrate consisting mainly of lymphocytes, macrophages and plamocytes. Interstitial cells are increased because of the presence of fat vacuoles, mild edema with congestion of septal capillaries [2]. About 80% cases present interstitial pneumonitis, with dry and unproductive cough, steroids grunts and bronchitis [1, 2, 11].

In the digestive tract, especially in the small intestine, there is proliferation of infected MPS cells by amastigotes forms, located at the duodenal-jejunal microvilli. The jejuna injury is characterized by inflammatory infiltrate composed of heavily parasite macrophages, plamocytes and lymphocytes; with an atrophy of crypts cells and loss of albumin responsible for the low plasma levels if this protein in the development of the disease and decrease of intestinal absorption may occur small ulceration in presence of the parasite. Also there is edema, enlargement of the villi and intestinal lymphatic vase, without changes on the structure of the organ. The appetite is normal or often increased [1, 2, 11].

The liver is always enlarged, with the surface slightly yellowed due to steatosis, it presents lightly hardened and with no nodules and can be felt up to 15 centimeters below the costal margin [1, 2, 11]. It can evolve a long-term liver fibrosis that compromises the permeability of sinusoidal wall. Also occurs hyperplasia and hypertrophy of Kupffer cells, which contain the parasite [2, 11]. Kidneys have interstitial nephritis, pinpoint hemorrhages, an interstitial inflammatory infiltrate because of macrophages and plasmocytes mainly around the vessels of renal cortex. The parasites are rare and found in the endothelium of glomerular capillaries. In chronic cases there are deposits of immunocomplexes in basement membrane of glomerulus and in fatal cases occur in hyaline thickening in glomerular mesangium [2, 11].

Visceral leishmaniasis in the spleen causes severe splenomegaly. Its increase is due to hyperplasia of the MPS and large amounts of amastigotes cells within the splenic pulp. There is a decrease in the number of T lymphocytes mainly in the T dependent areas [2, 11].

At the end of the first month the spleen reaches beyond the costal margin, may get to the lower abdomen [1]. The Malpighian corpuscles are sometimes hypertrophied, sometimes are atrophied, the follicular arterioles are thickened, its endothelium is full of parasites and the organ is congested. Although there is an increase of connective tissue, hyaline degeneration, fibrous nodules and granulomatous Gandygama type [11]. In the acute phase of infection, the capsule of the organ is smooth, soft and often its weight caries between 500 g and 2.0 kg. In the chronic phase the organ is hard, firm and painless to palpation, there is a process of fibrosis, atrophy of lymphoid follicles, myeloid metaplasia, and can occur anemic infarcts [11]. Anemia is responsible for the pale and its progressively worsening may lead to cardiorespiratory manifestations with syncope episodes, anemic cardiac murmur [11]. If leishmaniasis is not treated properly it can develop an aggravation of the disease as extreme cachexia, severe muscle hypertrophy, depletion, massive hepatosplenomegaly, severe hemorrhagic manifestations, delirium, hallucinations and infectious complications that generally leads to death [1].

In the last two decades, VL reappeared in the world to an alarming extent. In Brazil, urban epidemics were observed in several cities and the disease has been established as an opportunistic infection in patients with AIDS, as has been observed in the south of Europe [23]. Moreover, the expansion of the epidemic affecting young people or groups of individual with co-morbidities has caused high number of deaths.

The infectious complications and hemorrhage are the main risk factors to lead to death in VL. Early diagnosis is crucial to reduce mortality through the introduction of effective prophylactic and therapeutic measures [23].

FINAL CONSIDERATIONS

According to literature review, it is possible to see the need for an effective program to control the disease, so that we can avoid the risks of disease transmission and take prevention measures. For such health programs, efforts are needed from the population and government, especially regarding sanitation, control of dog populations, vaccine against canine VL, early diagnosis, among many other preventive measures. Only then we could think of eradicating VL from our country.

REFERENCES

[1] Marzochi, MCA; Coutinho, SG; Souza, WJS; Amendoeira, MRR. *Leishmaniose Visceral* (Calazar). Vol. 41 Nº 5, 1981.

[2] Dietze, R; Carvalho, SFG. *Medicina Tropical.* 1 ª ed. Capitulo 5; Editora Atheneu, São Paulo: 2003.

[3] WHO. World Health Organization. The 17th Programme Report of the UNICEF/UNDP/World Bank/ WHO Special Programme for Research and Training in Tropical Diseases 2006. [Acesso em 20/08/2008] Disponível: http://www.who.int/tdr/leishmaniasis.htm

[4] Desjeux, P. The increase risk for leishmaniasis worldwide. *Trans R. Soc. Trop. Med. Hyg.* 2001; 95: 239-241.

[5] Rey, L. Parasitologia. 3ª edição. Rio de janeiro: Guanabara Koogan; 2001

[6] Valente, SAS; Fraiha Neto, H; Leäo, RNQ; Miles, MAL. Doenças Infecciosas e Parasitárias – Enfoque Amazônico. – Belém: Cejup: UEPA: Instituto Evandro Chagas, 1997.

[7] Deane, LM. Leishmaniose Visceral no Brasil, Estudos sobre reservatórios e transmissores realizado no Estado do Ceará. Tese. Rio de Janeiro. *Serv. Nac. Ed.* Sanitária, 1956.

[8] Deane, LM ; Deane, MP. Encontro de leishmanias nas vísceras e na pele de uma raposa, em zona endêmica de calazar, nos arredores de Sobral, Ceará. O Hospital. Rio de Janeiro, v. 45, p. 419-421, 1954.

[9] Obe, RL. Revista Paraense de Medicina – Vol. 11 Nº 1, 1997.

[10] Silva, ES; Gontijo, CM; Pirmez, C; Fernandes, O; Brazil, RP. Short report: detection of Leishmania DNA by polimerase chain reaction on blood samples from dogs with Visceral Leishmaniasis. 2001. pp 896-898.

[11] Huggins, DW; Medeias, LB; Júnior, HAF. Leishmaniose Visceral- *Calazar Revista Brasileira de Medicina* – Vol. 50 Nº 9, p 1203-1222, 1993.

[12] Ministério da Saúde. Doenças Infecciosas e Parasitárias – Guia de Bolso. 3ª ed. Vol. II Série B. *Textos Básicos de Saúde.* Brasília, 2004.

[13] LAISON, R.; Shaw, JJ. New World leishmaniasis – the neotropical *Leishmania* species. Chapter 13, 1998.

[14] Ministério da Saúde. Guia de Vigilância Epidemiológica. 6ª ed. Série A. Normas e Manuais Técnicos. Brasília, 2005.

[15] Secretaria de Saúde do Estado de São Paulo. II informe técnico: Leishmaniose visceral Americana. Instituto Adolf Lutz, 2003

[16] Cupolillo, E; Junior, GG; Momen, H. A general classification of new world *Leishmania* using numerical zymotaxonomy. 1994. pp 296-311.

[17] Lainson, R; Dye, C; Shaw, JJ; Macdonald, DW; Courtenay, O; Souza, AA; Silveira, FT. Amazonian visceral leishmaniasis – distribution of the vector Lutzomyia longipalpis (Lutz and Neiva) in relation to the fox Cerdocyon thous (linn.) and the efficiency of this reservoir host as a source of infection. *Mem. Inst. Oswaldo Cruz.* 85(1): 135-7, 1990.

[18] Ministério da Saúde. Manual de Vigilância e Controle da Leishmaniose visceral. 1ª ed. Editora MS, Brasília, 2004.

[19] Ministério da Saúde. Doenças Infecciosas e Parasitárias – Guia de Bolso. 5ª ed. Ampliada Série B. Textos Básicos de Saúde. Brasília, 2005.

[20] Ministério da Saúde. Gráfico de evolução da leishmaniose visceral de 1980 a 2005. [Acesso em 22/08/2008]. Disponível: http://portal.saude.gov.br/portal/arquivos/pdf/leish_visceral_serie_2006.pdf

[21] Ministério da Saúde. Programa Nacional de Vigilância e Controle das Leishmanioses – Gerência Técnica das Leishmanioses COVEV/CGDT/DEVEP/SVS/MS. Secretaria de Vigilância em Saúde, 2008.

[22] Ministério da Saúde. Doenças Infecciosas e Parasitárias – Aspectos Clínicos, Vigilância Epidemiológica e Medidas de Controle - Guia de Bolso. 2ª ed. Revisada e Ampliada. 2000.

[23]Ministério da Saúde. Leishmaniose Visceral Grave – Normas e Condutas. Editora MS, Brasília: 2006.

In: Vertebrate and Invertebrate Animal Approaches
Editor: José Rosa Gomes
ISBN: 978-1-63117-102-4

Chapter 30

Black-tailed Prairie Dog Affinity for Sunlight and Its Potential Animal Welfare Management Implications

Adam E. M. Eltorai and Robert W. Sussman
Department of Anthropology
Washington University in St. Louis, Missouri, US

Abstract

Black-tailed prairie dogs *(Cynomys ludovicianus)* are socially complex, diurnal rodents that live in intricate cities. Prairie dogs are very popular among zoo visitors. Thus, optimal welfare conditions of these captive animals are of the utmost importance. As mammals, prairie dogs employ various strategies and adaptations in order to maintain ideal body temperatures. In this study, the distributional pattern of a captive population of black-tailed prairie dogs is studied. The prairie dogs' affinity for sunlight suggests that current captive conditions may need to be modified in order to optimize for thermoregulation needs.

Introduction

Proper environmental conditions enhance captive animals' welfare and performance of normal activity (Hosey, 2005). The experiences of zoo visitors are colored by the animal activity (Margulis et al., 2003). Thus, monitoring and maintaining ideal captive animal welfare conditions is essential. In order to properly assess the efficacy of current welfare management strategies, one must examine the animals' behavior. Temperature, and therefore sunlight patterns, play a major role in determining captive animals' behavior (Stoinski et al., 2002). To further expand on our earlier work (Eltorai and Sussman, 2010), we investigate a novel relationship between captive black-tailed prairie dogs *(Cynomys ludovicianus)* and their patterns of distribution with respect to sunlight.

METHODS

During the summer of 2009, observations were detailed at the Saint Louis Zoo (Saint Louis, Missouri, USA). The Saint Louis Zoo is home to 25 black-tailed prairie dogs. The prairie dog population consists of a mixture of adults and pups. Neither individual identity nor sex were able to be identified. A total of 901 instantaneous scan samples (as outlined in Lehner, 1996) were taken every 5 minutes to determine the number of individuals in sunlight. To account for the potentially confounding variable of normal changes in sunlight patterns due to time of day and to confirm that the prairie dogs were not simply distributed uniformly about the enclosure and were, in fact, following the sunlight, samples were taken at points throughout the day (9AM-7PM). More specifically, approximately equal numbers of observations were detailed in the morning (9AM-11:59AM), afternoon (12PM-2:59PM), and late afternoon (3PM-7PM). During each scan, the number of individuals in direct sunlight was counted in order to determine the percentage of the population in direct sunlight. To determine the percentage of enclosure with direct sunlight, the exhibit was evenly divided into five areas. If sunlight illuminated any part of an area, 20% of the enclosure was said to be covered by sunlight.

To statistically analyze the data, we used a *paired* t-test. To normalize the prairie dog distribution levels in sunlight, we divided the percentage of prairie dogs in sunlight by the percentage of the enclosure covered by sunlight. We were then able to compare the percentage of prairie dogs in sunlight with respect to the amount of sunlight available in the enclosure for each of the three times of day. The percentage data meets the assumptions of a parametric test, given the large number of data points collected in each of the three times of day, We acknowledge the potential limitation of analysis given the nature of the non-independence of some data points.

RESULTS

In the figure, we show the average number of regions within the enclosure—as a percentage of enclosure with direct sunlight—that experience direct sunlight during the morning, afternoon, and late afternoon. We observed a highly significant relationship between the distribution of prairie dogs in sunlight and that of the percent of the enclosure with direct sunlight ($p = 0.0073$, $t(2) = 11.6019$). In other words, throughout the day, the percent of the prairie dog population that moves to be in an area of direct sunlight exposure is unlikely due to chance. Because the way the data was collected rules out the possibility that the prairie dogs were simply distributed uniformly about the enclosure, our findings strongly suggest that the prairie dogs have an affinity for sunlight and therefore re-positioned themselves accordingly.

DISCUSSION

These data indicate that the prairie dogs' pattern of distribution within an enclosure parallels that of the sunshine patterns within the enclosure. These findings suggest that further research is required to establish the temperature regulation needs (i.e. ideal temperatures) of prairie dogs and, potentially, other mammals. Such investigations must be done in a controlled laboratory setting, as environmental factors such as wind, time of day, and time of year make the measurements of a species' ideal temperature all but impossible. Proper investigations into other optimal environmental conditions for captive animals will inform zoo keepers and researchers on how to best manage and monitor the welfare of animals.

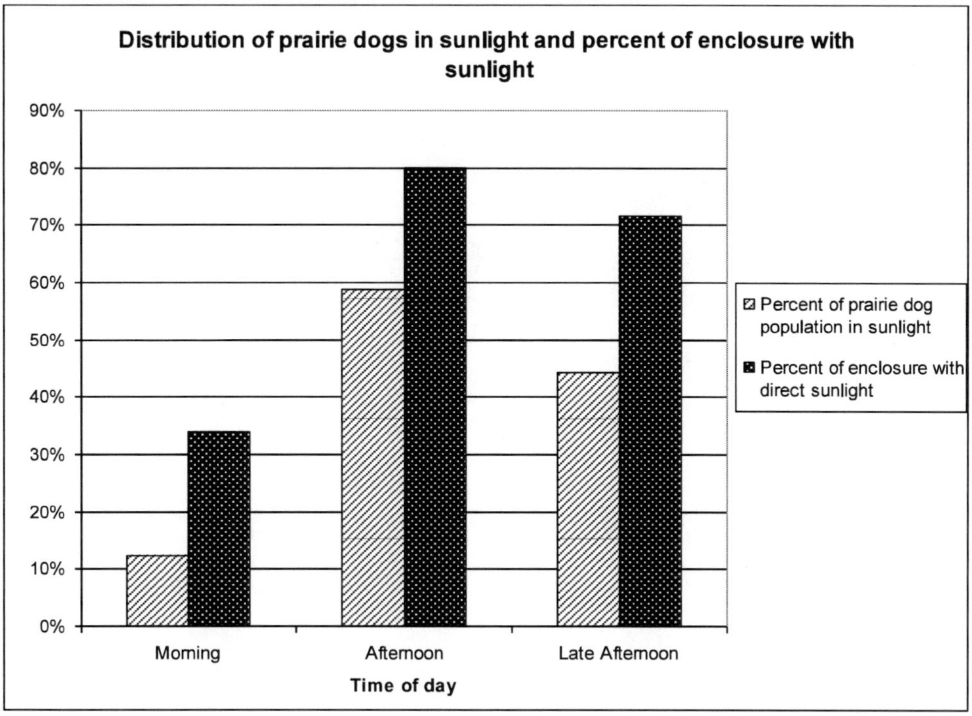

Side-by-side comparison of prairie dog distribution during the morning (9AM-11:59AM), afternoon (12PM-2:59PM), and late afternoon (3PM-7PM) with respect to sunlight, along with the percentage of enclosure with direct sunlight.

By improving the environmental conditions for captive animals, the animals will demonstrate more natural and appealing behaviors to zoo crowds. By enhancing the zoo visitor experience, visitors will be more inclined to return and support the zoo, which, therefore, will support further animal welfare research.

REFERENCES

Eltorai, A. E. M. and Sussman, R. W. (2010). The "Visitor Effect" and captive black-tailed prairie dog behavior. *Der Zoologische Garten, 79, 109-120.*

Hosey, G.R. (2005). How does the zoo environment affect the behavior of captive primates? *Applied Animal Behavior Science, 90, 107-129.*

Lehner, P.N. (1996). *Handbook of ethological methods.* 2nd Ed. UK: Cambridge University Press.

Margulis, S. W., Hoyos, C., and Anderson, M. (2003). Effect of felid activity on zoo visitor interest. *Zoo Biology, 22, 587–599.*

Stoinski T.S., Hoff M.P., Maple T.L. (2002). The effect of structural preferences, temperature and social factors on visibility in Western Lowland Gorillas (Gorilla G. Gorilla). *Environment and Behavior, 34, 493-507.*

In: Vertebrate and Invertebrate Animal Approaches
Editor: José Rosa Gomes
ISBN: 978-1-63117-102-4

Chapter 31

CLINICAL SPECTRUM OF HYPOTHYROIDISM IN WEST BENGAL, INDIA

Nisha Balmiki[#1], Srikanta Guria[#2], Biswabandhu Bankura[#1], Arup Kumar Pattanayak[1], Tapas Kumar Das[3], Anirban Sinha[4], Sudipta Chakrabarti[5], Subhankar Chowdhury[4] and Madhusudan Das[1*]

[1]Department of Zoology, University of Calcutta, Kolkata, West Bengal, India
[2]Department of Zoology, Parasitology and Medical Entomology Laboratory, Darjeeling Govt. College, Darjeeling, West Bengal, India
[3]Bagnan Rural Hospital, Bagnan, Howrah, West Bengal, India
[4]Department of Endocrinology, Institute of Post Graduate Medical Education and Research, Kolkata, West Bengal, India
[5]Institute of Life Sciences, Nalco Square, Chandra Sekharpur, Bhubneswar, India

ABSTRACT

Hypothyroidism is a progressive disorder of the thyroid gland. Thyroid hormones are extremely important and have diverse actions in the body. Disorders associated with altered thyroid hormone secretion are common and affect about 5% women and 0.5% men. It's estimated that more than 200 million people at minimum worldwide have thyroid disease. India too, is no exception. The present study is aimed to enumerate the clinical spectrum of hypothyroidism in the population of West Bengal to establish the etiology of the disease in this population. In our study a significant number of patients (52%) have gained body weight. Lethargy and cold intolerance are significantly different between cases and controls and associated with hypothyroidism. Hypothyroidism is

[#] Equal contributors.
[*] Corresponding author: e-mail: madhuzoo@yahoo.com; Telephone: 91-09831281756; Fax Number: 91-33-24614849.

associated with menstrual abnormality. Anti-TPO antibody positive have been found in 22% of patients in contrast to a prevalence of 7-20% in the general population. Urinary iodine levels indicate that iodine deficiency is currently not a public health problem in our study population.

Keywords: Thyroid Stimulating Hormone (TSH), Hypothyroidism, Anti-TPO antibody

INTRODUCTION

Hypothyroidism is a progressive disorder of the thyroid gland. The thyroid gland secretes thyroid hormones, primarily 3, 5, 3'5'-l-tetraiodothyronine or thyroxine (T_4), and a lesser quantity of 3, 5, 3'-l-triiodothyronine (T_3) (Nusse and Whitehead, 2001). Thyroid hormones are extremely important and have diverse actions in the body. Thyroid hormones are essential in all phases of life cycle, including foetal and neonatal neurological development, overall growth, development, physical and mental efficiency, energy production and reproduction (Hetzel, 1983). Disorders associated with altered thyroid hormone secretion are common and affect about 5% women and 0.5% men. It is estimated that more than 200 million people at minimum worldwide have thyroid disease. India too, is no exception. According to a projection from various studies on thyroid disease, it has been estimated that about 42 million people in India suffer from thyroid diseases (Unnikrishnan and Menon, 2011). The present study is aimed to enumerate the clinical spectrum of hypothyroidism in the population of West Bengal to establish the etiology of the disease in this population.

MATERIALS AND METHODS

Subject Population

100 hypothyroid patients (cases) and their corresponding sex and age matched 100 normal individuals (controls) were screened depending on their clinical manifestations, detailed familial history from the Institute of Post Graduate Medical Education and Research (IPGME and R), Kolkata.

Clinical Investigations

Clinical information includes complications like lethargy, cold intolerance, loss of memory, constipation and weight gain. In case of female we also recorded menarche, last child birth, abortion, cycle duration and periodicity.

TSH Assay

Quantitative sandwich immunoassay kit (Siemens, India) was used to assay serum TSH level.

T_3/FT_4 Assay

Serum T_3 or FT_4 concentrations were determined by radioimmunoassay (RIA). ^{125}I-labeled thyroid hormone (either T_3 or T_4) competes with thyroid hormone in the serum sample and for antibody sites on the tube, in the presence of blocking agents for thyroid hormone binding proteins.

Anti-TPO Antibody Assay

An Enzyme-linked immunosorbent assay (*ELISA*) assay was employed for the quantitative measurement of human TPO antibodies in serum. TPO antibody values of > 60 IU/mL were considered as positive (Maskari and Alnaqdy, 2006).

Urinary Iodine Assay

The iodine in the urine is measured by a modification of the traditional colorimetric method of Sandell and Kolthoff (1937). This was done using the Ammonium persulfate method as described by Pino et al., 1996.

RESULTS

Study Samples

Among the cases 12% male, 88% female and the ratio (male: female) is 1:7 (in world population the ratio is 1:6). The cases and controls were well balanced in terms of age and gender (Table 1). There is no statistically significant difference between case and control groups.

Table 1. Study characteristics of case and controls

	Case (n=100)	Control (n=100)	p-value
Sex			
Male	12 (12 %)	15 (15%)	0.68
Female	88 (88%)	85 (85%)	
Age (Mean ± SD)	33.65 ± 17.71	31.87 ± 15.76	0.45

Clinical Characteristics

Among case population we found familial history in 35% of female and 42% of male. The significant clinical manifestations of hypothyroidism were goiter (65%), lethargy (65%), muscle cramp (47%), loss of memory (32%), hoarseness of voice (40%), weight Gain (52%), cold intolerance (62%), constipation (49%).

In our investigation, among 88 female patients, we observed spontaneous abortion in 28%, aged 20-35 years. We observed 15 % patients having menstrual cycle abnormality.

Thyroid Hormone Levels

In cases, TSH level was significantly high and FT_4, T_3 level was significantly low as compared to controls (Figure 1 A, B and C).

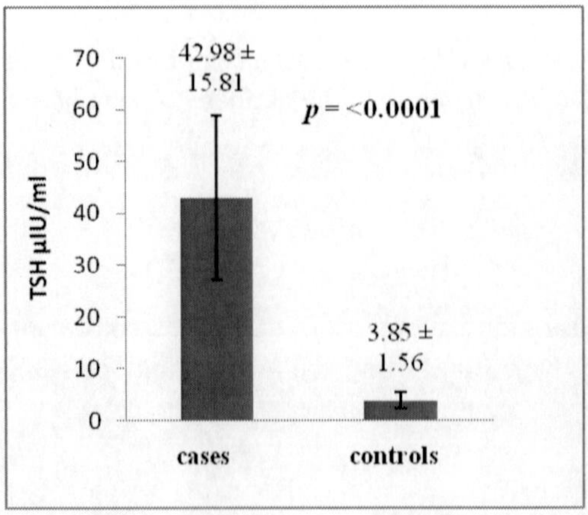

Figure 1. (A).

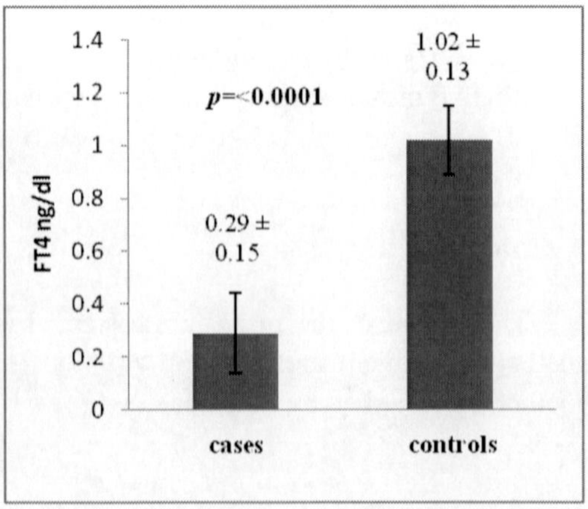

Figure 1. (B).

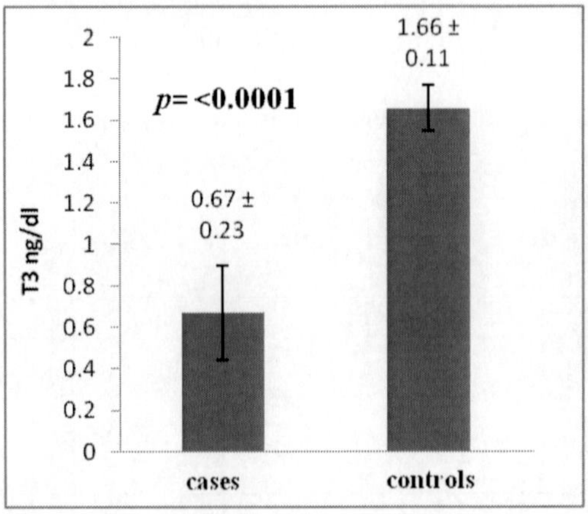

Figure 1. (C).

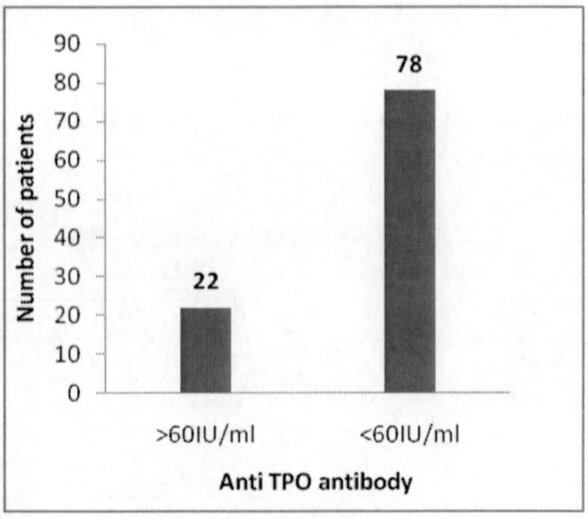

Figure 1. (D).

Anti TPO Antibody

In the present investigation both cases and controls were assayed for serum anti TPO Antibody. We found 22 patients were anti TPO antibody positive (Figure 1D).

Urinary Iodine

In our study sample we observed 93 (93%) case and 95 (95%) control individuals were within the level of 10 –20 µg/dL UIE which was in the range of optimal iodine nutrition (10 – 20 µg/dL) (WHO, UNICEF and ICCIDD, 2001).

Table 2. Clinical Characteristics of the study population

	Cases (N=100)	Controls (N=100) p-value
Family history#		
Male	42	0
Female	35	0
Goiter#	65	0
Weight Gain #		
Yes	52	31 0.004*
No	48	69
Loss of memory		
Yes	32	26 0.43
No	68	74
Lethargy #		
Yes	65	39 <0.0001*
No	35	61
Muscle cramp #		
Yes	47	34 0.08
No	53	66
Cold intolerance #		
Yes	62	41 0.003*
No	38	59
Constipation #		
Yes	49	45 0.67
No	51	55

At diagnosis.
*P-Value < 0.05 is considered to be statistically significant.

DISCUSSION

The present study is aimed to investigate the spectrum of clinical manifestations in hypothyroid patients of West Bengal, India. We compared those clinical presentations with control individuals in the same population. The patient samples exhibited significant increase in the level of serum TSH then that of controls. As expected patients with hypothyroidism showed lower levels of FT4 andT3 as compared to controls (Figure 1B and C). In our study a significant number of patients (52%) have gained body weight. In normal condition whenever our body needs energy, hypothalamus sends hunger signals to the brain to have food especially carbohydrates.

Once the hunger is satisfied, it is sensed by the hypothalamus that releases serotonin and thus we stop having food. This system is hampered in people with hypothyroidism. If there is decreased levels of thyroid hormones then the cells require less energy and the extra energy is then stored in the body leading to weight gain (Caroline et al., 2008). Additionally, lethargy and cold intolerance are significantly different between cases and controls and associated with hypothyroidism.

It is because thyroid hormone deficiency generally decreased oxygen consumption and defects specific to individual organs (Silva, 2003) and decrease in tissue calorigenesis results in cold intolerance.

It is interesting to note that in our study population 28 hypothyroid females were spontaneously aborted and 15 have menstrual abnormailty. Therefore, thyroid dysfunction might be one of the reasons behind menstrual abnormailty. Earlier research revealed relationship between menstruation and thyroid disease (Truong et al., 2005). Hypothyroidism is associated with menorrhagia, dysmenorrhea. Infertility, premenstrual syndrome, polymenorrhea and menorrhagia have also been reported (Brayshaw et al., 1986) in hypothyroidism.

Autoantibodies to TPO have been widely used to show the population at most risk for the development of autoimmune thyroiditis (AITD). Anti -TPO autoantibodies positive have been found in 22% of patients in contrast to a prevalence of 7-20% in the general population (Tomer and Davies, 2003 and Huber et al., 2008). Therefore, presence of serum anti -TPO antibody is an important contributor for hypothyroidism. It is important to note that urinary iodine levels of both cases and controls are in normal range. Therefore, urinary iodine levels indicate that iodine deficiency is currently not a public health problem in our study population.

Acknowledgments

This work was supported by grant from the Department of Biotechnology, Government of India (Ref. No. BT/PR9342/Med/12/346/2007). The authors would like to thank the families who gave their consent and collaborated in this study.

References

Abramowicz, MJ; Tarkovnik, HM; Varela, V et al. Identification of a mutation in the coding sequence of the human thyroid peroxidase gene causing congenital goiter. *J. Clin. Invest.*, 1992. 90, 1200–1204.

Asakawa, H. Thyroid peroxidase (TPO) gene and pathogenic TPO mutation. *Nippon. Rinsho.*, 1994. 52(4), 864-868.

Baker, JR; Arscott, P; Johnson, J. An analysis of the structure and antigenecity of different forms of human thyroid peroxidase. *Thyroid.*, 1994. 4(2), 173-178.

Bakker, B; Bikker, H;Vulsma, T; De Randamie, JSE; Wiedijk, BM and De Vijlder, JJM. Two decades of screening for congenital hypothyroidism in the Netherlands: TPO gene mutations in total iodide organification defects (an update). *J. Clin. Endocr. Metab.*, 2000. 85, 3708-3712.

Banga, JP; Mahadevan, D; Barton, GJ; Sutton, BJ; Saldanha, JW; Odell, E; McGregor, AM. Prediction of domain organisation and secondary structure of thyroid peroxidase, a human autoantigen involved in destructive thyroiditis. *FEBS Lett.*, 1990.266, 133-141.

Bavadam, L. Imposing iodine. *Frontline.,* 2006. 13-14.

Caroline S. Fox; Michael J. Pencina; Ralph B. D'Agostino; Joanne M. Murabito; Ellen W. Seely; Elizabeth N. Pearce; Ramachandran S. Vasan. Relations of Thyroid Function to Body Weight: Cross-sectional and Longitudinal Observations in a Community-Based Sample. *Archintern med.*, 2008. 168 (6), 587.

Fugazzola, L; Mannavola, D; Vigone, MC; Cirello, V; Weber, G and Beck-Peccoz, P. Total Iodide Organification Defect: Clinical and Molecular Characterization an Italian Family. *Thyroid.,* 2005.15(9), 1085- 1088.

Kimura, S; Hong, YS; Kotani, T; Othaki, S; Kikkawa, F. Structure of the human thyroid peroxidase gene: comparison and relationship to the human myeloperoxidase gene. *Biochemistry.,* 1989. 28, 4481– 4489.

Kimura, S; Kotani, T; McBride, OW; Umeki, K; Hirai, K; Nakayama, T; Ohtaki, S. Human thyroid peroxidase: complete cDNA and protein sequence, chromosome mapping, and identification of two alternately spliced mRNAs. *Proc. Nat. Acad. Sci.,* 1987. 84, 5555-5559.

Kopp, P. Thyroid hormone synthesis: thyroid iodine metabolism. In: Braverman L, Utiger R, eds. Werner and Ingbar's the thyroid: a fundamental and clinical text. 9th ed. Philadelphia: Lippincott, Williams, Wilkins. 2005. 52–76.

Libert, F; Ruel, J; Ludgate, M et al. Thyroperoxidase, an autoantigen with a mosaic structure made of nuclear and mitochondrial gene modules. *EMBO J.,* 1987. 6, 4193– 4196.

Maskari, AM; Alnaqdy, A. Frequency of Thyroid Microsomal and Thyroid Peroxidase Antibody Levels in a Selected Group of Omani Patients with Graves' Disease. *Kuwait Med. J.,* 2006. 38 (1), 10- 13.

Pino, S; Fang, SL; Braverman, LE. Ammonium persulfate: a safe alternative oxidizing reagent for measuring urinary iodide. *Clin. Chem.,* 1996. 42, 239–43.

Rivolta, CM; Esperante, SA; Gruneiro-Papendieck, L; Chiesa, A; Moya, CM; Domene, S; Varela, V and Targovnik, HM. Five novel inactivating mutations in the thyroid peroxidase gene responsible for congenital goiter and iodide organification defect. *Hum. Mutat.*, 2003. 22(3), 259.

Sandell, EB; Kolthoff, IM. Micro determination of iodide by a catalytic method. *Mikrochim. Acta.*, 1937. 1, 9– 25

Silva, JE. The Thermogenic Effect of Thyroid Hormone and Its Clinical Implications. *Ann. Intern. Med.,* 2003. 139, 205-213.

Tajima, T; Tsubaki, J and Fujieda, K. Two novel mutations in the thyroid peroxidase gene with goitrous hypothyroidism. *Endocrine Journal.*, 2005. 52(5), 643-645.

Wu, JY; Shu, SG; Yang, CF; Lee, CC and Tsai, FJ. Mutation analysis of thyroid peroxidase gene in Chinese patients with total iodide organification defect: identification of five novel mutations. *Journal of Endocrinology.*, 2002. 172, 627–635.

INDEX

A

abiotic parameter, 271
Abraham, 148
access, 5, 83, 118
acclimatization, 118, 131, 214
acetic acid, 84, 95, 99, 161, 258
acetone, 111, 112
acetonitrile, 96
acetylcholine, 197
acetylcholinesterase, 185, 186, 188, 189, 193, 196, 198, 199
acid, 3, 4, 5, 6, 9, 11, 83, 84, 89, 92, 93, 94, 95, 96, 99, 100, 101, 103, 113, 124, 132, 145, 160, 161, 186, 187, 188, 199, 200, 223, 258
activity level, 83, 129, 134
actuality, 24
adaptation, 190, 222, 223, 266, 274, 275, 285
adaptations, 224, 291
adductor, 141
adenopathy, 287
adenosine, 94
adenosine triphosphate, 94
adhesion, 139
adipose, 34
adulthood, 124, 245, 247, 249, 285
adults, 20, 122, 249, 292
adverse conditions, 286
adverse effects, 185
aerobic capacity, 214
affective disorder, 100, 101
Africa, 81, 122, 180, 266
age, 16, 17, 20, 21, 22, 38, 100, 110, 113, 114, 226, 228, 235, 237, 238, 239, 240, 241, 242, 247, 250, 253, 296, 297
aggregation, 145, 226, 234
aggression, 106, 112, 116
aggressive behavior, 106, 111
agricultural sector, 159
agriculture, 118, 151, 152
AIDS, 288
air pollutants, 88
alanine, 26, 29, 81, 83
alanine aminotransferase, 81, 83
albumin, 83, 153, 288
alcohol consumption, 94
alcohol dependence, 103
alcoholism, 92, 93, 94, 101, 102
alcohols, 40, 105, 107, 109, 111, 112
aldehydes, 82, 86, 105, 109, 111
algae, 118
alkalinity, 186
allele, 94
allergy, 88
ALT, 25, 26, 29, 81, 83, 85, 86, 87
aluminium, 236
alveolar macrophage, 51, 59, 69, 70, 79
alveoli, 67
amines, 102
amino, 3, 4, 5, 6, 9, 10, 11, 26, 29, 30, 38, 43, 50, 59, 93, 94, 99, 101, 160, 190, 199
amino acid(s), 3, 4, 5, 6, 9, 10, 11, 93, 94, 99, 101, 160, 190, 199
ammonia, 185, 186, 187, 189, 190, 191, 192, 193, 194, 195, 196, 197, 198, 199, 200
ammonium, 185, 187, 188, 190, 269, 271
amphibia, 117, 118, 121, 122, 263, 264, 276
amphibians, 117, 118, 122
anatomy, 157
anemia, 287
anesthetics, 107
animal behavior, 121
animal welfare, 291, 293
annotation, 4, 6
ANOVA, 15, 18, 96, 189, 216, 217, 219, 220, 221, 271

antibody, 70, 79, 287, 296, 297, 298, 300
antidepressant(s), 94, 100, 103
antioxidant, 22, 81, 82, 85, 86, 87, 129, 130, 136, 137, 147, 168, 186, 190, 198, 199, 200, 214, 218, 222, 223
apex, 181
apoptosis, 15, 16, 17, 19, 20, 21, 22, 23, 24, 29, 30, 45, 50, 58, 70, 76, 79, 103
apoptotic pathways, 23
appetite, 197, 288
apples, 236
applications, 88
aquaculture, 146, 185, 197, 198, 213
aquaria, 119, 131, 204
aquarium, 119, 204
aquatic systems, 186, 198
Argentina, 15, 21, 70
arithmetic, 15, 18
aromatic hydrocarbons, 82, 86, 137
arsenic, 61, 62, 63, 64, 65, 66, 67, 69, 70, 140
arsenic poisoning, 61, 62, 69
arterioles, 288
arthropods, 172
ascorbic acid, 200
Asia, 148, 168, 214, 234, 266
aspartate, 81, 92, 101
aspirate, 46, 47, 54, 55, 62, 63, 72
assessment, 17, 51, 59, 70, 79, 84, 102, 137, 146, 233
asthma, 82, 87, 88
asthmatic children, 87
asymptomatic, 287
atherogenesis, 101
atoms, 113
atopy, 88
atrophy, 36, 37, 43, 78, 94, 162, 288
attachment, 48, 159, 163
attractant, 106
authors, 21, 257, 262
autoantibodies, 300
Autoantibodies, 300
autoimmune disease(s), 282
autopsy, 26, 34, 40, 63, 73
azadirachtin, 129, 130, 131, 132, 133, 134, 135, 136, 140

B

bacteria, 148
bacterial infection, 137
bacterial speciation, 4
Bangladesh, 140, 146, 160, 167, 282
barbiturates, 21
barriers, 283
basal lamina, 157
base, 153, 176, 178, 180, 223
basement membrane, 157, 159, 162, 163, 288
basophils, 203, 204, 210
behavioral change, 102
behaviors, 106, 112, 293
benefits, 22
benzene, 82, 86, 89, 167
binding, 30, 88
bioaccumulation, 121
bioassay, 198
biochemical processes, ix
biochemistry, 88, 122, 137, 167
bioinformatics, 3, 7, 13
biological activities, 152, 156
biological processes, ix
biomarkers, 130, 137, 147, 148, 157, 167, 168, 198, 224
biomass, 117, 121
biosynthesis, 43, 100, 115
biotic, 117, 118
birth rate, 276
bleeding, 281
blood, 29, 72, 73, 74, 78, 82, 86, 95, 103, 139, 140, 141, 160, 161, 186, 197, 200, 204, 205, 209, 210, 212, 222, 224, 282, 285, 286, 287, 289
blood smear, 204
blood transfusion, 286
blood vessels, 282
bloodstream, 29
body fluid, 100
body size, 233, 275
body weight, 15, 17, 25, 26, 31, 34, 38, 40, 43, 46, 51, 54, 59, 72, 119, 120, 121, 190, 226, 235, 242, 295, 299
bone, 10, 103, 282, 286, 287
bone marrow, 282, 286, 287
bowel, 15, 20
brain, 37, 94, 101, 102, 103, 186, 190, 197, 198, 199, 299
Brazil, viii, 107, 115, 148, 257, 258, 259, 263, 281, 282, 283, 284, 285, 286, 288, 289
breathing, 186, 187, 198, 204, 223, 286
breeding, 83, 122, 213, 221, 226, 235, 241, 265, 266, 267, 269, 271, 272, 273, 274, 276
Britain, 254
Brno, 212
bronchial asthma, 87, 88
bronchitis, 288
buffer, 83, 84
Burma, 182
burning, 82, 86, 88

C

Ca^{2+}, 158
cabbage, 236
cachexia, 287, 288
cadmium, 137, 158, 168, 223
Cairo, 235, 236, 243, 245, 246
calcium, 10, 101, 154, 166
cancer, 7, 37, 62, 70, 82, 86
capillary, 69, 107
capsule, 288
carbohydrates, 285, 299
carbon, 94, 99, 113, 119
carbon atoms, 113
carbon dioxide, 119
carcinogenesis, 82
carcinoma, 22
Caribbean, 234
caries, 288
carnivores, 158
catabolism, 43, 94
catabolized, 93
catalysis, 186, 198
catalytic activity, 190, 197
catecholamines, 102
category a, 6
category b, 160
catfish, 186, 187, 189, 192, 193, 194, 195, 196, 197, 199, 210, 214, 215, 219, 220, 221, 222, 223, 224
CBS, 92, 93
cDNA, 301
cell, 15, 16, 17, 20, 21, 22, 23, 26, 29, 30, 82, 258
cell death, 15, 16, 20, 21, 22, 23
cell division, 23
cell surface, 144, 145, 210
central nervous system, 82, 94, 99, 197
centromere, 260, 261
cerebrospinal fluid, 92, 94, 99, 100
cerebrovascular disease, 91, 99
character, 262
cheilosis, 287
chemical(s), 6, 11, 62, 70, 72, 86, 95, 106, 107, 108, 113, 114, 116, 117, 121, 130, 134, 145, 165, 186, 187, 214, 215, 222
chemical characteristics, 6, 11
chemical properties, 6, 11
chemotaxis, 56
chemotherapy, 21, 22
Chicago, 96
chicken, 5
children, 87, 88
Chile, 70
chimpanzee, 9, 10, 12
China, 182
chitosan, 211
chloroform, 26, 34, 40, 47, 63, 73
choline, 94, 101, 102, 118, 197
chromatography, 92, 96, 102, 105, 106
chromosome, 4, 257, 258, 260, 261, 262, 263, 301
chromosome map, 301
chronic obstructive pulmonary disease, 88
cilia, 163
circadian rhythm(s), 16, 20, 23
circadian rhythmicity, 16
circulation, 203, 210, 282
cities, 285, 288, 291
classes, 5, 6, 105, 107, 109, 110, 111, 112, 139
classification, 172, 290
climate(s), 121, 214, 266, 283
climate change, 214
clinical presentation, 299
clinical symptoms, 37
clusters, 116
CO2, 82, 86, 120
cobalamin, 93
coding, 183, 300
collagen, 34, 35, 37, 39, 40, 42, 43, 59, 82, 156
Colombia, 285
colon, 15, 17
colonization, 172, 186
color, 84, 174, 178, 285
combined effect, 186
commercial, 83, 117, 159
communication, 106, 114, 124
communities, 118, 121
community, 88, 121, 122, 156
comparative analysis, 3, 4, 6, 13
compensation, 214, 224
competition, 115, 243
complement, 261, 262, 287
complex computational models, 4
complexity, 12, 214
complications, 281, 288, 289, 296
composition, 105, 106, 107, 109, 110, 111, 112, 114, 198, 204
compositional pattern, 109
compounds, 15, 16, 70, 82, 86, 89, 105, 106, 107, 108, 109, 110, 111, 112, 131
computational biology tools, 3, 6, 9, 10, 12, 13
computer, 96
concentration, 83, 84
conception, 17
conditioning, 186
conductivity, 269, 271
configuration, 101
conformity, 143

congestion, 16, 21
conjugation, 130, 136
connective tissue, 74, 155, 156, 157, 161, 162, 163, 164, 166, 288
consent, 300
conservation, 3, 4, 5, 6, 10, 11, 12, 13, 146, 211, 234
conserving, 100
constipation, 296, 297
constituents, 85, 87, 222
consumption, 87, 94, 95, 100
contact dermatitis, 86
containers, 131, 141, 152, 161
contaminant, 155
contamination, 129, 130, 151, 160, 165, 271
control, 17, 20, 23, 25, 26, 28, 83, 85, 86
control group, 26, 28, 34, 36, 40, 46, 48, 54, 56, 62, 63, 72, 74, 85, 92, 96, 99, 189, 190, 191, 297
conversion, 30
cooling, 84
cooperation, 13, 101
copper, 168, 204
copulation, 100, 241, 246, 248, 249, 253, 285
correlation(s), 92, 96, 102, 213, 216, 217, 229, 239, 241, 242, 252, 266
correlation coefficient, 216, 229, 239, 241, 242, 252
cortex, 288
cost, 168
cotton, 124, 160
cough, 288
covering, 208
crabs, 137
creatine, 223
creatine phosphokinase, 223
creatinine, 160
creep, 283
cretinism, 34
crop(s), 130, 157
crowds, 293
CRP, 103
CSF, 92, 94, 99, 100, 103
CT, 164
cues, 118
culture, 124
cure, 140, 287
CV, 205, 206, 207, 209, 210
cycles, 83, 100, 243, 270, 282
cycling, 172
cystathionine, 92, 93
cysteine, 87, 93
cytoarchitecture, 156, 157
cytochrome, 137, 266, 275
cytogenetics, 257, 263
cytometry, 147
cytoplasm, 17, 42, 63, 73, 142, 145, 166, 206, 207, 208, 210, 220, 282
cytotoxicity, 79, 139, 145, 146

D

damages, 166, 186, 197, 218
dance, 116
database, 3, 4, 5, 9, 10, 11, 203
death(s), 15, 16, 23, 25, 29, 30, 284, 288
death rate, 283
decomposition, 86, 130, 172, 186
defects, 79, 300
defence, 79, 211
defense, 82, 87
deficiency, 50, 51, 59, 70, 79, 88, 94, 99, 100, 101, 296, 300
deficit, 78
definition, 17
degradation, 29, 147, 167
dehiscence, 43
dehydration, 153
delirium, 288
denaturation, 270
dependent variable, 237, 247
deposition, 82, 122, 222
deposits, 288
depression, 91, 93, 94, 99, 100, 101, 102, 103, 146, 174, 176, 180
depth, 17, 236
dermatitis, 86, 88, 287
dermatology, 24
dermatoses, 21
dermis, 39, 282
destruction, 26, 46, 71, 76, 121, 157, 160, 197
detachment, 157, 161
detection, 95, 102, 289
deterrence, 130
detoxification, 86, 130, 134, 136, 137, 146, 152, 154, 156, 159, 166
deviation, 110, 141, 142, 143, 144, 145
diabetes, 43, 71, 72, 78, 79, 80
diabetic patients, 78, 79
diet, 26, 34, 40, 46, 54, 62, 72, 83, 91, 95, 100, 152
dietary intake, 100, 101, 102
differentiation, 23
digestion, 153, 154, 158, 166
digestive enzymes, 154
diluent, 96
dimorphism, 15, 16, 20
diploid, 261, 262
discs, 124

diseases, ix, 16, 21, 39, 43, 44, 45, 53, 82, 86, 91, 99, 140, 148, 186, 198, 200, 282, 296
dislocation, 83
disorder, 46, 54, 72, 102, 103, 295, 296
disposition, 130
dissociation, 22
dissolved oxygen, 186, 269
distilled water, 84, 107, 187, 215, 258
distribution, 22, 146, 167, 172, 181, 182, 198, 203, 210, 212, 224, 232, 275, 276, 282, 283, 285, 290, 291, 292, 293
divergence, 5, 13, 14, 263, 266
diversity, ix, 6, 111, 172, 283
division, 22, 23
division of labor, 106
dizziness, 82
DNA, 4, 16, 17, 18, 26, 46, 82, 86, 89, 101, 103, 130, 166, 265, 266, 267, 270, 274, 276, 289
DNA damage, 86, 103
DNA polymerase, 265, 267, 270
DNA strand breaks, 86
dogs, 283, 285, 286, 289, 291, 292
dominance, 111
dopamine, 100, 102
dopaminergic, 101
dosage, 91, 95
dosing, 16, 21
drinking water, 70
DRS, 146, 157
drug delivery, 16, 21
drugs, ix, 51, 79
duodenum, 15, 17, 20
dysmenorrhea, 300

E

earthworms, 182
East Asia, 266
ecology, 121, 137, 157, 158, 173, 233, 234
ecosystem, 117, 118, 129, 140, 151, 152, 160
ectothermic, 222
edema, 162, 288
editors, 263
egg, 114, 123, 124, 126, 127, 227, 234, 235, 236, 237, 238, 242, 245, 246, 248, 249, 250, 253, 285, 286
Egypt, 199, 235, 236, 245, 246
elaboration, 13
election, 266, 274, 275
electric conductivity, 271
electron, 171, 180, 181, 205, 206, 207, 208, 209, 210, 212
electron microscopy, 171, 181, 212
electrophoresis, 266, 270
ELISA, 297
elongation, 4, 270
e-mail, 3, 9, 265, 295
emission, 86
encapsulation, 139, 145
encoding, 82
endangered, 146
endocrine, 20, 25, 39, 53, 121, 200
endothelial cells, 69
endothelium, 288
energy, 146, 190, 227, 285, 296, 299
England, 157, 242
enlargement, 27, 29, 154, 167, 288
environment(s), 30, 50, 82, 88, 111, 131, 137, 141, 152, 168, 186, 204, 211, 215, 227, 229, 282, 285, 293
environmental change, 223
environmental conditions, ix, 83, 106, 118, 226, 229, 291, 293
environmental effects, 70
environmental factors, 82, 136, 190, 198, 214, 293
Environmental Protection Agency, 160
environmental stress, 167, 200
enzymatic activity, 83, 188
enzyme(s), 3, 4, 6, 16, 29, 81, 82, 84, 85, 86, 87, 88, 93, 100, 129, 130, 131, 132, 136, 137, 145, 148, 154, 160, 167, 168, 186, 188, 189, 190, 197, 198, 199, 214, 216, 222, 223, 267, 274
eosinophils, 203, 204, 207, 210
epidemic, 288
epidemiology, 88, 283
epidermis, 39, 40, 41, 42, 286
epilepsy, 100
epinephrine, 84, 89
epithelial cells, 23, 161, 162, 164, 165, 166, 287
epithelium, 16, 21, 22, 23, 24, 153, 154, 155, 156, 157, 161, 162, 163, 165, 166, 258, 286
ERS, 127
erythrocytes, 138, 218, 222
erythropoietin, 102
esophagus, 286
ester, 111
ethanol, 91, 94, 95, 98, 99, 100, 101, 153, 173, 204, 259
ethyl alcohol, 124
etiology, 295, 296
euchromatin, 264
Europe, 241, 245, 254, 266, 274, 275, 276, 288
evidence, 12, 62, 87, 99, 101, 111, 115, 122, 264, 286, 287
evolution, ix, 4, 12, 13, 114, 182, 258, 263, 266, 275, 276

excitotoxicity, 103
excretion, 156, 197
experimental condition, 10, 198, 243
exploitation, 232
exposure, 15, 16, 62, 64, 65, 70, 82, 83, 87, 88, 89, 117, 129, 130, 131, 132, 133, 134, 136, 137, 139, 140, 141, 142, 143, 144, 145, 146, 149, 151, 153, 154, 156, 157, 159, 160, 162, 166, 167, 168, 186, 187, 197, 198, 200, 223, 271, 292
extinction, 84
extracellular matrix, 145
extraction, 96
extracts, 107, 108, 111, 112, 137, 254

F

families, 4, 182, 241, 300
family, 258
farmers, 130, 140
farms, 140
fasting, 16, 22, 72, 95
fat, 287
fatty acids, 85
fauna, 171, 172, 182
females, 15, 17, 19, 20
fertility, 171
fertilization, 232, 235, 237, 241, 243, 247, 253
fever, 21, 287
fiber, 36, 37, 140, 161
fibroblasts, 34, 38, 43, 69
fibrosis, 159, 161, 163, 288
filtration, 129, 136, 153, 155, 156, 165
financial, 102, 113, 127, 197, 211
financial support, 113, 197
Finland, 169
fish, 5, 16, 22, 119, 140, 166, 168, 169, 185, 186, 187, 189, 190, 191, 197, 198, 199, 200, 204, 210, 211, 212, 213, 214, 215, 218, 219, 221, 222, 223, 224, 225, 226, 227, 228, 229, 232, 233, 234
fish oil, 16, 22
fisheries, 198, 233
fitness, 118, 122, 166
fixation, 31, 38, 43, 46, 54, 63, 70, 73, 80, 141
flora, 186, 271
fluctuations, 214, 218
fluid, 46, 54, 63, 73, 92, 94, 99, 100, 153
fluorescence, 46
fluoxetine, 103
folate, 91, 93, 94, 95, 96, 98, 99, 100, 101, 102, 103
folic acid, 100
follicle, 39
follicles, 39, 40, 41, 43, 288
food, 17, 74, 75, 95, 106, 114, 115, 116, 119, 124, 146, 152, 153, 154, 155, 156, 157, 159, 165, 166, 169, 199, 211, 214, 218, 229, 233, 234, 236, 299
food chain, 146
Ford, 146
formaldehyde, 161, 212
formation, 16, 21, 75, 79, 93, 100, 131, 155, 218
formula, 55, 226, 227, 236, 246, 247
fragments, 83, 172
free radicals, 82, 87
freshwater, 129, 130, 131, 137, 139, 140, 146, 147, 148, 151, 152, 153, 156, 157, 158, 159, 160, 166, 167, 168, 169, 185, 186, 187, 198, 213, 214, 215, 219, 220, 221, 222, 223, 224, 233
freshwater species, 153
fruits, 160, 235, 236
functional changes, 145, 165
fungi, 235
fungus, 235
fusion, 59

G

gallium, 70
gastrointestinal tract, 23
gel, 270
gender, 15, 16, 18, 21
gene duplication, 4, 12
gene expression, 20
gene transfer, 4
generation, 87
gene(s), 13, 20, 30, 51, 82, 183, 274
genetic marker, 274
genetics, 266
genome, 4, 12
genotype, 4
genus, 81, 171, 180, 181, 258, 262, 263, 264, 266, 275, 276
geography, 182
Germany, 83, 94, 95, 161, 183, 188, 216, 243, 270
gigantism, 172, 183
gill, 129, 131, 132, 133, 134, 135, 136, 137, 151, 152, 153, 154, 156, 157, 158, 159, 161, 162, 163, 165, 166, 167, 168, 185, 186, 188, 189, 190, 191, 192, 193, 194, 195, 196, 197
gland, 10, 25, 39, 53, 105, 106, 107, 108, 109, 110, 111, 112, 113, 114, 115, 116, 129, 131, 132, 133, 134, 135, 136, 151, 152, 153, 154, 155, 156, 157, 158, 159, 160, 161, 162, 164, 165, 166, 167, 168, 169, 224, 254, 295, 296
glomerulus, 288
glucocorticoid, 43

glucose, 26, 29, 72, 73, 74, 78, 138, 186, 200, 214, 223
glucose tolerance, 29, 72, 74
glucose tolerance test, 72, 74
glucose-induced insulin secretion, 29
glutamate, 103, 185, 189, 190, 191, 194, 198
glutamine, 190
glutathione, 81, 83, 84, 85, 87, 88, 89, 92, 93, 129, 130, 136, 137, 138, 214
glycine, 92, 93
glycogen, 160
glycolysis, 190
goblet cells, 156
goiter, 297, 300, 301
Golgi bodies, 207
gonads, 226
gout, 21
gracilis, 262
grades, 153, 204
granules, 142, 205, 206, 207, 210
grass, 186, 190, 198
gray matter, 186
grids, 204
groups, 17, 25, 83, 85, 88, 262, 264
growth, 10, 13, 14, 22, 23, 39, 43, 45, 50, 53, 117, 118, 121, 130, 185, 186, 190, 197, 199, 200, 214, 227, 233, 235, 236, 237, 238, 239, 240, 241, 242, 243, 245, 246, 250, 253, 271, 276, 296
growth factor, 14, 22, 50
growth hormone, 10, 13, 14, 50
growth rate, 190, 233, 236, 237, 238, 239, 240, 241, 242, 245, 246, 250, 253
Gulf of Mexico, 233

H

habitat, 121, 129, 139, 140, 146, 151, 152, 155, 156, 157, 160, 166, 211, 232, 266, 274, 275
hair, 39, 40, 43, 124, 176, 178, 180, 285
hair follicle, 39, 40, 43
hallucinations, 288
haplotypes, 276
hardness, 119, 120, 186, 199
harmful effects, 81, 146
healing, 40, 43, 145
health, 61, 71, 72, 81, 82, 87, 88, 89, 136, 137, 141, 146, 198, 285, 289, 296
health care, 198
health effects, 82, 89
health risks, 146
health status, 141
heavy metals, 156, 157
height, 96, 124, 246, 247, 249, 250, 251, 252, 253
helium, 107
hemorrhage, 289
hepatocytes, 86, 100, 197
hepatomegaly, 287
hepatosplenomegaly, 287, 288
hepatotoxicity, 103
herbicide, 121
heterochromatin, 208, 209, 260, 262, 264
histology, 27, 28, 29, 36, 37, 147, 156, 158, 166
history, 127, 128, 233, 246, 254, 270, 277, 283, 296, 297, 299
HM, 300, 301
homeostasis, 16, 26, 29, 50, 130, 139, 146
homocysteine, 91, 92, 93, 95, 96, 97, 99, 100, 101, 102, 103
homovanillic acid, 92, 100
honey bees, 114, 115
hookworm, 282
horizontal gene transfer, 4
hormone(s), 10, 11, 12, 13, 25, 26, 27, 30, 31, 33, 34, 38, 39, 43, 40, 44, 45, 47, 50, 51, 53, 55, 58, 59, 106, 218, 295, 296, 297, 299, 300, 301
host, 59, 123, 124, 127, 212, 246, 282, 285, 286, 290
hub, 6, 13
human, ix, 5, 9, 10, 12, 13, 16, 22, 24, 59, 70, 82, 86, 88, 89, 100, 101, 102, 140, 146, 152, 160, 168, 187, 276, 282, 283, 285, 286, 297, 300, 301
human body, 100
human health, 82, 88
humidity, 124, 126, 242, 249, 283, 286
Hunter, 242, 243
hyaline, 288
hydrocarbons, 82, 86, 105, 109, 111, 112, 137
hydrogen, 82, 84, 86, 130, 132, 136, 214
hydrogen peroxide, 82, 84, 86, 130, 132, 136, 214
hydrolysis, 197
hydroperoxides, 86
hydroxide, 187
hydroxyl, 82, 89, 130, 136
hyperglycemia, 78
hyperplasia, 10, 69, 287, 288
hypersensitivity, 22, 103
hyperthyroidism, 26, 29, 37, 42, 43, 50, 51, 53, 58, 59
hypertrophy, 38, 287, 288
hypothalamus, 10, 299
hypothesis, 89, 91, 99, 100, 101, 102, 121, 229, 266, 267, 274, 275
hypothyroidism, 25, 26, 29, 30, 31, 33, 34, 37, 38, 39, 42, 43, 45, 48, 50, 51, 59, 295, 296, 297, 299, 300, 301
hypoxia, 16, 22, 82, 197, 204

I

ideal, 291, 292
identification, 6, 7, 88, 112, 127, 172, 204, 301
identity, 3, 5, 6, 9, 10, 11, 12, 13, 106, 112, 204, 292
illumination, 17, 95, 131
immersion, 17, 83, 161, 205, 211
immune function, 26, 50, 58, 69, 71, 78, 197
immune modulation, 45, 53
immune reaction, 80
immune response, 45, 50, 53, 58, 61, 69, 78, 148, 198, 204
immune system, 10, 50, 51, 59, 62, 70, 72, 86
immunity, 51, 59, 82, 140, 153, 287
immunodeficiency, 287
immunomarkers, 148
immunomodulation, 146
immunomodulatory, 51, 62
immunosuppression, 282, 287
implementation, 30, 87
impregnation, 153, 259, 260
in vitro, 15, 20, 30, 51, 54, 78, 103, 197, 222
in vivo, 15, 20, 21, 22, 50, 59, 78, 86, 133, 134, 135, 148, 282
incidence, 22, 78, 186, 282, 285
incubation period, 126, 235, 236, 237, 242, 246, 249
incubator, 124
indentation, 178
independence, 292
indices, 15, 16, 18, 19, 20
individuals, 20, 82, 105, 106, 107, 109, 112, 118, 119, 131, 152, 173, 187, 219, 220, 221, 228, 236, 241, 245, 246, 247, 248, 253, 259, 267, 287, 292, 296, 299
induction, 29, 37, 59, 72, 189
infants, 38
infarction, 287
infection, 79, 137, 281, 282, 283, 285, 286, 287, 288, 290
inferences, 222
inflammation, 37
ingestion, 61, 101, 165
inguinal, 287
inhibition, 30, 79, 87, 88, 116, 132, 133, 135, 136, 137, 197, 198
inhibitor, 15, 20, 100
initiation, 4
injury, 22, 23, 30, 70, 82, 103, 186, 236, 288
innate immunity, 140
insecticide, 117, 118, 119, 120, 121, 122, 130
insects, 106, 112, 168, 285, 286
insertion, 6, 11
insulin, 14, 26, 29, 30, 38, 50, 78, 79
integration, 106
integrity, 122, 190
intensive farming, 185
interaction, 26, 86
intercourse, 286
interface, 152
interference, 214
interstitial nephritis, 288
interstitial pneumonitis, 288
intestine, 16, 19, 20, 21, 22, 23, 94, 213, 215, 216, 217, 218, 219, 220, 221, 222, 223, 258, 288
intoxication, 21, 70, 156
invertebrates, ix, 3, 4, 6, 129, 198
investment, 227
iodine, 296, 297, 299, 300, 301
ions, 190
Iran, 91, 101
Ireland, 79
iron, 130
isoleucine, 11
isozymes, 274
Israel, 254, 265, 266, 267, 269, 273, 274, 275, 276, 277
issues, 158, 221

J

Japan, 169, 226, 234
jejunum, 20
joints, 178, 180
Jordan, 16, 22, 148, 168
juveniles, 167, 186, 197, 198, 199, 236, 237, 246, 249

K

karyotype, 257, 258, 259, 260, 261, 262
keratinocytes, 43
kidney(s), 29, 30, 70, 82, 94, 144, 160, 168, 186, 203, 204, 208, 210, 213, 215, 216, 217, 218, 219, 220, 221, 222, 282
kill, 118
killing, 17
kinetics, 22
kinship, 10
KOH, 173
Korea, 103
Krebs cycle, 190
Kuwait, 233, 301

L

laboratory studies, 118, 241
lactate dehydrogenase, 223, 224
lactic acid, 124
lakes, 140, 200, 223, 233
lamella, 154, 161, 174, 176
large intestine, 16, 19, 20
larva, 124, 126, 127, 285
larvae, 117, 121, 122, 124, 186, 190, 198, 200, 265, 266, 267, 269, 270, 271, 272, 273, 274, 275, 276, 285
larval development, 137
larval stages, 200
lead, 87, 100, 139, 151, 157, 159, 166, 204, 212, 288, 289
legs, 126, 172, 173, 174, 176, 178, 180, 181
leishmaniasis, 282, 283, 287, 288, 289, 290
lesions, 145, 165
lethal factor, 190
lethargy, 296, 297, 300
leucine, 11
leucocyte, 204
leukocytes, 70, 79, 212
leukopenia, 287
life cycle, 124, 126, 127, 237, 239, 243, 246, 249, 250, 285, 296
light, 15, 17, 22, 26, 28, 34, 36, 40, 46, 48, 54, 56, 62, 63, 72, 73, 74, 83, 86, 95, 119, 124, 141, 142, 153, 178, 183, 206, 207, 236, 271, 285
light conditions, 22
line, 263
lipid peroxidation, 81, 82, 84, 85, 86, 87, 89, 132, 186, 200, 214
lipids, 130, 224
liquid chromatography, 92, 95, 102
liver, 29, 30, 38, 62, 81, 82, 83, 84, 85, 86, 87, 94, 100, 103, 154, 169, 185, 186, 188, 189, 190, 191, 192, 193, 194, 195, 196, 197, 198, 199, 213, 214, 215, 216, 217, 218, 219, 220, 221, 222, 223, 282, 286, 288
liver cells, 100, 154
liver damage, 85
localization, 70, 198, 212
loci, 267
longevity, 125, 126, 127, 286
lumen, 17, 155, 156, 159, 162, 164, 166
lung cancer, 62, 70
lung disease, 82, 88
lymph, 282, 286, 287
lymph node, 282, 286, 287
lymphocytes, 168, 204, 208, 210, 211, 288
lymphoid, 50, 59, 203, 288
lymphoid tissue, 50, 59
lysis, 270, 287
lysosome, 167

M

machinery, 136, 246
macrophages, 28, 29, 45, 46, 48, 49, 50, 51, 53, 54, 55, 56, 57, 58, 59, 61, 62, 63, 64, 65, 66, 69, 70, 71, 72, 73, 74, 75, 76, 77, 78, 79, 204, 209, 210, 211, 286, 287, 288
magnitude, 136, 151, 152, 153
major depression, 94
major depressive disorder, 103
majority, 226, 262, 286
malaria, 282
malate dehydrogenase, 214, 222
males, 15, 17, 19, 20, 258, 262
malnutrition, 282, 287
mammal(s), 10, 12, 39, 45, 53, 54, 61, 140, 160, 210, 286, 291, 293
mammalian brain, 101
man, 282, 285, 286
management, 21, 160, 225, 291
mandates, 21
mandible, 175, 177, 179
mantle, 153, 161
manufacturer, 17
marine environment, 137
marrow, 282, 286, 287
marsh, 236, 242
mass, 10, 105, 106, 107, 153, 161, 188, 191, 192, 193, 194, 195, 196, 216, 219, 237, 242, 287
mass spectrometry, 106
materials, 130, 154
matrix, 145, 154, 156, 161, 172, 270
matter, 155, 171, 172, 178, 186, 204, 285
measurement(s), 101, 102, 168, 223, 226, 249, 270, 293, 297
measures, 87
median, 172, 176
medicine, 14, 88, 140
Mediterranean, 21, 24, 168, 233, 234, 245
meiosis, 182
melatonin, 30, 88
mellitus, 43, 71, 72, 79
membranes, 63, 73, 85, 94, 222
memory, 296, 297, 299
menarche, 296
menorrhagia, 300
menstruation, 300
mental retardation, 94
mercury, 186, 198, 199

messengers, 94
Metabolic, 102, 222
metabolic changes, 224
metabolic disorder, 72
metabolism, 30, 38, 51, 59, 86, 93, 94, 99, 100, 102, 130, 137, 152, 154, 169, 186, 199, 214, 218, 222, 301
metabolites, 86, 100, 103
metacentric chromosome, 259, 261
metals, 156, 157, 167
metamorphosis, 121, 265, 271
metaphase, 261
meter, 187
methanol, 46, 54, 63, 73, 258
methionine, 11, 92, 93, 99, 100, 101
methyl group(s), 94
methylation, 94, 99, 101, 102
Mexico, 182, 233
mice, 15, 16, 17, 19, 20, 22, 23, 30, 31, 43, 51, 79, 80, 103, 168
microorganism(s), 151, 172
microscope, 28, 36, 46, 54, 63, 73, 124, 141, 142, 161, 173, 205, 259
microscopic investigations, 180
microscopy, 40, 48, 56, 63, 74, 171, 181, 212
microsomes, 158
microtome, 153, 161
Middle East, 266
migration, 232, 282
millipede, 182, 183
mineralization, 10
Missouri, 292
mitochondria, 203, 206, 208, 210, 214, 223, 224, 266
mitochondrial DNA, 266, 274, 275, 276
mitosis, 16, 23
mitotic index, 16
mitral stenosis, 103
models, ix, 4, 50, 79
modifications, 161, 188, 214, 258
modules, 301
moisture, 236, 246
molecular oxygen, 136
molecular weight, 109
molecules, 6, 10, 12, 16, 106
mollusks, 146, 199
monolayer, 141, 162
Montenegro, 287
Moon, 223
morbidity, 153
Morocco, 276
morphogenesis, 59
morphological variations, 140, 257, 258
morphology, ix, 5, 6, 29, 30, 38, 40, 43, 46, 49, 50, 56, 59, 64, 74, 76, 77, 79, 106, 112, 140, 151, 172, 203, 204, 261, 262, 274, 275
morphometric, 226
mortality, 62, 63, 66, 67, 69, 70, 73, 77, 78, 79, 120, 147, 153, 186, 200, 237, 238, 243, 245, 249, 284, 289
mortality rate, 284
mosaic, 301
Moscow, 182
mRNAs, 301
MTS, 137
mucosa, 22
mucus, 161, 163, 165, 248
multicellular organisms, 16
multiplication, 287
murmur, 288
muscle mass, 10
muscles, 29
mussels, 157, 167, 169
mutation(s), 271, 300, 301
Myanmar, 140, 160
myeloid metaplasia, 288
myopathy, 33, 37, 38

N

NaCl, 46, 54, 62, 63, 72, 141, 161
NADH, 83, 85
naphthalene, 138
National Academy of Sciences, 276
native species, 214
natural selection, 266
nausea, 82
necrosis, 21, 51, 72, 76, 78, 197
negative effects, 117
Nepal, 282
nephritis, 288
nephropathy, 282
nerve, 37, 156
nervous system, 82, 94, 99, 197
Netherlands, 204, 300
neurodegenerative diseases, 91, 99
neurodegenerative disorders, 21
neurogenesis, 23
neurological disease, 148
neuronal systems, 101
neurons, 103
neurosecretory, 10
neurotoxicity, 99, 101, 103
neurotransmitter, 99, 100
neurotransmitters, 99, 100, 140
neutral, 11, 83, 185

neutrophils, 82, 204, 205, 206, 210
New Zealand, 91, 95, 223, 243
nickel, 167
Nigeria, 59
Nile, 186, 190, 198, 200, 233
nitrite, 131, 186, 199, 200
nitrogen, 30, 38, 43, 50, 59, 185, 186, 190, 199
N-N, 205, 206, 207
nodes, 282, 286, 287
nodules, 177, 288
norepinephrine, 100, 101
North Africa, 266, 276
North America, 223, 241
nuclei, 10, 15, 17, 28, 36, 42, 49, 143, 156, 161, 206, 210
nucleic acid, 100, 132, 160
nucleolus, 206, 208, 209, 264
nucleus, 17, 76, 143, 144, 203, 205, 206, 207, 208, 209, 210
nurses, 112
nutrient(s), 154, 172
nutrition, 102, 148, 166, 187, 299

O

observations, 20
obsessive-compulsive disorder, 102
obstructive lung disease, 82, 88
octane, 96
oil, 16, 17, 22, 158, 161, 166, 167, 205
oils, 82
oocyte, 233
OPA, 270
opportunities, 11
optical density, 84
order, 17
organ, 15, 19, 21, 23, 30, 38, 39, 43, 50, 53, 59, 61, 151, 152, 153, 154, 156, 157, 160, 162, 166, 181, 186, 217, 218, 219, 288
organic chemicals, 214
organic compounds, 82, 86
organic matter, 171, 172, 178, 186, 285
organism, 136, 151, 152, 159, 160, 190, 200, 271
organs, 19, 21, 25, 33, 136, 137, 151, 152, 153, 156, 158, 160, 161, 165, 171, 216, 217, 237, 287, 300
ornamental plants, 236, 246
osmium, 204
osmolality, 212
ovaries, 225, 226, 227
ovum, 225, 234
oxidation, 82, 188, 190, 223
oxidative damage, 87, 89, 130, 218, 223
oxidative stress, 50, 82, 86, 87, 101, 130, 136, 137, 147, 186, 190, 198, 199, 200, 218
oxygen, 22, 72, 78, 82, 86, 129, 130, 132, 136, 140, 146, 161, 186, 199, 200, 204, 214, 218, 223, 265, 269, 271, 300
oxygen consumption, 300
oyster, 146, 147, 148

P

Pacific, 147
Pakistan, 233
palpation, 288
pancreas, 25, 27, 28, 29, 30, 74, 94
pancreatitis, 29, 30
parallel, 17, 29, 154, 161
parameters, 85
parasite, 285, 286, 287, 288
parasites, 287, 288
pastures, 243
pathogens, 211
pathology, 101, 148, 211
pathophysiology, 94
pathways, 21, 23, 93
PCR, 265, 267, 270, 271, 273, 274, 275
pellagra, 287
peptide(s), 6, 9, 10, 11, 287
peptide hormone, 10, 11
periodicity, 17, 22, 296
peripheral blood, 212, 286
peritoneum, 46, 54, 62, 72
permeability, 63, 73, 288
permission, 182
permit, 20, 259
peroxidation, 81, 82, 84, 85, 86, 87, 89, 132, 186, 200, 214
peroxide, 82, 84, 86, 87, 130, 132, 136, 214, 222
pesticide, 118, 121, 129, 131, 135, 136, 139, 140, 147, 151, 152, 153, 159, 160, 167
pests, 118, 124, 130, 140, 160, 254
petroleum, 166, 167
pH, vii, 46, 54, 62, 73, 83, 84, 96, 119, 120, 131, 141, 161, 185, 186, 187, 188, 189, 190, 191, 192, 193, 197, 199, 200, 204, 212, 215, 216, 269, 271
phagocytic cells, 59, 287
phagocytosis, 23, 45, 46, 48, 50, 53, 55, 56, 58, 59, 79, 139, 145, 154
pharmaceutical, 102
pharmacology, 102, 148
pharynx, 286
phenol, 85, 145, 223
phenotype, 4
phenylalaline, 11

phenylalanine, 100
Pheromones, 106, 113
Philadelphia, 301
phosphate, 46, 54, 62, 73, 83, 84, 132, 138, 188, 204, 214, 215, 216
phospholipids, 100
phosphorylation, 214
phylogenetic tree, 270, 274
physical characteristics, 6
Physiological, 167, 200
physiology, 16, 82, 122
pigmentation, 126
pituitary gland, 10, 224
placebo, 44, 100
plant diseases, 140
plants, 123, 124, 127, 235, 236, 246
plaque, 44
plasma, 87
plasma levels, 288
plasma membrane, 75, 79, 208
plasma proteins, 274
plasticity, 111
playing, 106
pneumonitis, 288
polar, 236
polarization, 262
policy, 21
pollen, 115
pollutants, 82, 88, 140, 156, 160, 186, 222
pollution, 81, 82, 130, 132, 136, 137, 148, 157, 166, 168, 198, 214
polycyclic aromatic hydrocarbon, 82, 86
polymerase, 265, 267, 270
polymerase chain reaction, 265, 267
polymerization, 22
polymorphisms, 266
polypeptide, 10
polypropylene, 26, 34, 40, 46, 54, 62, 72
polyunsaturated fat, 85
polyunsaturated fatty acids, 85
ponds, 140, 152, 160, 265, 267, 269, 271, 275
pools, 118, 266
population, 20, 62, 112, 118, 122, 139, 140, 144, 145, 151, 152, 158, 159, 160, 211, 226, 254, 257, 258, 262, 270, 274, 276, 286, 289, 291, 292, 295, 296, 297, 299, 300
population density, 151, 286
positive correlation, 217, 241, 252
potassium, 83, 84, 132, 188
potato, 236
poultry, 140
poverty, 285
precipitation, 96
predators, 128, 286
premenstrual syndrome, 300
preparation, 156, 187, 213, 215, 218, 219, 222
prevention, 21, 136, 289
primate, 12, 13
principles, 24, 88, 199
pro-inflammatory, 102
project, 87, 174, 177, 180, 197
prolactin, 50, 224
proliferation, 16, 21, 22, 23, 43, 70, 222, 288
promoter, 82
prophylactic, 289
proptosis, 37
prostate gland, 167
protection, 87, 130, 211, 218
protective mechanisms, 130
protein sequence, 3, 4, 5, 10, 11, 13, 301
protein synthesis, 10, 214, 221, 223
proteinase, 270
proteins, 4, 11, 12, 83, 130, 214, 222, 274, 297
protocol, 87, 259
pseudopodia, 74, 144, 145, 203, 210
psoriasis, 44
puberty, 149
public health, 61, 87, 146, 296, 300
pulp, 288
pupa, 285
P-value, 74
pyridoxine, 93

Q

Queensland, 234

R

radiation, 7, 16, 20, 21, 22, 23, 82
radicals, 78, 82, 87, 130, 136
rainfall, 275
rainforest, 234
rate of change, 247
reactions, 80, 94, 99, 130, 132
reactive oxygen, 22, 72, 78, 130, 136, 218, 223
reactivity, 287
reason, 29
reception, 23
receptors, 78, 79, 80, 103
recognition, 106, 112, 113, 157, 243
reflection, 29
reform, 147
regeneration, 10, 225
region, 260, 261

regions of the world, 282
regression, 229, 237, 239, 243, 247, 250
regression analysis, 237, 243, 247
regression equation, 229
regulation, 89
relationship, 30
relaxation, 34
relevance, 6
reliability, 17
repair, 145
repellent, 106
replication, 4
reproduction, 10, 152, 156, 197, 214, 225, 227, 243, 254, 276, 296
researchers, ix, 4, 293
residues, 5, 6, 9, 11, 12, 129, 145
resources, 106
respiration, 152, 153, 166
respiratory, 86
response, 24, 42, 45, 50, 53, 59, 61, 65, 69, 70, 74, 77, 78, 103, 114, 115, 116, 130, 132, 135, 136, 146, 166, 186, 190, 197, 198, 199, 200, 204, 287
retardation, 94, 239
reticulum, 103
retina, 33, 34, 35, 36, 37
reverse transcriptase, 4
rhinitis, 88
Rhizopus, 59
rhythm, 20, 21, 23, 24
rhythmicity, 16
risk(s), 82, 86, 87, 88, 99, 101, 129, 137, 146, 151, 214, 281, 282, 289, 300
risk assessment, 137
risk factors, 282, 289
RNA, 4, 13, 38, 160, 166
rodents, 291
room temperature, 17, 127, 128, 140, 153, 161
root(s), 13, 235, 236
rowing, 274
Royal Society, 157, 254
rubber, 236, 246
rules, 292
runoff, 140, 151, 152, 160, 186
rural areas, 283

S

safety, 146
salinity, 214
saliva, 286
salmon, 200, 222, 223
SAP, 146, 157
SAS, 247, 254, 289
saturation, 78, 214
Saudi Arabia, 81, 88, 225, 226, 233
scanning electron microscopy, 171, 181
scavengers, 87
scent, 111, 114
school, 88
science, ix, 147, 243
scleroderma, 21
scope, 246
seasonal changes, 213, 214, 216, 218, 222, 226
seasonal flu, 218
seasonality, 214, 234
secretion, 10, 26, 29, 30, 38, 78, 105, 106, 107, 109, 110, 111, 112, 114, 154, 295, 296
sediment(s), 140, 160, 186
seedlings, 236
segregation, 226
self-fertilization, 235, 237, 241, 247, 253
Senegalese sole, 199
senescence, 4, 242
sensation, 153
sensitivity, 16, 20, 23, 103, 136, 157
serotonin, 92, 94, 99, 100, 101, 299
serum, 25, 26, 27, 29, 31, 34, 37, 40, 42, 47, 50, 55, 58, 83, 85, 92, 94, 95, 96, 98, 99, 103, 224, 267, 287, 296, 297, 298, 299, 300
serum albumin, 83
serum glutamic pyruvic transaminase, 26
sewage, 198
sex, 16, 38, 99, 103, 106, 109, 110, 113, 213, 215, 216, 217, 218, 219, 222, 225, 226, 227, 232, 233, 262, 292, 296
sex differences, 106
sex ratio, 225, 226, 227, 232, 233
sex reversal, 232
sex steroid, 99, 103
sexual dimorphism, 15, 16, 20
sexual intercourse, 286
SGPT, 25, 26, 27, 29
shape, 29, 43, 166, 178, 203, 209, 237, 249
sheep, 30
showing, 5, 17, 67, 94, 106, 153, 206, 207, 208, 209, 210, 213, 218, 225
shrimp, 137, 140
shrubs, 283
signalling, 51
signals, 113, 299
signs, 21, 34, 247, 287
silica, 270
silver, 199, 259, 264
sinuses, 156
skeletal muscle, 185, 188, 189, 190, 191, 192, 193, 194, 195, 196, 197, 214, 223

skin, 22, 39, 40, 41, 42, 43, 62, 282, 286, 287
skin diseases, 43
slow basal rate, 11
small intestine, 16, 20, 21, 23, 288
smoke, 81, 82, 85, 86, 87, 88, 89
smoking, 82
social integration, 106
social regulation, 106
society, 146
sodium, 61, 62, 64, 65, 83, 84, 94, 137, 157, 167, 168, 187, 188, 215, 216, 258
sodium hydroxide, 187
software, 259, 270, 276
Solea senegalensis, 198, 199, 200
solution, 63, 72, 73, 83, 131, 141, 161, 187, 204, 226, 258, 270
somatic cell, 4
source populations, 276
South Africa, 113, 180
South America, 258, 276
SP, 107, 165, 257
Spain, 224
speciation, 4, 13, 276
specifications, 17
spectrum, 82
spiders, 137
spindle, 16, 22, 143
spine, 176, 180, 181
spleen, 25, 26, 28, 29, 30, 61, 63, 64, 65, 66, 67, 69, 282, 286, 288
splenomegaly, 282, 287, 288
spontaneous abortion, 298
Spring, 267, 269
Sri Lanka, 171
SSS, 161
standard deviation, 141, 142, 143, 144, 145
standard error, 18, 85, 229
standards, 21, 24
starvation, 169, 224
state(s), 23, 27, 34, 40, 47, 55, 95, 100, 106, 113, 136, 139, 257, 258, 282, 283, 284
statistics, 113, 254
stem cells, 29
stenosis, 103
sterile, 62, 131, 141, 161
steroids, 99, 103, 288
stimulation, 70, 100, 129, 190, 254
stimulus, 10
stock, 124, 204, 233
stomatitis, 287
storage, 154, 188
strain, 20
strategic planning, 213
stress, 31, 38, 43, 50, 51, 59, 82, 86, 87, 101, 103, 107, 130, 132, 136, 137, 140, 146, 147, 157, 166, 167, 169, 185, 186, 190, 197, 198, 199, 200, 214, 218, 220
stress response, 166, 186, 190, 198, 199, 200
stressors, 121
structure, 6, 16, 74, 152, 153, 158, 161, 167, 169, 176, 190, 210, 233, 264, 275, 288, 300, 301
subacute, 153, 154
substitution(s), 5, 11, 287
substrate, 100, 124, 130, 132, 188, 216
sucrose, 131, 188
Sudan, 282
suicide, 94, 103
suicide attempts, 94, 103
sulfate, 84
sulfur, 94
Sun, 30, 186, 190, 198, 200
supplementation, 100
suppression, 62, 70, 79, 139, 190
surface area, 165, 172
surveillance, 152, 199
survival, 23, 117, 118, 121, 127, 140, 159, 166, 186, 197, 199, 200, 226, 275
survival value, 159
susceptibility, 72, 79, 121, 186, 198
suspensions, 258
suture, 178, 180
swelling, 37, 162
symptoms, 34, 37, 82, 88, 281, 283, 287
syndrome, 16, 21
synthesis, 10, 16, 43, 87, 88, 94, 100, 141, 149, 190, 214, 221, 222, 223, 254, 301

T

T lymphocytes, 288
Taiwan, 198
tanks, 118, 140, 161, 215
target, 86, 106, 151, 152, 160
target organs, 151, 152
taxa, 11, 118, 276
taxonomy, 4, 6, 12, 182, 262
teams, 79
technical assistance, 21
teeth, 174, 176, 179, 181
telephone, 3, 9
telomerase components, 4
telomerase protein, 3, 4, 5, 6, 13
telomere, 4
TEM, 203, 204
temperature, 17, 83, 95, 119, 124, 126, 127, 128, 131, 140, 141, 153, 161, 185, 186, 187, 188, 189,

190, 197, 198, 199, 200, 213, 214, 215, 216, 217, 218, 222, 223, 233, 242, 249, 266, 269, 271, 274, 275, 286, 292, 294
temporal variation, 16, 22
tendons, 114
tensile strength, 43
tension, 204
testicle, 258
testis, 258
testosterone, 149
Thailand, 89
therapeutic effects, 99
therapeutics, 88
therapy, 88, 101
thermoregulation, 291
thrombus, 103
thyroid, 25, 30, 31, 33, 34, 37, 38, 39, 40, 43, 44, 45, 46, 50, 51, 53, 54, 59, 295, 296, 297, 299, 300, 301
thyroid cancer, 37
thyroid gland, 25, 39, 53, 295, 296
thyroiditis, 37, 300
thyrotropin, 224
tibia, 174, 178, 180, 181
time periods, 271
tissue, 10, 22, 33, 34, 35, 37, 39, 50, 59, 69, 74, 82, 84, 99, 102, 131, 152, 154, 155, 156, 157, 159, 160, 161, 162, 163, 164, 166, 186, 188, 190, 197, 204, 211, 216, 223, 236, 267, 270, 287, 288, 300
TNF-α, 78
toluene, 82, 86
tooth, 174, 178, 180
toxic effect, 85, 87, 132, 140, 166
toxic substances, 130
toxicity, 61, 64, 69, 70, 99, 117, 118, 122, 131, 132, 136, 137, 138, 140, 146, 148, 152, 153, 155, 159, 160, 161, 166, 167, 168, 169, 185, 186, 189, 190, 191, 192, 193, 194, 195, 196, 197, 198, 199, 200, 218
toxicology, 89, 118
toxin, 129, 157, 166
transaminases, 31
transcription, 16, 30, 38
transformation, 23, 214, 264, 287
transfusion, 286
transmission, 282, 283, 285, 286, 289
Transmission Electron Microscopy (TEM), 203
transport, 122, 145, 153, 156, 159, 165, 223
transverse section, 153, 156
treatment, 16, 20, 21, 23, 24, 26, 29, 30, 34, 38, 50, 91, 94, 95, 96, 97, 99, 100, 101, 103, 153, 156, 187, 281
trees, 81, 82
trial, 44
tricyclic antidepressant(s), 100
triggers, 160
triiodothyronine, 43, 296
tryptophan, 100
TSH, 38, 296, 298, 299
tumor, 21, 23, 51
tumor necrosis factor, 21, 51
tumours, 62
turnover, 16, 21
type 1 collagen, 59

U

UK, 204, 293
ultrastructure, 79, 107, 114, 204, 210, 212
UNDP, 289
uniform, 36, 131, 153, 205
United, 83, 141, 161
urban, 156, 283, 285, 288
urban areas, 285
urbanization, 281, 282, 283, 285
urea, 72, 160, 186, 197, 200
uric acid, 160
urine, 297
USA, 17, 26, 34, 94, 115, 122, 141, 243, 270, 276, 292
UV, 121, 188, 216

V

vaccine, 289
vacuole, 74, 75, 165, 286
valuation, 86
valve, 95
variability, 264
variables, 16, 21
variance, 15, 18
variations, 11, 14, 15, 16, 20, 22, 64, 65, 107, 140, 186, 214, 215, 222, 223, 224, 227, 257, 258, 264, 266, 267, 270, 274, 276, 277
vascular system, 211
vasomotor, 88
vector, 118, 282, 283, 285, 286, 287, 290
vegetables, 160, 236
vegetation, 235, 283
vein, 95
vertebrates, ix, 3, 4, 5, 6, 10, 286
vessels, 74, 156, 282, 288
viscera, 282
vision, 37
vitamin B1, 91, 94, 95, 96, 98, 99, 100, 101, 103

vitamin B12, 91, 94, 95, 96, 98, 99, 100, 101, 103
vitamin B12 deficiency, 94
vitamin B6, 100
Vitamin C, 200
vitamins, 99, 100
volatile organic compounds, 82, 86
volatility, 113
vulnerability, 121, 153
vulva, 178, 181

W

walking, 187, 214
Washington, 291
waste, 186
wastewater, 199
water quality, 120, 160, 185, 197
wealth, 3, 4
weight gain, 31, 38, 43, 51, 59, 250, 296, 300
weight loss, 287
welfare, 291, 293
Western Australia, 245
Western Europe, 245
wetlands, 118, 122, 131, 203
wheezing, 87
windows, 96
wood, 88, 161
workers, 86, 88, 89, 105, 106, 107, 108, 109, 110, 111, 112, 114, 115, 162, 232
World Bank, 289
World Health Organization (WHO), 282, 289, 299
worldwide, 72, 282, 289, 295, 296
wound healing, 40, 43, 145

Y

yeast, 53, 54, 55, 56, 57, 58, 59
yield, 20, 185, 188
yolk, 227
young people, 288

Z

zinc, 167
zooplankton, 118